Introduction to Differential and Difference Equations through Modeling

This book presents an opportunity to learn difference and differential equations through a modeling-first approach. The text is meant as an introduction to those equations and not as a text only for modeling courses. No previous exposure to these equations is expected. Modeling in *Introduction to Differential and Difference Equations through Modeling* is presented as the vehicle for learning difference and differential equations.

Although the topics in difference and differential equations are consistent with those in other textbooks, this approach differs. The presentation starts with a model (or several models) and offers the solution with minor discussions. Then, methods to obtain those solutions are presented and show these same models and others again in more detail.

This approach is designed to focus on the use of difference and differential equations to solve real-world problems and to learn not only these primary topics, but how to apply these through modeling.

The authors begin with a review of matrix algebra, then an introduction to modeling. The text progresses to discrete dynamical systems, and then to the standard organization of most differential equation texts, making the alignment with a current syllabus easier.

Technology is a significant modeling component. Excel®, Python®, and Maple® are presented as methods to solving the models. This material has been class tested at the US Military Academy at West Point, Marian University, the College of William & Mary, and the Naval Postgraduate School with great success.

Textbooks in Mathematics
Series editors: Al Boggess, Kenneth H. Rosen

Numerical Methods and Analysis with Mathematical Modelling
William P. Fox and Richard D. West

Business Process Analytics
Modeling, Simulation, and Design
Manuel Laguna and Johan Marklund

Quantitative Literacy Through Games and Gambling
Mark Hunacek

Measure Theory and Integral
Theory and Practice
John Srdjan Petrovic

Contemporary Abstract Algebra, Eleventh Edition
Joseph A. Gallian

Student Solutions Manual for Gallian's Contemporary Abstract Algebra, Eleventh Edition
Joseph A. Gallian

Algebra, Second Edition
Groups, Rings, and Fields
Louis Halle Rowen and Uzi Vishne

Functional Analysis for the Applied Mathematician
Todd Arbogast and Jerry L. Bona

Exploring Linear Algebra, Second Edition
Labs and Projects with Mathematica ®
Crista Arangala

Measure Theory and Fine Properties of Functions, Second Edition
Lawrence Craig Evans and Ronald R. Gariepy

Set Theory
An Introduction to Axiomatic Reasoning
Robert André

Introduction to Differential and Difference Equations through Modeling
William P. Fox and Robert E. Burks

https://www.routledge.com/Textbooks-in-Mathematics/book-series/CANDHTEXBOOMTH

Introduction to Differential and Difference Equations through Modeling

William P. Fox and Robert E. Burks, Jr.

CRC Press
Taylor & Francis Group
Boca Raton London New York

CRC Press is an imprint of the
Taylor & Francis Group, an **informa** business
A CHAPMAN & HALL BOOK

First edition published 2025
by CRC Press
2385 Executive Center Drive, Suite 320, Boca Raton, FL 33431

and by CRC Press
4 Park Square, Milton Park, Abingdon, Oxon, OX14 4RN

CRC Press is an imprint of Taylor & Francis Group, LLC

Library of Congress Cataloging-in-Publication Data
Names: Fox, William P., 1949- author | Burks, Robert author
Title: Introduction to differential and difference equations through
modeling / William P. Fox and Robert E. Burks.
Description: First edition. | Boca Raton, FL : CRC Press, 2025. | Includes
bibliographical references and index.
Identifiers: LCCN 2024052725 (print) | LCCN 2024052726 (ebook) |
ISBN 9781032949000 hardback | ISBN 9781032947341 paperback |
ISBN 9781003582274 ebook
Subjects: LCSH: Differential equations--Mathematical models | Difference
equations--Mathematical models | Mathematical models
Classification: LCC QA372 .F693 2025 (print) | LCC QA372 (ebook) |
DDC 515/.35--dc23/eng/20250214
LC record available at https://lccn.loc.gov/2024052725
LC ebook record available at https://lccn.loc.gov/2024052726

ISBN: 9781032949000 (hbk)
ISBN: 9781032947341 (pbk)
ISBN: 9781003582274 (ebk)

DOI: 10.1201/9781003582274

Typeset in Palatino
by KnowledgeWorks Global Ltd.

To our wives, Hamilton and Angie, and our families

Contents

Preface

Mathematical modeling is both a skill and an art. You have to practice it to maintain and enhance the use of those skills and develop the craft of modeling. The more practice you get, the better you become in this craft known as modeling. The topics covered in this book are the ones we have taught in our mathematical modeling courses over the past thirty years. Each chapter starts with a problem designed to engage you in modeling. The problem highlights "what" the issue or challenge is that needs solving. From there, we apply the principles of mathematical modeling, walking through the steps to develop a model. The main focus is the model itself, and we introduce technology as a tool to either solve the model or perform sensitivity analysis. We rely on technology to build, calculate, or implement the model, and then we dive into the model's analysis. Over the past decade, we've taught a three-course sequence on mathematical modeling. We selected Excel as the preferred software due to its availability, ease of use, and student familiarity with it. Plus, it remains useful in their future careers.

In this book, we define a mathematical model as a description of a real-world system using mathematical concepts to facilitate the explanation of change in the system or to study the effects of different elements of the system and to understand changes in the patterns of behavior.

Mathematical models can take many forms, including but not limited to dynamical systems, statistical models, differential equations, or game theoretic models. These and other types of models can overlap, with a given model involving a variety of abstract structures. In general, mathematical models may include logical models, as long as logic is taken as a part of mathematics. In many cases, the quality of a scientific field depends on how well the mathematical models developed on the theoretical side agreeing with the results of repeatable experiments. Lack of agreement between theoretical mathematical models and experimental measurements often leads to important advances as better theories are developed.

Audience

The length of this book prevents us from addressing every potential nuance in modeling real-world problems. We attempt to provide a useful set of models and potentially appropriate techniques to obtain useful results for common problems the reader may encounter as a guide. We do assume a basic or fundamental background in mathematical modeling and therefore only spend a little time establishing the procedure before we return to providing examples and solution techniques. The book is an introduction to mathematical modeling and would be best used for individuals or groups who have already had some introductory mathematical courses and have been introduced to technology at some level. *Introduction to Differential and Difference Equations through Modeling* would be of interest to mathematics departments that offer mathematical modeling courses focused on discrete modeling or modeling for decision making.

The following groups would benefit from using this book:

- Undergraduate students in quantitative methods courses in business, operations research, industrial engineering, management sciences, industrial engineering, or applied mathematics.
- Graduate students in discrete mathematical modeling courses covering topics from business, operations research, industrial engineering, management sciences, industrial engineering, or applied mathematics.

Objectives

The objective of *Introduction to Differential and Difference Equations through Modeling* is to present and illustrate advanced applied mathematical modeling techniques that are accessible to students from many disciplines. The goal is to support the development of a desire for lifelong mathematical learning, habits of mind, and competent and confident problem solvers for the 21st century. The goal is to be illustrative in nature. While the focus of this book is on utilization in a mathematical modeling course, it does have applications for decision makers in any discipline. The book provides an overview to the decision makers of a wide range of applications of quantitative approaches to aid in the decision-making process. As we constantly remind our students, mathematical modeling is not designed to tell you what to do but it does provide insights and is designed to support critical thinking and the decision-making process. We view the mathematical modeling process as a framework for decision makers. This framework consists of four key components: the formulation process, the solution process, the interpretation of the solution in the context of the actual problem, and sensitivity analysis. The users of mathematical modeling should question the procedures and techniques and assumptions used in the analysis during every step in the process. At a minimum, you should always ask yourself, "Did I use an appropriate technique" to obtain a solution and why were other techniques discounted during the process? Another question could be, "Did I oversimplify the modeling process" so that any solution I develop will not really apply in this situation? This book addresses a vital need for students who plan to use technology in solving complex problems, whether through physical models or creative mathematical modeling like discrete dynamical systems.

Technology

We've selected Maple, Excel, and R® to support the mathematical modeling process due to their wide accessibility. It's important for readers to have a fundamental understanding of one of these three technologies to get the most from this book. We provide relevant code for these technologies, where applicable, in each chapter. This allows you to follow the general discussion and then apply the technology of your choice to illustrate the models. However, we do not cover the basic setup for these technologies.

Technology – whether Excel, Maple, or R – is simply a means to an end. Building and interpreting the model are the core elements, and technology is just a tool to carry out the necessary calculations. Beyond using Excel's built-in features, including the Solver and Data Analysis Tool Pack, we demonstrate custom features that can be built within Excel. Many of the figures in this book were generated using technology. It's important to note that one limitation of Excel is its weaker graphics capabilities, especially in 3D. While we've used every feature available in Excel, there were times when we had to develop custom templates, macros, and programs to solve certain problems. These materials will be provided upon request. Both Maple and R complement Excel well, offering better graphics and many integrated commands for mathematical modeling. Maple templates and example solutions are also available upon request from the authors.

Organization

This book is well applied when solving problems with differences and differential equations. But what makes it unique is practical mathematical modeling. We will see how these equations apply directly to real scenarios. We start with a thorough review of linear algebra. This provides the necessary foundation for the solutions to various equations. From there we proceed to basic solution schemes and numerical methods so that you can begin to tackle difference and differential equations in a substantive way. Basic examples are woven throughout for demonstration purposes. At the heart of this text are hands-on modeling projects. Each chapter begins with a model, followed by detailed guidance on how to interpret and solve it. The scenarios present problems solved later in the chapter using the techniques introduced. Often, solving these models requires combining multiple methods covered in earlier sections.

The text includes fundamental exercises for practice. Multiple modeling scenarios demonstrate the wide application of the methods. Every chapter offers several examples where the techniques are applied step by step.

This book contains information that could easily be covered in an advanced semester course focused on mathematical modeling or a semester survey course on the topics. The book is designed to provide instructors the flexibility to pick and select material to support their courses. Chapters 2–11 provide material and solution techniques to address common real-world problems.

- **Chapter 1: Matrix Algebra Review.** This chapter provides a review of basic matrix algebra concepts essential for solving systems of linear equations and ordinary differential equations (ODE). Determinants, inverses, and basic operations are introduced. It emphasizes understanding matrix properties. Practical exercises include matrix manipulation using Maple, which is a helpful computational tool.
- **Chapter 2: Mathematical Modeling and Technology for Difference and Differential Equations**. Focusing on the process of mathematical modeling, this chapter covers the use of technology in solving modeling equations. It

explains model fitting with data transformations and proportionality. Readers will explore the applications of Excel, Maple, and Python in creating and solving accurate mathematical models.

- **Chapter 3: Modeling Discrete Dynamical Systems (DDS).** This chapter introduces DDS and their applications, highlighting equilibrium and stability. Real-world examples like drug dosage models are used to illustrate these systems. The chapter will provide examples applying DDS to analyze long-term behaviors.
- **Chapter 4: Systems of First Order Difference Equations.** This chapter focuses on an examination of systems of equations where each variable has a time index $t = 0, 1, 2, \ldots$ and variables of different time{periods are connected in some manner. These systems are called systems of difference equations and are useful to describe dynamical systems with discrete time. The chapter introduces the reader to the use of matrix algebra to solve these systems of equations.
- **Chapter 5: Introduction, Basic Concepts, and Techniques in Modeling First Order Ordinary Differential Equations.** This chapter provides the reader with an introduction to modeling first order differential equations where the system contains variables where the rate of change is dependent upon the current state of the variable. This chapter expands on the techniques presented in the previous chapters to allow for an examination of more complex problems.
- **Chapter 6: Modeling with Numerical Solutions to Differential Equations: IVP for ODEs with Technology.** In this chapter, numerical methods such as Euler's and Runge-Kutta methods are explored. Delving into higher-order equations, this chapter applies concepts like Hooke's law to scenarios such as bungee jumping to help better understand these real-world systems. Real-world examples show how physical forces like gravity and air resistance shape solutions. Detailed models, including velocity and safety factors, enrich the understanding of the content for the reader. The emphasis is on solving complex equations that lack analytical solutions. Examples include solving motion equations and predicting system behaviors under various conditions.
- **Chapter 7: +++++Numerical Output for Analysis: Graphical and Percent Error.** This chapter offers an exploration on finding general solutions for applying ordinary differential equations to more complex systems. Solutions are both graphical and analytical, focusing on second order linear equations. Readers are provided with opportunities to solve real-world problems through specific exercises. The focus of this chapter is on applying Euler's method to higher order differential equations. The exercises take students through step-by-step applications, illustrating how Euler's method solves physical systems. The chapter also addresses the accuracy of these numerical methods in real scenarios.
- **Chapter 8: Higher Order Differential Equations.** This chapter dives into systems of both linear and nonlinear differential equations, with applications in mechanics, vibrations, and circuits. Nonhomogeneous equations are introduced to the reader, along with methods to solve them using undetermined coefficients. Using matrix notation to solve these systems, it introduces readers to concepts analyzing complex systems efficiently and accurately.

- **Chapter 9: Higher Order Numerical Methods to Solve IVP and BVP.** This chapter introduces the reader to how to tackle ordinary differential equations arising in real-world applications that need to be solved exactly. The chapter focuses on applying numerical techniques to help qualitatively analyze these problems. Real-life examples, like mechanical vibrations, are modeled to demonstrate the use of these techniques. The chapter emphasizes practical applications through structured exercises.
- **Chapter 10: System of Linear and Nonlinear ODEs.** This chapter covers the analysis of systems of linear and nonlinear ordinary differential equations. It applies these systems to real-world scenarios such as brine tank systems and diffusion problems. Readers will have the opportunity to explore solutions to both theoretical and practical problems.
- **Chapter 11: Numerical Solutions to Systems of ODEs.** A variety of numerical methods are introduced for solving systems of ODEs, with a particular focus on competition and predator-prey models. The chapter also discusses the use of Python in implementing these numerical methods, providing practical tools for readers.
- **Chapter 12: Modeling Using Laplace Transforms.** Laplace transforms are applied to solving complex differential systems in this chapter. Real-world applications such as electrical circuits and population models are explored. The chapter provides exercises to practice both forward and inverse Laplace transforms, solidifying comprehension.

About the Authors

Dr. William P. Fox is currently a visiting professor of computational operations research at the College of William & Mary. He is an emeritus professor in the Department of Defense Analysis at the Naval Postgraduate School and teaches a three-course sequence in mathematical modeling for decision making. He received his PhD in industrial engineering from Clemson University. He has taught at the United States Military Academy for twelve years until retiring, and at Francis Marion University where he was the chair of mathematics for eight years. He has many publications and scholarly activities including twenty plus books and one hundred and fifty journal articles.

Colonel (R) Robert E. Burks, Jr. is an associate professor in the Defense Analysis Department of the Naval Postgraduate School (NPS), and Director of the NPS' Wargaming Center. He holds a PhD in operations research from the Air Force Institute of Technology. A retired logistics army colonel with more than thirty years of military experience in leadership, advanced analytics, decision modeling, and logistics operations, he served as an army operations research analyst at the NPS, TRADOC Analysis Center, United States Military Academy, and the United States Army Recruiting Command.

William P. Fox and Robert E. Burks also wrote:
Advanced Mathematical Modeling with Technology, 2021, CRC Press.

Other books by William P. Fox:

Decision Analysis through Modeling and Game Theory, 2024, CRC Press.

Modeling Change and Uncertainty Machine Learning and Other Techniques, 2022, CRC Press.

Mathematical Modeling in the Age of the Pandemic, 2021, CRC Press.

Advanced Problem Solving Using Maple: Applied Mathematics, Operations Research, Business Analytics, and Decision Analysis (with William Bauldry), 2020, CRC Press.

Mathematical Modeling with Excel (with Brian Albright), 2020, CRC Press.

Nonlinear Optimization: Models and Applications, 2020, CRC Press.

Advanced Problem Solving with Maple: A First Course (with William Bauldry), 2019, CRC Press.

Mathematical Modeling for Business Analytics, 2018, CRC Press.

Acknowledgments

We thank all the mathematical modeling students we have had over the last thirty years as well as all the colleagues who have taught mathematical modeling with us during this adventure. We particularly single out the following who helped in our three-course mathematical modeling sequence at the Naval Postgraduate School over the years: Bard Mansger, Steve Horton, Patrick Driscoll, and Greg Mislick. We are especially appreciative of the mentorship of Frank R. Giordano over the past thirty plus years.

1

Matrix Algebra Review

We have designed this chapter to provide a quick refresher to matrix algebra. The intent is not to teach matrix algebra in this chapter but to allow the reader to refresh their memory of the language and structure of matrix algebra. This refresher will make for a smoother transition to some of the later chapters in this book.

System of Equations

A system of equations is a collection of two or more equations that share the same set of variables, where the solution represents values that satisfy all the equations simultaneously.

In the real world, systems of equations often model situations where multiple conditions must be met at the same time. For example:

- Business: A company wants to determine how many units of two different products to produce in order to meet both production capacity limits and profit goals.
- Science: In chemistry, systems of equations can model reactions where several substances interact under specific constraints of temperature and pressure.
- Finance: Determining how to allocate resources, like investing in different assets to meet both risk and return goals.

In each case, the variables (like quantities of products or resources) need to satisfy several conditions (the equations), and solving the system helps find the optimal or feasible solution for the scenario.

Example 1.1

Consider a population of a species of animals where there are three age groups. Let $a(n)$ be the 0–1 year age group, $b(n)$ be the 1–2 year age group (young adults), and c(n) be the 2–3 year age group. Suppose that the survival rates of each age group are: the 0–1 year age group is 50% and the 1–2 age group is 50%. Also, each animal from the 2–3 year age group will give birth to 3 offspring, and each animal from the 1–2 age group will give birth to one offspring.

DOI: 10.1201/9781003582274-1

The system of equations that may model this population with three age groups is as follows:

$$\begin{cases} a(n+1)=b(n)+3c(n) \\ b(n+1)=\dfrac{1}{2}a(n) \\ c(n+1)=\dfrac{1}{2}b(n) \end{cases}$$

These equation constitutes a system discrete dynamical systems as we will see in Chapter 4. In Chapter 6, we will present a system of differential equations such as:

$$\begin{aligned} x'(t) &= 2x(t)-3y(t) \\ y'(t) &= x(t)+y(t) \end{aligned}$$

We will learn to make some calculations that will make the solution process much easier for this system of dynamical systems as well as the system of differential equations. To make the calculations easier to follow, we will learn to write this system in "matrix notation," using only the coefficients of the respective variables.

A matrix is an array of numbers with m rows, each row containing n numbers. We call this an m by n ($m \times n$) matrix.

$$\begin{bmatrix} a_{11} & a_{12} & \cdots & a_{1n} \\ a_{21} & a_{21} & \cdots & a_{2n} \\ \cdot & \cdot & \cdot & \cdot \\ \cdot & \cdot & \cdot & \cdot \\ \cdot & \cdot & \cdot & \cdot \\ a_{m1} & a_{m2} & \cdots & a_{mn} \end{bmatrix}$$

The first matrix element, a_{11}, represents the number in the first row and the first column.

Likewise, a_{12} represents the number in the first row and the second column. The element a_{mn} represents the number in the mth row (last row) and n^{th} column (last column).

A *square matrix* is a matrix that has the same number of rows and columns, i.e., a $m \times m$ matrix.

Matrix Form for a System of Equations

Using our population model, we write it in matrix form as follows:

$$\left.\begin{aligned} a(n+1) &= b(n)+3c(n) \\ b(n+1) &= \frac{1}{2}a(n) \\ c(n+1) &= \frac{1}{2}b(n) \end{aligned}\right\} \Rightarrow \begin{bmatrix} 0 & 1 & 3 \\ \frac{1}{2} & 0 & 0 \\ 0 & \frac{1}{2} & 0 \end{bmatrix} \begin{bmatrix} a(n) \\ b(n) \\ c(n) \end{bmatrix} = \begin{bmatrix} a(n+1) \\ b(n+1) \\ c(n+1) \end{bmatrix}$$

Or for the other differential equation system, we could write as

$$X'(t) = \begin{bmatrix} 2 & -3 \\ 1 & 4 \end{bmatrix} \begin{bmatrix} x(t) \\ y(t) \end{bmatrix}$$

We will come back to this, but first we must consider some basic matrix operations in order to fully understand the matrix equation.

Finding the Determinant of a Matrix

The determinant of a matrix is a number associated with every square matrix. The value of the determinant depends on the particular elements in the rows and columns (of the associated matrix). If the matrix is formed as a result of modeling a physical system, then knowing the determinant of the matrix gives us useful information about the properties of that physical system.

Although all square matrices have a determinant, in this course we will mostly be concerned with finding the determinant of 2×2 and 3×3 matrices. Consider the 2×2 matrix, $\boldsymbol{A}$, below:

$$\boldsymbol{A} = \begin{bmatrix} a_{11} & a_{12} \\ a_{21} & a_{22} \end{bmatrix}$$

The subscripts on the elements of $\boldsymbol{A}$ represent the row and column, respectively, of the element's position in the matrix. In general, a_{ij} is the element in the ith row and the jth column position of a matrix.

We will use primarily two symbols to denote the determinant of a matrix:

1. The determinant of $\boldsymbol{A} = \det(\boldsymbol{A})$
2. The determinant of $\boldsymbol{A} = \begin{vmatrix} a_{11} & a_{12} \\ a_{21} & a_{22} \end{vmatrix}$

The determinant of a 2×2 matrix is especially easy to compute and it will be our building block for the calculation of a 3×3 determinant. Using the same matrix $\boldsymbol{A}$ from above:

$$\det(\boldsymbol{A}) = \begin{vmatrix} a_{11} & a_{12} \\ a_{21} & a_{22} \end{vmatrix} = a_{11}a_{22} - a_{21}a_{12}$$

For a numerical example consider the matrix $\boldsymbol{B}$ shown below:

$$\boldsymbol{B} = \begin{bmatrix} 3 & 5 \\ 2 & 3 \end{bmatrix}$$

Then $\det(\boldsymbol{B}) = 3(3) - (2)(5) = -1$.

A 3 × 3 determinant is not much harder after you understand the pattern. For the 3 × 3 matrix C,

$$C = \begin{bmatrix} a_{11} & a_{12} & a_{13} \\ a_{21} & a_{22} & a_{23} \\ a_{31} & a_{32} & a_{33} \end{bmatrix}$$

the determinant is

$$\det(C) = a_{11}\begin{vmatrix} a_{22} & a_{23} \\ a_{32} & a_{33} \end{vmatrix} - a_{12}\begin{vmatrix} a_{21} & a_{23} \\ a_{31} & a_{33} \end{vmatrix} + a_{13}\begin{vmatrix} a_{21} & a_{22} \\ a_{31} & a_{32} \end{vmatrix} \quad (1.1)$$

Study Equation 1.1 until you discover some patterns; several should become evident.

First, notice that computing a 3 × 3 determinant involves computing three 2 × 2 determinants. Each 2 × 2 determinant in Equation 1.1 is called a **minor, $\boldsymbol{M}_{ij}$**. The minor $\boldsymbol{M}_{ij}$ associated with element a_{ij} can be found by crossing out the ith row and the jth column of the matrix. For instance, the minor $\boldsymbol{M}_{11}$ associated with element a_{11} is found by crossing out the first row and the first column of the matrix. The other minors are found in the same manner. Using the concept of minors, we could rewrite Equation 1.1 as Equation 1.2:

$$\det(C) = a_{11}\boldsymbol{M}_{11} - a_{12}\boldsymbol{M}_{12} + a_{13}\boldsymbol{M}_{13}. \quad (1.2)$$

Second, notice that Equation 1.2 has a sign pattern: +, −, +. This pattern is obtained by designating the a_{11} position as "+" and then alternating the sign pattern across the top row.

If the entries in a square matrix are all real numbers then the determinant can be computed using your calculator or the computer. If the matrix involves symbols then it must be computed by hand or by using your computer. You should become proficient in using all of these methods.

Review of Matrix Notation, Terminology, and Operations

An $m \times n$ matrix **A** is a rectangular array of m rows and n columns shown in Equation 1.3.

$$A = \begin{bmatrix} a_{1,1} & a_{1,2} & a_{1,3} & \cdots \\ a_{2,1} & a_{2,2} & a_{2,3} & \cdots \\ a_{3,1} & a_{3,2} & a_{3,3} & \cdots \\ . & . & . & . \\ . & . & . & . \\ . & . & . & . \end{bmatrix} \quad (1.3)$$

We will use bold capital letters to denote a matrix such as **A**.

If we need to add matrices they need to be of the same dimension, $m \times n$. We then add like entries.

Example 1.2: Addition of matrices

$$A=\begin{bmatrix} 1 & -4 \\ 5 & 6 \end{bmatrix},\ B=\begin{bmatrix} -6 & 7 \\ 8 & 1 \end{bmatrix} \text{ then}$$

$$A+B=\begin{bmatrix} 1+-6 & -4+7 \\ 5+8 & 6+1 \end{bmatrix}=\begin{bmatrix} -5 & 3 \\ 13 & 7 \end{bmatrix}$$

For multiplication by a scalar, c, we multiply c by each term.
Example 1.2. Multiplication by a scalar - 4**A**

$$4A=4\begin{bmatrix} 1 & -4 \\ 5 & 6 \end{bmatrix}=\begin{bmatrix} 4 & -16 \\ 20 & 24 \end{bmatrix}$$

Subtraction is defined as **A – B = A + (–1*B)**

Common properties are:

$$\begin{aligned} &\mathbf{A}+\mathbf{0}=\mathbf{0}+\mathbf{A}=\mathbf{A} \\ &\mathbf{A}-\mathbf{A}=\mathbf{0} \\ &\mathbf{A}+\mathbf{B}=\mathbf{B}+\mathbf{A} \\ &\mathbf{A}+(\mathbf{B}+\mathbf{C})=(\mathbf{A}+\mathbf{B})+\mathbf{C} \\ &\mathbf{C}(\mathbf{A}+\mathbf{B})=\mathbf{cA}+\mathbf{cB} \\ &(\mathbf{c}+\mathbf{d})\mathbf{A}=\mathbf{cA}+\mathbf{dA} \end{aligned}$$

The transpose of a matrix, A^T of a $m \times n$ matrix yields a $n \times m$ matrix.

Example 1.3: If $B=\begin{bmatrix} 1 \\ 2 \\ 3 \end{bmatrix}$ **then** $B^T=\begin{bmatrix} 1 & 2 & 3 \end{bmatrix}$

Matrix multiplication is a bit trickier. If **A** is $m \times n$ then to perform matrix multiplication of A*B then B must be a $n \times p$ matrix.

Example 1.4: Given $A=\begin{bmatrix} 1 & -4 \\ 5 & 6 \end{bmatrix},\ B=\begin{bmatrix} -6 & 7 \\ 8 & 1 \end{bmatrix},\ C=\begin{bmatrix} 1 & 2 & 3 \\ 4 & 5 & 6 \end{bmatrix}$

Then $\mathbf{AB}=\begin{bmatrix} 1 & -4 \\ 5 & 6 \end{bmatrix}\begin{bmatrix} -6 & 7 \\ 8 & 1 \end{bmatrix}=\begin{bmatrix} (1*-6)+(-4*8) & 1*7+(-4*1) \\ 5*-6+6*8 & 5*7+6*1 \end{bmatrix}=\begin{bmatrix} -38 & 3 \\ 18 & 41 \end{bmatrix}$

$$AC=\begin{bmatrix} -15 & -18 & -21 \\ 29 & 40 & 51 \end{bmatrix}$$

We note that in multiplication that in general, $\mathbf{AB} \neq \mathbf{BA}$

We denote the matrix **I** as a square matrix ($m \times m$) of order m whose main diagonal is all **1.**

$$\mathbf{AI} = \mathbf{A} = \mathbf{IA}$$

If **A** is a square matrix, then the inverse of **A** is a square matrix **B** such that **AB=I=BA.** We will also refer to **B** as $\mathbf{A}^{-1}$. In order for $\mathbf{A}^{-1}$ to exist then the determinant of **A**, det(**A**) must be nonzero. If so, the matrix **A** is said to be nonsingular. If the determinant of **A** is zero, then the matrix **A** is singular.

There are many ways to compute the determinant, we assume the student has already been exposed to these methods.

Example 1.5: Given $A = \begin{bmatrix} 1 & -4 \\ 5 & 6 \end{bmatrix}$, **compute** $\mathbf{A}^{-1}$

First we find the determinant: $|\mathbf{A}| = (1)(6) - (-4)(5) = 26$

$$A^{-1} = \begin{bmatrix} 3/13 & 2/13 \\ -5/26 & 1/26 \end{bmatrix}$$

Example 1.6: Show $AA^{-1} = I$

$$AA^{-1} = \begin{bmatrix} 1\left(\frac{3}{13}\right)+(-4)\left(-\frac{5}{26}\right) & 1\left(\frac{2}{13}\right)+(-4)\left(\frac{1}{26}\right) \\ 5\left(\frac{3}{13}\right)+6\left(-\frac{5}{26}\right) & 5\left(\frac{2}{13}\right)+6\left(\frac{1}{26}\right) \end{bmatrix} = \begin{bmatrix} 1 & 0 \\ 0 & 1 \end{bmatrix}$$

If **A** is a matrix valued function, then we might differentiate the matrix-valued function one element at a time.

Example 1.7: Given $x(t) = \begin{bmatrix} t^2 \\ t+3 \\ e^{-2t} \end{bmatrix}$ **and** $A(t) = \begin{bmatrix} \cos(t) & t \\ 2t & \sin(t) \end{bmatrix}$

Then $x'(t) = \begin{bmatrix} 2t^2 \\ 1 \\ -2e^{-2t} \end{bmatrix}$ **and** $A'(t) = \begin{bmatrix} -\sin(t) & 1 \\ 2 & \cos(t) \end{bmatrix}$

Matrix Algebra Using Maple

When can use the popular mathematical modeling software Maple to help us soft Matrix Algebra problems. We must open the linear algebra section of Maple.

```
> with(LinearAlgebra):
```

Method 1: Matrices can be authored as expressions of the form [[a_1,b_1,…,z_1],[a_2,b_2,…, z_2],…,[a_z,b_z,…,z_z]]. Note: the brackets are important! The following example demonstrates the author format for the matrix

$$\begin{bmatrix} 1 & 2 & 3 \\ 4 & 5 & 6 \\ 7 & 8 & 9 \end{bmatrix}$$

A:=Matrix(3,3, **[[1,2,3],[4,5,6],[7,8,9]]**;

$$\mathbf{A} := \begin{bmatrix} \mathbf{1} & \mathbf{2} & \mathbf{3} \\ \mathbf{4} & \mathbf{5} & \mathbf{6} \\ \mathbf{7} & \mathbf{8} & \mathbf{0} \end{bmatrix}$$

Matrix Operations

Addition and Scalar Multiplication of Matrices

To add two matrices in Maple, first author each matrix, or declare each matrix, then use the normal [+] operator to add the two matrices. They can only be added if they are both of the same dimensions, $m \times n$. The [−] operator will subtract two matrices. Use the [*] operator to multiply a matrix by a scalar, but use the period, [.], to multiply two matrices.

1. Example of Scalar Multiplication:

$$\mathbf{3 * [2}x\mathbf{, -5, }x^2\mathbf{]} \text{ simplifies to } \mathbf{[6}x\mathbf{, -15, 3}x^2\mathbf{]}.$$

2. Example of Matrix Multiplication in Maple:

$$\mathbf{A} := [[\mathbf{1},\ \mathbf{2}],[\mathbf{2},\mathbf{3}]]$$
$$\mathbf{B} := [[-\mathbf{1},-\mathbf{1}],[-\mathbf{1},\mathbf{2}]]$$
$$\mathbf{MatrixMatrixMultiply(A.B)} = [[-\mathbf{3},\mathbf{3}],[-\mathbf{5},\mathbf{4}]]].$$

3. The Dot Product: Derive will compute the dot product of two vectors using the [.] operator. For example:

$$[\mathbf{1},\mathbf{2},\ -\mathbf{4}]\cdot[\mathbf{2},\ -\mathbf{4},\mathbf{1}] = -\mathbf{10}$$

   Notice that the answer to a dot product is a scalar.

4. The Matrix Inverse: In Maple, we obtain the inverse of a matrix by first authoring or declaring the matrix and then raising it to the –1 power, with the [^] key. For example, the Maple code for finding the inverse would be: IA:=Inverse(A)

A:= Matrix ([[1, 2], [3, 4]]);

$$A := \begin{bmatrix} 1 & 2 \\ 3 & 4 \end{bmatrix}$$

A^{-1} is IA:= Matrix Inverse(A);

$$A^2 := \begin{bmatrix} -2 & 1 \\ 1.5 & -0.5 \end{bmatrix}$$

5. The Determinant of a Matrix: If your matrix ***M*** is a square matrix, the function Determinant(M) returns the determinant of ***M***. For example, the expression with(LinearAlgebra):

   M := Matrix([[0, –75, –55], [–75, 0, 65], [–55, 65, 0]]);

$$M := \begin{bmatrix} 0 & -75 & -55 \\ -75 & 0 & 65 \\ -55 & 65 & 0 \end{bmatrix}$$

   Determinant(M);

$$536250$$

   This is one way that you can tell if a matrix is invertible. If the Determinant (*A*) = 0, then the inverse of A,A^{-1}, does not exist, and you either find no solution to your system of equations or infinitely many solutions.

6. Eigenvalues: If ***A*** is a square matrix and v is a variable, the Eigenvalues(A) return the eigenvalues of matrix ***A***. The eigenvalues of the matrix A are the zeros of its characteristic polynomial. If you want to see the characteristic polynomial for a matrix, we use the Maple command > Characterisitc Polynomail(M, x)

   with(Linear Algebra):

   M := Matrix([[1, 1, 0], [0, 1, 3], [0, 0, 2]]);

$$M := \begin{bmatrix} 1 & 1 & 0 \\ 0 & 1 & 3 \\ 0 & 0 & 2 \end{bmatrix}$$

   Characteristic Polynomial(M, x);

$$x^3 - 4x^2 + 5x - 2$$

   solve(Characteristic Polynomial(M, x), x);

$$(1,2)$$

   Eigenvalues(M);

$$(1.0, 1.0, 2.0)$$

7. **Eigenvectors:** To find eigenvectors, we use the Maple code >Eigenvectors(A)

A := Matrix ([[1, 2], [3, 4]]);

$$A := \begin{bmatrix} 1 & 2 \\ 3 & 4 \end{bmatrix}$$

Eigenvalues(A);

$$(-0.372, 5.372)$$

Evalf(Eigenvectors(A));

$$v_1 = \begin{bmatrix} -0.825 \\ 0.566 \end{bmatrix}$$

$$v_2 = \begin{bmatrix} -0.416 \\ -0.909 \end{bmatrix}$$

We note that use the *evalf*() to simplify the result.

Solving Simultaneous Linear Equations

A system of 3 linear, independent equations, such as

$$\begin{cases} 5x + 3y - 7z = 4 \\ 2x - 8y + z = 6 \\ -x + 9y + 4z = 5 \end{cases}$$

is equivalent to the matrix equator

$$\begin{bmatrix} 5 & 3 & -7 \\ 2 & -8 & 1 \\ -1 & 9 & 4 \end{bmatrix} \cdot \begin{bmatrix} x \\ y \\ z \end{bmatrix} = \begin{bmatrix} 4 \\ 6 \\ 5 \end{bmatrix}$$

Maple can solve systems of linearly independent equations easily using matrix algebra.

B := Transpose([[5, 3, –7], [2, –8, 1], [–1, 9, 4]]);

$$M := \begin{bmatrix} 5 & 3 & -7 \\ 2 & -8 & 1 \\ -1 & 9 & 4 \end{bmatrix}$$

C := [[4, 6, 5]];

$$C := \begin{bmatrix} 4 \\ 6 \\ 5 \end{bmatrix}$$

MD: = MatrixInverse(B);

$$M := \begin{bmatrix} \frac{41}{302} & \frac{75}{302} & \frac{53}{302} \\ \frac{9}{302} & -\frac{13}{302} & \frac{19}{302} \\ -\frac{5}{151} & \frac{24}{151} & \frac{23}{151} \end{bmatrix}$$

Evalf(MatrixMatrixMultiply(MD, C));

$$\begin{bmatrix} 2.9106 \\ 0.1755 \\ 1.5828 \end{bmatrix}$$

Exercises

1.1 Clear your worksheet and Declare a Constant, a, to be [[1, 2, 3], [4, 5, 6], [7, 8, 0]]. Also, Declare a Constant, b, to be [[1, 3, 1], [2, – 1, 1], [3, 2, 2]]. Using Maple create the matrices for the following expressions and Simplify:

i. a + b ii. a*b iii. a^2 iv. a^(--1) v. b^(–1)

What happened in the above Problem?

1.2 Given the matrix $A = \begin{bmatrix} 2 & 1 & 1 \\ 2 & 3 & 2 \\ 1 & 1 & 2 \end{bmatrix}$

a. Is it nonsingular (invertible)? What is A^{-1}?
b. What are its eigenvalues? Its eigenvectors?

References and Additional Readings

Horn, R. A., & Johnson, C. R. (2012). *Matrix Analysis* (2nd ed.). Cambridge University Press.
Lay, D. C., Lay, S. R., & McDonald, J. J. (2014). *Linear Algebra and Its Applications* (5th ed.). Pearson.
Strang, G. (2016). *Introduction to Linear Algebra* (5th ed.). Wellesley-Cambridge Press.

2

Mathematical Modeling and Technology for Difference and Differential Equations

Mathematical modeling is a powerful process that allows its users to understand, predict, and optimize real-world systems and problems. It is an excellent problem-solving skill that provides a structured framework for analyzing and improving systems. Mathematical modeling is particularly effective in helping the user to simplify complexity, inform decision making, enhance critical thinking, and quantify uncertainty.

A **mathematical model** is a description of a system using mathematical concepts and language. The process of developing a mathematical model is termed **mathematical modeling.** Mathematical models are used not only in the natural sciences (such as physics, biology, earth science, meteorology) and engineering disciplines (e.g. computer science, artificial intelligence) but also in the social sciences (such as business, economics, psychology, sociology, and political science); physicists, engineers, statisticians, operations research analysts, and economists use mathematical models most extensively. A model may help to explain a system, and to study the effects of different components, and make *predictions* about behavior.

Mathematical models can take many forms, including but not limited to dynamical systems, statistical models, differential equations, or game theoretic models. These and other types of models can overlap, with a given model involving a variety of abstract structures. In general, mathematical models may include logical models, as long as logic is taken as a part of mathematics. In many cases, the quality of a scientific field depends on how well the mathematical models developed on the theoretical side agree with the results of repeatable experiments. Lack of agreement between theoretical mathematical models and experimental measurements often leads to important advances as better theories are developed.

Overview and the Process of Mathematical Modeling

Mathematical modeling is about using math to make sense of real-world problems. Whether you're studying how diseases spread, traffic patterns, or figuring out the best way to manage resources, a mathematical model helps simplify these complex systems so we can understand them better. The following are several classic mathematical examples from a variety of fields.

Ecological Population Dynamic

In a wildlife reserve, the population of a species (e.g., rabbits) is tracked yearly. The population in year n is represented by P_n, and it is affected by factors like birth rate, death rate, and limited resources.

DOI: 10.1201/9781003582274-2

A common model in this case would be the Logistic Growth Model. It assumes that the population grows exponentially at first but is limited by the carrying capacity of the environment.

$$P_{n+1} = P_n + r * P_n * (1 - P_n / K)$$

where:

P_n = population at time n
r = intrinsic growth rate
K = carrying capacity of the environment

This model helps wildlife managers predict population growth, decide how much to invest in habitat restoration, and understand the long-term viability of the species under current environmental pressures.

Highway Traffic Flow

Traffic flow on a congested highway can be modeled using discrete dynamical systems to predict the buildup of traffic over time and the impact of lane changes, traffic lights, or accidents.

A simple model uses a cellular automaton where each section of the highway is represented as a cell, and vehicles move between cells at discrete time steps:

$$C_{n+1} = C_n + V - L - A$$

where:

C_n = current vehicle count on the highway at time step n
V = number of vehicles entering the highway
L = number of vehicles leaving the highway
A = number of vehicles involved in accidents or blockages

This model helps traffic engineers predict traffic patterns, optimize traffic light timings, and manage road closures. It can also be used to simulate the effects of adding lanes or improving public transit options to reduce congestion.

Predator Prey Model (Biology)

In a natural ecosystem, populations of predators (e.g., wolves) and prey (e.g., deer) interact with one another. Discrete dynamical systems are used to model how the population sizes of both groups evolve over time.

A classic model for this is the Lotka–Volterra predator-prey model:

$$P_{n+1} = P_n + r * P_n - \alpha * P_n * Q_n$$
$$Q_{n+1} = Q_n - s * Q_n + \beta * P_n * Q_n$$

where:

P_n = prey population at time n
Q_n = predator population at time n
r = prey growth rate
α = predation rate
s = predator death rate
β = rate of predator reproduction based on prey availability

This model helps ecologists understand the long-term dynamics of predator-prey interactions and develop conservation strategies to maintain a balanced ecosystem.

Bridge Jumping (Bungee Jumping)

In the sport of bridge jumping or bungee jumping, a willing participant attaches one end of bungee cord to himself, attaches the other end to a bridge railing, and then drops off a bridge. Is this a safe sport? Can we describe these?

A classic model for this is a damped harmonic oscillator due to the elastic cord's restoring force and the air resistance acting on the jumper. In this system, the dynamics of the jumper's motion are governed by the forces acting on them: gravity pulling them down, the elastic force of the bungee cord pulling them back up, and air resistance slowing them down.

The motion of a bungee jumper can be described by the following differential equation for a damped harmonic oscillator:

$$m * d^2x/dt^2 = -k * x - c * dx/dt + m * g$$

where:

m = mass of the jumper
k = spring constant of the bungee cord
x = displacement from the equilibrium position
c = damping coefficient (representing air resistance)
g = acceleration due to gravity
dx/dt = velocity, and d^2x/dt^2 is the acceleration of the jumper

These are all events that we can model using mathematics. This textbook will help you understand what a mathematical modeler might do for you as a confident problem solver using the techniques of mathematical modeling. As a decision maker, understanding the possibilities and asking the key questions will enable better decisions to be made.

The Modeling Process

In this chapter, we turn our attention to the process of modeling and examine many different scenarios in which mathematical modeling can play a role.

Mathematical modeling requires as much art as it does science. Thus, modeling is more of an *art* than a science. Modelers must be creative – willing to be more artistic or original in their approach to the problem. They must be inquisitive – questioning their assumptions, variables, and hypothesized relationships. Modelers must also think outside the box in order to analyze the models and their results. Modelers must ensure their model and results pass the commonsense test. Science is very important and understanding science enables one to be more creative in viewing and modeling a problem. Creativity is extremely advantageous in problem solving with mathematical modeling.

To gain insight, we should consider one framework that will enable the modeler to address the largest number of problems. The key is that there is something *changing for which we want to know the effects*. We call this the **system** under analysis. The real-world system can be very complicated or very simplistic. This requires a process that allows for both types of real-world systems to be modeled within the same process.

Consider striking a golf ball with a golf club from a tee. Our first inclination is to use the equations about distance and velocity that we used in high school mathematics class. These equations are very simplistic and ignore many factors that could impact the fall of the ball such as wind speed, air resistance, mass of the ball, and other factors. As we add more factors we can improve the precision of the model. Adding these additional factors makes the model more realistic and more complicated to produce. Understanding this model might be a first start in building a model for such situations or similar situations such as a bungee jumper or bridge swinger. These systems are similar for part of the model: the free fall portion has similar characteristics.

Figure 2.1 shows a closed-loop process for modeling. Given a real-world situation like the one above, we collect data in order to formulate a mathematical model. This mathematical model can be one we derive or select from a collection of already built mathematical models depending on the level of sophistication required. Then we analyze the model that we used and reach mathematical conclusions about it. Next, we interpret the model and either make predictions about what has occurred or offer an explanation as to why something has occurred. Finally, we test our conclusion about the real-world system with new data. We may refine or improve the model to improve its ability to predict or explain the phenomena. We might even reformulate a new mathematical model.

Mathematical Modeling

We will build some mathematical models describing change in the real world. We will solve these models and analyze how good our resulting mathematical explanations and predictions are. The solution techniques that we employ in subsequent chapters take advantage of certain characteristics that the various models enjoy. Consequently, after building the models, we will **classify** the models based on their mathematical structure.

When we observe change, we are often interested in understanding why change occurs the way it does, perhaps to analyze the effects of different conditions, or perhaps to predict what will happen in the future. Often, a mathematical model can help us understand behavior better while allowing us to experiment mathematically with different conditions. For our purposes, we will consider a mathematical model to be a mathematical construct designed to study a particular real-world system or behavior. The model allows us to use mathematical operations to reach mathematical conclusions about the model as illustrated in Figure 2.1.

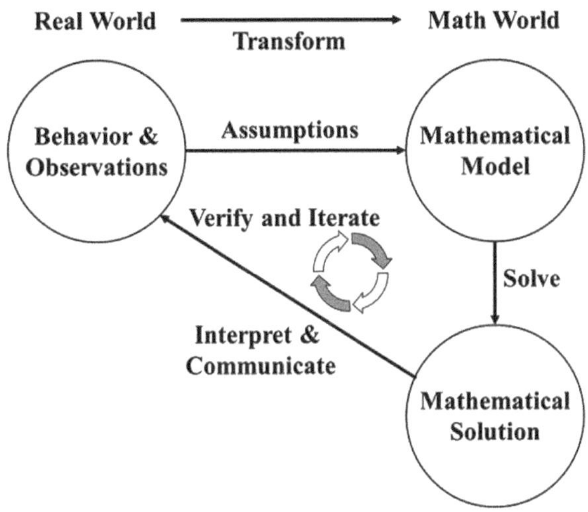

FIGURE 2.1
Modeling real-world systems with mathematics (Fox and Burks, 2022).

Models and Real-World Systems

A system is an assemblage of objects joined by some regular interaction or interdependence. Examples include sending a module to Mars, handling the United States debt, a fish population living in a lake, a TV satellite orbiting the earth, delivering mail, and locations of service facilities – all are examples of a system. The person modeling is interested in understanding not only how a system works but also what interactions cause change and how sensitive the system is to changes in these inputs. Perhaps the person modeling is also interested in predicting or explaining what changes will occur in the system as well as when these changes might occur.

A possible basic technique used in constructing a mathematical model of some system is a combined mathematical-physical analysis. In this approach, we start with some known physical principles or reasonable assumptions about the system. Then we reason logically to obtain conclusions. Sometimes we have data and let's help us come up with a reasonable model. Modelers must be open to many avenues to solve problems.

Figure 2.2 suggests how we can obtain real-world conclusions from a mathematical model. First, observations identify the factors that seem to be involved in the behavior of interest. Often we cannot consider, or even identify, all the relevant factors, so we make simplifying assumptions excluding some of them. Next, we conjecture tentative relationships among the identified factors we have retained, thereby creating a rough "model" of the behavior. We then apply mathematical reasoning that leads to conclusions about the model. These conclusions apply only to the model and may or may not apply to the actual real-world system in question. Simplifications were made in constructing the model and the observations upon which the model is based invariably contain errors and limitations. Thus, we must carefully account for these anomalies and test the conclusions of the model against real-world observations. If the model is

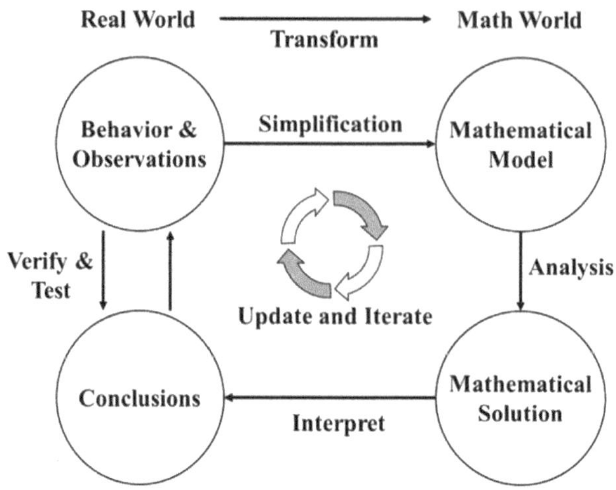

FIGURE 2.2
In reaching conclusions about a real-world behavior, the modeling process is a closed system (adapted from Giordano et al., 2014).

reasonably valid, we can then draw inferences about the real-world behavior from the conclusions drawn from the model. In summary, we have the following procedure for investigating real-world behavior:

Step 1. Observation the system or hypothesis the system (if one does yet exist), identify the key factors involved in the real-world behavior, simplify initially, and refine later as necessary.

Step 2. Conjecture or guess the possible relationships or inter-relationships among the factors and variables identified in step 1.

Step 3. Solve the model.

Step 4. Interpret the mathematical conclusions in terms of the real-world system.

Step 5. Test the model conclusions against real-world observations-the common sense rule.

Step 6. Perform model testing or sensitivity analysis.

There are various kinds of models that we will introduce as well as methods or techniques to solve these models in the subsequent chapters. An efficient process would be to build a library of models and then be able to recognize various real-world situations to which they apply. Another task is to formulate and analyze new models. Still, another task is to learn to solve an equation or system in order to find more revealing or useful expressions relating to the variables. Through these activities, we hope to develop a strong sense of the mathematical aspects of the problem, its physical underpinnings, and the powerful interplay between them.

Most models do simplify reality. Generally, models can only approximate real-world behavior. Next, let's summarize a ***process*** for formulating a model.

Model Construction

Let's focus our attention on the process of model construction. An outline is presented as a procedure to help construct mathematical models. In the next section, we will illustrate this procedure with a few examples.

Our nine steps, modified from a six-step approach by Giordano (Giordano, et al., 2014), act as a guide for thinking about the problem and getting started in the modeling process.

Step 1. Understand the problem or the question asked.

Step 2. Make simplifying assumptions. Justify your assumptions.

Step 3. Define all variables and provide units.

Step 4. Construct a model.

Step 5. Solve and interpret the model.

Step 6. Verify the model.

Step 7. Identify the strengths and weaknesses of your model.

Step 8. Sensitivity Analysis or Model Testing of the model. Do the results pass the "common sense" test?

Step 9. Implement and maintain the model for future use.

We will start with a discussion of each step in more depth.

Step 1. Understand the problem or the question asked:

Identifying the problem to study is usually difficult. In real life, no one walks up to you and hands you an equation to be solved. Usually, it is a comment like, "we need to make more money," or "we need to improve our efficiency." We need to be precise in our formulation of the mathematics to describe the situation.

Step 2. Make simplifying assumptions:

Start by brainstorming the situation. Make a list of as many factors, or variables, as you can. Realize we usually cannot capture all these factors influencing a problem. The task is simplified by reducing the number of factors under consideration. We do this by making simplifying assumptions about the factors, such as holding certain factors as constants. We might then examine to see if relationships exist between the remaining factors (or variables). Assuming simple relationships might reduce the complexity of the problem. Once you have a shorter list of variables, classify them as independent variables, dependent variables, or neither.

Step 3. Define all variables:

It is critical to define all your variables and provide the mathematical notation to be used for each.

Step 4. Select the modeling approach and formulate the model:

Using the tools in this text and your own creativity build a model that describes the situation and whose solution helps to answer important questions.

Step 5. Solve and interpret the model:

We take the model we constructed in Steps 1–4 and solve it. Often this model might be too complex or unwieldy so we cannot solve it or interpret it. If this happens, we return to Steps 2–4 and simplify the model further.

Step 6. Verify the model:

Before we use the model, we should test it out. There are several questions we must ask. Does the model directly answer the question or does the model allow for the answer to the questions to be answered? Is the model useable in a practical sense (can we obtain data to use the model)? Does the model pass the commonsense test?

We like to say that we corroborate the reasonableness of our model rather than verify or validate the model.

Step 7. Strengths and weaknesses:

No model is complete with self-reflection of the modeling process. We need to consider not only what we did correctly, but what we did that might be suspect as well as what we could do better. This reflection also helps in refining models.

Step 8. Sensitivity analysis and model testing:

A modeler wants to know how the inputs affect the ultimate output for any system. Passing the common sense is essential. One of the authors once had a class model Hooke's law with springs and weights. He asked them all to use their model to see how far the spring would stretch using their weight. They all provided the numerical answers, but none said that the spring would break under their weight.

Step 9. Refine, implement, and maintain the model:

A model is pointless if we do not use it. The more user-friendly the model the more it will be used. Sometimes the ease of obtaining data for the model can dictate its success or failure. The model must also remain current. Often this entails updating parameters used in the model.

Making Modeling Assumptions

In its simplest form, the modeling process is built upon a set of necessary assumptions:

Make assumptions

Do some "math"

Derive and interpret conclusions

We say that one cannot question the math but can question the assumptions used to get to the model used. Assumptions drive the modeling as well as the analysis. According to Albright, every model is based upon some set of assumptions. These can be trivial or more complex depending on what we know or can observe about the problem. It also might be affected by the available data.

That is why we say we can question the assumptions and ensure they are justified.

Illustrative Examples. We now demonstrate the modeling process that was presented in the previous section. Emphasis is placed on problem identification and choosing appropriate (useable) variables. We do not build the models as these modeling examples are repeated later in the book and the models are completed and discussed there.

Example 2.1: Prescribed Drug Dosage

Scenario: Consider a patient that needs to take a newly marketed prescribed drug. To prescribe a safe and effective regimen for treating the disease, one must maintain a blood concentration above some effective level and below an unsafe level.

Understanding the Problem: Our goal is a mathematical model that relates dosage and time between dosages to the level of the drug in the bloodstream. What is the relationship between the amount of drug taken and the amount in the blood after time, *t*? By answering this question, we are empowered to examine other facets of the problem of taking a prescribed drug.

Assumptions: We should choose or know the disease in question and the type (name) of the drug that is to be taken. We will assume the drug is Rythmol, a drug taken to control the heart rate and is called an antiarrhythmic. We need to know or to find the decaying rate of Rythmol in the bloodstream. This might be found from data that has been previously collected. We need to find the safe and unsafe levels of Rythmol based on the drug's "effects" within the body. This will serve as bounds for our model. Initially, we might assume that the patient's size and weight have no effect on the drug's decay rate. We might assume that all patients are about the same size and weight. All are in good health and no one takes other drugs that affect the prescribed drug. We assume all internal organs are functioning properly. We might assume that we can model this using a discrete time period even though the absorption rate is a continuous function. These assumptions help simplify the model.

Example 2.2: Population Growth or Decay

Let's assume we are interested in building a model that models the growth or decay of a population of bacteria being developed in a petri dish. The collected over 10 hours is shown in Table 2.1. We might want to know the size of the bacteria population after 24 or 48 hours.

TABLE 2.1

Data Bacteria Growth over Time

Hour	0	1	2	3	4	5	6	7	8	9	10
Bacteria	1	2	4	8	16	32	64	128	256	512	1024

Problem Identification: Find a relationship between time and bacteria growth

Assumptions: We will assume that there is no outside interference. We further assume that the time is measured in hours. We assume that the data was collected the same way for time period. We might assume a proportionality relationship.

Proportionality and Geometric Similarity

You are with a famous archeologist on a dig in Southern California. The archeologist finds a femur bone of what appears to be a somewhat prehistoric bird-like creature. Of initial concern to the archeologists is to determine the size of this creature based on the bone found. They turn to you as the mathematical scientist on this expedition. You tell them that there are modeling tools that will help determine the size of this creature. One of these tools is using proportionality with geometric similarity.

In our previous modeling work, we examined the concept of proportionality as a simplifying assumption. In this chapter, we formalize the concept of proportionality and use it to test relationships among variables chosen in our modeling process. Then, we will present another simplifying argument called *geometric similarity*.

Proportionality

Let's begin with a formal definition of proportionality. Two positive quantities x and y are said to be proportional (to each other) if one quantity is a constant multiple of the other. This definition implies that, $y = kx$ for some constant k. We will write $y \propto x$ to indicate that the quantity y is proportional to the quantity x. Thus,

$$y \propto x \text{ if and only if } y = kx \text{ for some nonzero constant k.}$$

In this chapter, we will restrict ourselves to models where k is *positive* (k > 0).

It makes sense that if $y \propto x$ that $x \propto y$ because the positive constants of proportionality are reciprocals of each other. Thus, if $y = kx$ then $x = (1/k)y$.

Let's visualize the geometric interpretation of proportionality. In Figure 2.3, we plot the geometric interpretation of $y \propto x$. The constant k represents the slope of the straight line that transverses the origin. The slope, k, is defined to be the $\frac{rise}{run} = \frac{\Delta y}{\Delta x}$.

Examples of proportionalities include:

$$y \propto x^2 \text{ if and only if } y = k_1 x^2 \text{ for } k_1 > 0.$$
$$y \propto e^x \text{ if and only if } y = k_2 e^x \text{ for } k_2 > 0.$$

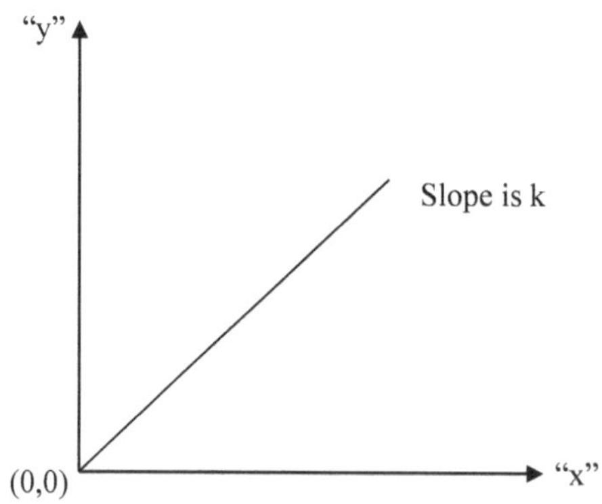

FIGURE 2.3
Graphical interpretation of proportionality.

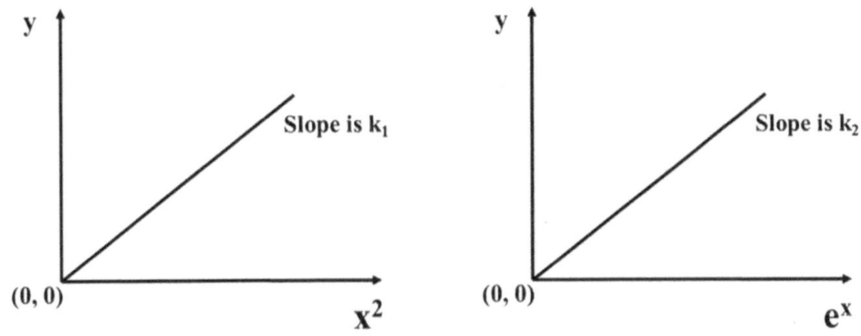

FIGURE 2.4
Geometric interpretation of proportionality.

Graphically, we can present the proportionality by altering the "x" axes to the quantity representing "x" in the proportionality argument. These are illustrated in Figure 2.4.

Since in model $y = kx^2$, $k > 0$ then $x \propto y^{1/2}$ because $x = \left(\frac{1}{\sqrt{k}}\right)y^{\frac{1}{2}}$. This leads us to consider a *transitive property for proportionality*:

If $y \propto x$ and $x \propto z$, *then* $y \propto z$. Thus, any variables proportional to the same variable are proportional to one another.

There have been many scientific discoveries that are proportionality arguments. Here are a couple of the most popular:

- Hooke's Law, $F = kS$
  where F is the restoring force in a spring stretched or compressed at a distance S.
- Newton's 2nd Law: $F = ma$ or $a = 1/m\ F$
  where a is acceleration of a mass m subjected to a net external force F.
- Omh's Law, $E = iR$
  where E is voltage, i represents the induced current, and R is the resistance.
- Boyle's Law $V = k/P$
  where under a constant temperature k the volume V is inversely proportional to the pressure, P.
- Kepler's 3rd law $T = cR^{3/2}$
  where T is the period in any measured dimension and R is the mean distance to the sun.
- Frick's Law: Rate of Diffusion $\propto$ Surface area x Concentration gradient/Thickness of membrane

Testing and Estimating Proportionality Arguments

The simplest method of constructing a model is by applying proportionality arguments to a set of data. By plotting the transformed data set for the proposed proportionality model, we can obtain an initial estimation of proportionality constant directly from the graph by finding the slope of the line that makes that line pass through the origin. The procedure for determining a proportionality constant for a model is presented below.

Procedure to estimate a proportionality argument is as follows:

1. Enter the data observed for the dependent and independent variables.
2. Plot the raw data points to check for trends and to identify potential data outliers.
3. Perform the transformations supporting a hypothesized proportionality.
4. Plot the transformed data to test the hypothesized proportionality.
5. Estimate the constant of proportionality, i.e., find the slope using least squares.

In this chapter, we demonstrate the Maple commands necessary to perform this method of "proportionality" model construction. First, we present the graphical plotting methods to check the proposed proportionality and then we present the commands for estimating the constants of proportionality. Next, several modeling applications are investigated using proportionality.

This description of proportionality graphically depicts a line through the origin. Thus, a quick test of a proportionality model is the plot of the data, which will approximate a straight line projected through the origin. Other types of proportionality models involve transformations, such as $y \propto x^2$ *or* $y \propto e^x$. Maple can quickly transform the original data, x, into many such functions, here x^2 and e^x, as we have previously seen in our introduction to Maple. A submodel involving a proportionality argument must reasonably approximate a line and it **must** pass through the origin.

Once a proportionality argument has been successfully identified, the next step is determining the slope of the line that "best" fits the data graphically. Initially, a large plot is advantageous to determine if the data approximates a line. Quite often this plot will not include the origin, which necessitates a second plot to verify that the "best" lines do pass through the origin.

Consequently, two plots may be required to accurately plot the transformed data. This procedure is illustrated by several examples. Since the line passes through the origin, we recommend using (0, 0) as one of these two points.

Example 2.3: Kepler's Law as a Proportionality Model

Let's consider Kepler's 3rd Law and use the data typically available for planetary motion (Table 2.2).

TABLE 2.2

Kepler's Law as a Proportionality Model

Body	Period (s)	Distance from Sun (m)
Mercury	7.60×10^6	5.79×10^{10}
Venus	1.94×10^7	1.08×10^{11}
Earth	3.16×10^7	1.5×10^{11}
Mars	5.94×10^7	2.28×10^{11}
Jupiter	3.74×10^8	7.79×10^{11}
Saturn	9.35×10^8	1.43×10^{12}
Uranus	2.64×10^9	2.87×10^{12}
Neptune	5.22×10^9	4.5×10^{12}

Because of the data's magnitudes, we decide to compute the period in days (we divide the period by the product of 60 sec/min * 60 min/.hr * 24 hrs/ day) and to scale the distance in millions of miles (divide by 1,000,000,000).

We will now model Example 2.3 using Maple.

Step 1. Enter the data raw data in Maple

```
>Xdata := [7.6, 19.4, 31.6, 59.4, 374, 935, 2640, 5220];
>Ydata := [5.79, 10.8, 15, 22.8, 77.9, 143, 287, 450];
```

Step 2. Plot the original data (Figure 2.5).

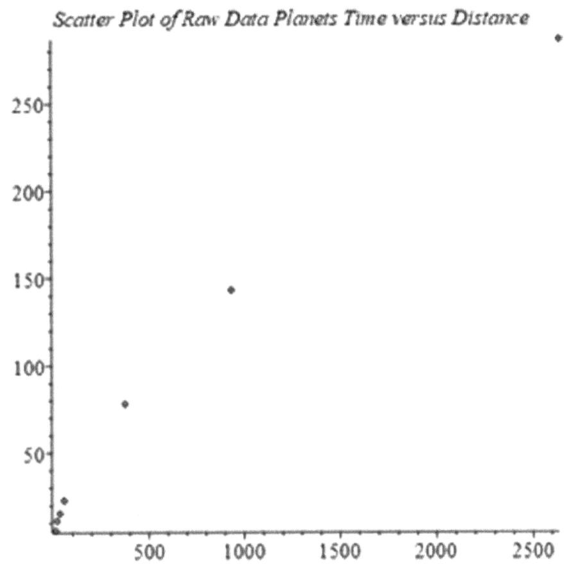

FIGURE 2.5
Maple output for Example 2.3.

We see that the data trends are increasing and concave down.

Step 3. We use Kepler's 3rd law $T \propto R^{3/2}$ and transform the data. Period remains the same but distance is transformed to distance raised to the 3/2 power. In Maple, we will use the command: d1 := *map(x- > x^3/2, distance1)* to achieve this transformation.

Step 4. We plot the transformed data to see if it is a reasonable line through the origin (Figures 2.6 and 2.7).

```
>d1 := map(x → x^(3/2), distance1);
```

```
d1 := [440.5729667, 1122.368925, 1837.117305, 3442.724502,
21742.33517, 5405.93734, 1.537527333x10^5, 3.0189691769x10^5,
4.543402591x10^5]
```

```
>pointploy({seq(period_days[i], d1[i], i = 1…9)})
```

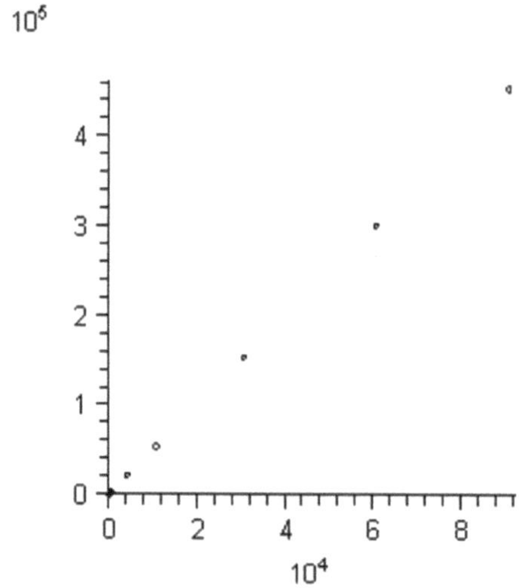

FIGURE 2.6
Maple transformed data output for Example 2.3.

```
> pointplot({seq([period_days[i],d1[i]],i=1..9)},style=point,
title='Proportionality_Test');
```

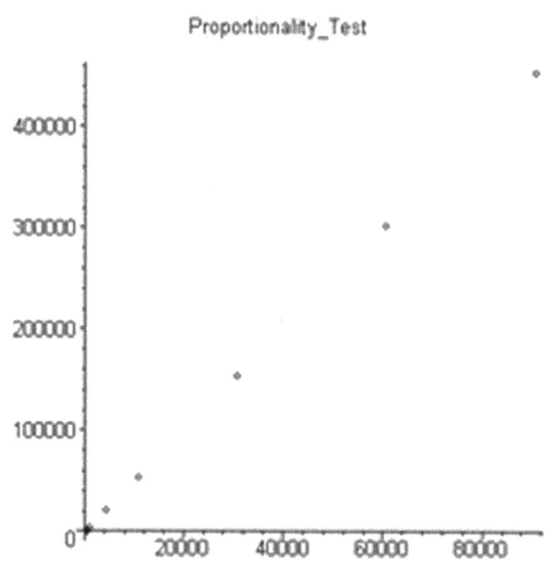

FIGURE 2.7
Proportionality test of maple transformed data output for Example 2.3.

The plot does look like it is a line through the origin.

Step 5. Obtain the slope using linear regression and Step 5 Visualization (Figure 2.8)

```
Slope by use of least squares:
```

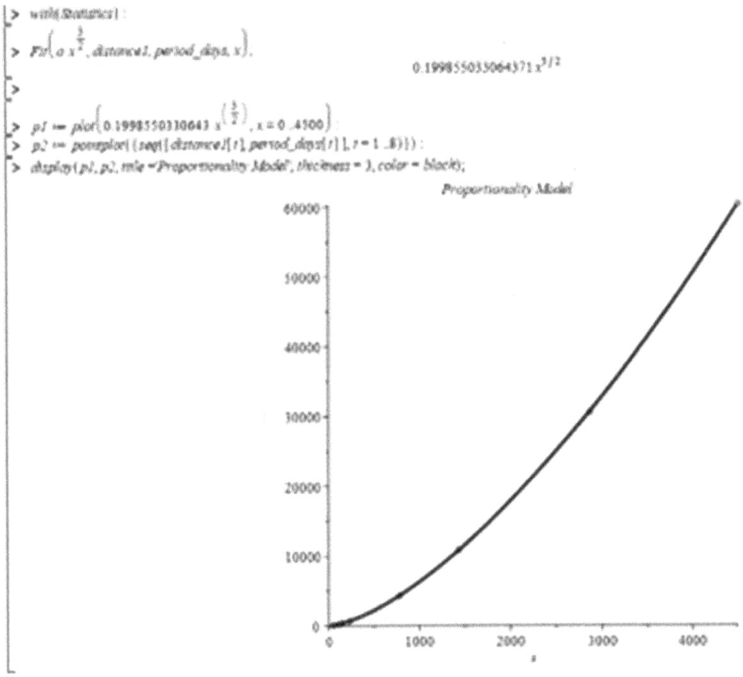

FIGURE 2.8
Maple linear regression fit of Example 2.3 data.

Using the least squares method, we see the fitted lines is a pretty good fit with our proportionality model.

Modeling Fitting Methods with Transformations

Often we might try a ln-ln transformation of a proposed model to check for linearity.

Example 2.4: Population Growth or Decay

Let's assume we are interested in building a model that models the growth or decay of a population of bacteria being developed in a petri dish. The collected over 10 hours is shown in Table 2.3. We might want to know the size of the bacteria population after 24 or 48 hours.

TABLE 2.3
Data Bacteria Growth over Time

Hour	0	1	2	3	4	5	6	7	8	9	10
Bacteria	1	2	4	8	16	32	64	128	256	512	1024

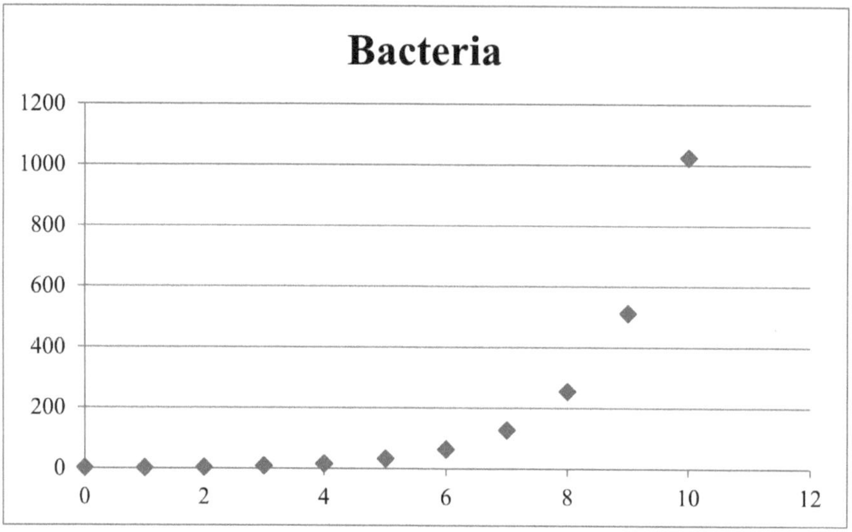

FIGURE 2.9
Plot of Example 2.4 data using Excel.

We plot the data. It is certainly a curve as seen in Figure 2.9.

We decide to try a ln-ln transformation. However, we cannot because one of the times is 0, and ln(0) is not defined. We decide the curve might look exponential, like $y = ae^{bx}$.

We now do a ln-ln of the model form $y = ae^{bx}$ transformation and obtain

$$ln(y) = ln(a) + b * x$$

We plot this as seen in Figure 2.10, which looks like a straight line.

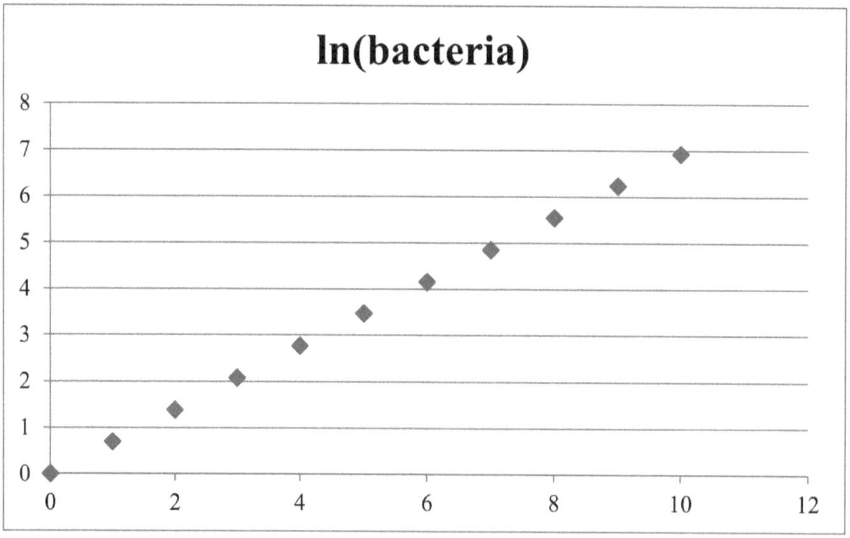

FIGURE 2.10
Plot of hour versus ln(bacteria) using Excel.

We use least squares to obtain this linear model. Ln(bacteria) = 0.68706 hour + 0.0507. We must transform this model back to its original x-y space. We do this by raising each side as exponents of the exponential, *e*. We obtain as our model"

$$Bacteria = 1.05\, e^{(.68706\, hour)}$$

Example 2.5: Nuclide Decay

A British physicist Lord Ernest Rutherford (1871–1937) established that an atom consists of a nucleus surrounded by electrons. The nucleus itself consists of protons and neutrons and two nuclei with identical numbers of protons and identical numbers of neutrons belonging to the same nuclear species or **nuclide**.

The accidental discovery of radioactivity by Henri Becquerel in 1896 followed quite closely by that of x-rays. Becquerel had left a substance containing uranium on a photographic plate wrapped in black paper. After the plate was developed, it showed images of crystals of uranium compounds. The uranium had emitted penetrating rays, and Becquerel coined the word radioactive to describe this property. Later research by Becquerel, Madam Currie, and Rutherford led to the discovery that there are three different kinds of radioactive emissions: alpha, beta, and gamma radiation. Ordinarily, it is of interest or concern to know how long it will take for a certain radioactive substance to lose some fraction of its total mass through the emission process. Let's construct a model to provide these answers.

Assumptions: Assume we have an object that if exposed to a certain substance will emit radiation. Assume that the emissions are all the same kind and that they all decay by the same process (alpha, beta, gamma). The decaying process is such that once a particular nucleus decays, it cannot repeat the process again. Moreover, the probability of its decaying in any time interval is the same. Suppose that at time t there are N undecayed nuclei present. Then the number of nuclei that decay during the time interval from t to $t + \Delta t$ must be proportional to the product of N and Δt. That is,

$$\Delta N = -kN\Delta t$$

where $k > 0$ is the proportionality coefficient, called the decay constant. Let us assume that the number of undecayed nuclei is a continuous function of time. Then dividing both sides of the last equation by Δt and passing to the limit as Δt → 0 results in the following differential equation.

$$\frac{dN}{dt} = -kN.$$

For a given radioactive element, the number k is determined from experimental data. We desire information concerning the radioactive decay of an element.

Problem Identification: Determine a relationship between time and the amount of radiation emitting from a specific substance after exposure.

Assumptions: We assume the rate in which an item decays is proportional to the amount of that item at any time, *t*. The model is accurate over a short period of time. Half-life is the measure of stability of a radioactive substance. The half-life is simply the time, t_i, takes one-half of the atoms to disintegrate. The longer the half-life, the more stable the item is.

TABLE 2.4

Example 2.5 Nuclide Decay Data

Time	Number of Nuclei, N
0	100
1	58.5
2	34.2
3	20
4	11.7
5	7
6	4
7	2.3
8	1.4

Model Construction: Ordinarily, it is of interest to know how long it takes an object emitting radiation to lose some fraction of its total mass or how much of its mass remains after a specified amount of time. Many radioactive materials disintegrate at a rate proportional to the amount present. For example, if x is the radioactive material and $Q(t)$ is the amount of radioactive material present at time, t, then the rate of change of $Q(t)$ with respect to t is given by $\frac{dQ}{dt} = -kQ$, where k represents the proportionality constant and $k > 0$. Let us refer to $Q(0) = Q_0$ as the initial quantity of radioactive material at any arbitrary time, $t = 0$.

Finding the proportionality constant k. from the data (Table 2.4):

Estimation using the law of radioactive decay. The law of radioactive decay leads to solution

$$Q(t) = Q_o e^{-kt}.$$

From the data, we know initially that Q(0) = 100 grams. After 8 hours there is 1.4 grams remaining. So

$$\begin{aligned} Q(8) = 1.4 &= 100e^{-t(8)} \\ 0.014 &= e^{(-8k)} \\ ln(.014)/-8 &= k \\ k &= 0.5335 \end{aligned}$$

If we use this method then we might use the model, $Q(t) = Q_o\, e^{-.5335kt}$.

If we use all the data in Table 2.5, then we must create the change in the number of nuclei and then plot versus N. The slope (using linear regression without an intercept) will be our value for k.

We plot the difference versus N and find the slope as −0.4152. If we use this approach our model is

$$Q(t) = Q_0\, e^{(-0.4152\, t)}.$$

From a modeling perspective, if you have all the data or a substantial amount of data, build the proportionality model to find the slope. If you only have a few data points, then use the model form to estimate the value of k.

TABLE 2.5

Difference in Example 2.5 Nuclide Decay

Are	Number of Nuclei, N	Difference
0	100	–41.5
1	58.5	–24.3
2	34.2	–14.2
3	20	–8.3
4	11.7	–4.7
5	7	–3
6	4	–1.7
7	2.3	–0.9
8	1.4	

As already demonstrated, technology will be critical to solving many of the problems we will present in the later chapters.

Technology

Technology plays an important part in executing the steps of the mathematical process. The partnering of technology with modeling is both key and essential to good modeling principles and practices. Whether you are solving models, running simulations, or visualizing data, having the right technological tools will make a difference. Technology will allow you to use larger datasets and explore more sophisticated real-world situations. In this book, we illustrate three powerful mathematical modeling technologies: Excel, Maple, and Python. Excel is great for managing and analyzing data, Maple excels at handling symbolic mathematics and complex algebraic operations, and Python provides versatility through programming, allowing you to build custom models and simulations. Each of these tools brings a unique set of capabilities allowing the modeler to better bridge the gap between theory and practical application.

Excel

Although Excel might not be the go-to technology for mathematicians or academicians it is a "go-to" tool for the real world. The more we can empower students in math, science, and engineering to use Excel properly then the better solutions will be for solving future problems.

Maple

Maple is an excellent technology for mathematics and operations research majors. Its power and graphical interface in two and three dimensions make it an excellent tool. Maple is a computer algebra system so it can find many closed form analytical solutions. Maple also has an excellent Numerical Analysis package that uses many if not all of the algorithms from Burden and Faires.

Python

Python is a free downloadable software tool for statisticians, engineers, and scientists. Python will be illustrated, as appropriate, within this text. Python code will also be provided within this text. Jupiter Notebook (which is also free) will be our Python interface.

Exercises

2.1 How would you approach a problem concerning a drug dosage? Do you always assume the doctor is right?

2.2 In modeling the size of any prehistoric creature, what information would you like to be able to obtain? What additional assumptions might be required?

2.3 In the oil rig problem, what other factors might be critical in obtaining a "good" model that predicts reasonably well? What variables could be important that were not considered?

2.4 For the model in Example 2.3, are the assumptions about the data reasonable? How would you collect data to build the model? What other variables would you consider?

References and Additional Readings

Albright, B. (2010). *Mathematical Modeling With Excel*. Jones & Bartlett.

Albright, B., and Fox, W. (2020). *Mathematical Modeling With Excel*, 2nd ed. Boca Raton, Fl: Taylor and Francis Publishers.

Burden, R., and Faires, D. (1997). *Numerical Analysis*. Pacific Grove, CA: Brooks-Cole.

COMAP, Modeling Competition Sites found at www.comap.com/contests

Fox, W., and Burks, R. (2022). *Modeling Change and Uncertainty*. New York, NY: Chapman and Hall/CRC.

Fox, W. (2018). *Mathematical Modeling for Business Analytics*. Boca Raton, Fl: Taylor and Francis Publishers.

Giordano, F., Fox, W., and Horton, S. (2014). *A First Course in Mathematical Modeling*, 5th ed. Boston, MA: Cengage Publishers.

3

Modeling Discrete Dynamical Systems (DDS)

Consider the growth of bacteria population. We often model population growth by assuming that the change in population is directly proportional to the current size of the given population. This produces a simple, first-order *discrete dynamical system* (DDS) similar to those seen earlier. It might appear reasonable at first examination, but the long-term behavior of growth without bound is disturbing. Why would growth without bound of a yeast culture in a jar (or controlled space) be alarming?

There are certain factors that affect population growth. Things include resources (food, oxygen, space, etc.). These resources can support some maximum population. As this number is approached, the change (or growth rate) should decrease and the population should never exceed its resource supported amount.

We model this phenomena as:

$$y_{n+1} = y_n + ky_n * (M - y_n)$$

where

y_n = population size after period n
n = time period measured in hours
k = constant of proportionality
M = carrying capacity of our system

Using proportionality arguments from Chapter 1, we find the proportionality slope is 0.0008.

We find the constant slope is 0.0008. We also find the equilibrium value, *ev*, of 621.

This model is

$$y_{n+1} = y_n + 0.0008y_n * (621 - y_n)$$

Again, this is nonlinear because of the $y^2(n)$ term. There is no closed form analytical solution for this equation, however, we may obtain a solution through iteration and graphing the iterated values in Excel (Figure 3.1).

The model shows stability in that the population (biomass) of the yeast culture approaches 621 as n gets large. Thus, the population is eventually stable at approximately 621 units. We present this model in a more complete form later in this chapter.

DOI: 10.1201/9781003582274-3

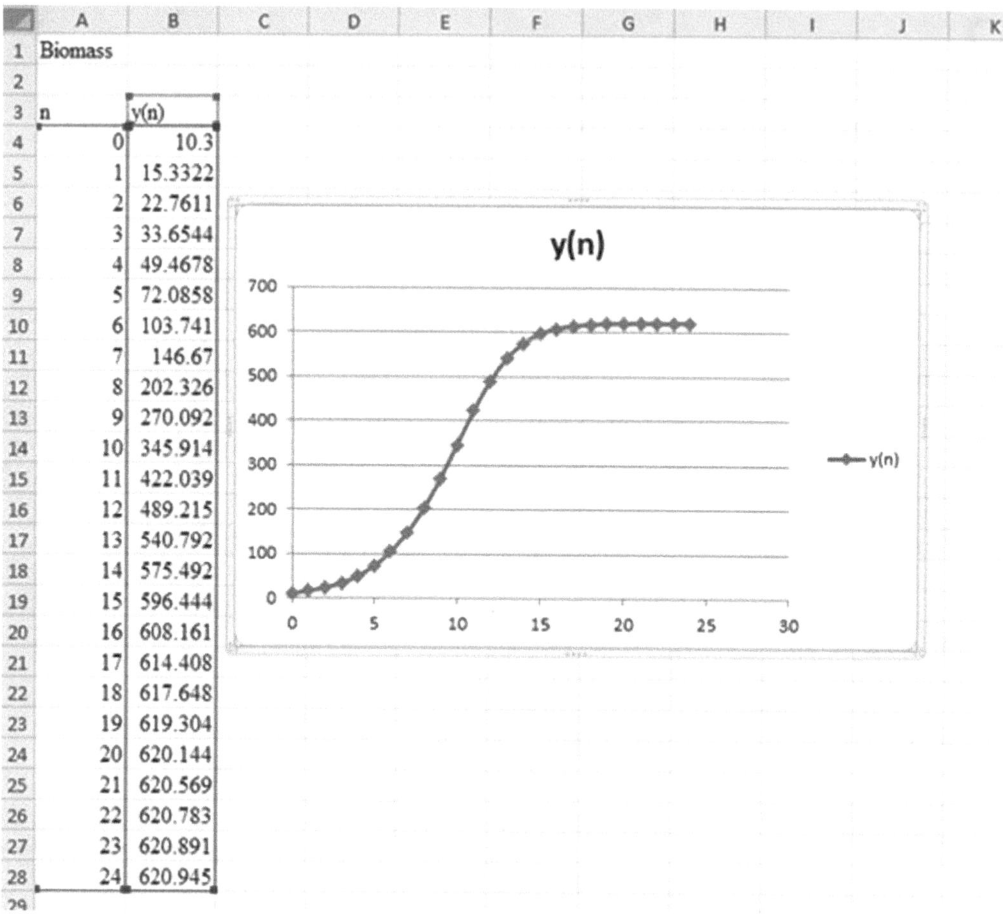

Biomass

n	y(n)
0	10.3
1	15.3322
2	22.7611
3	33.6544
4	49.4678
5	72.0858
6	103.741
7	146.67
8	202.326
9	270.092
10	345.914
11	422.039
12	489.215
13	540.792
14	575.492
15	596.444
16	608.161
17	614.408
18	617.648
19	619.304
20	620.144
21	620.569
22	620.783
23	620.891
24	620.945

FIGURE 3.1
Stability of growth of bacteria using Excel.

Introduction Modeling with Discrete Dynamical Systems

Consider car options today, should we buy or lease a car? How do we go about financing a new car? Once you have looked at the makes and models to determine what type of car you like, it is time to consider the "costs" and finance packages that lure potential buyers into the car dealerships. This process can be modeled as a dynamical system. Payments are made typically at the end of each month. The amount owed is predetermined as we will see later in the chapter.

What if your company is faced with a decision as to buying or leasing new computers for the company use for the next short-term horizon? How could this decision be analyzed? We might employ DDS to model and analyze the possible decisions.

A *DDS* (is a system that changes over time. In this case, the time interval is discrete such as minutes, hours, days, weeks, or years. A DDS is easy to model but may be difficult or even impossible to solve in closed form. In this chapter, although we present some easy closed form solutions, we concentrate more on building the models and obtaining numerical and graphical solutions.

Let's define the dynamical system *A(n)* to be the amount of antibiotic in our bloodstream after *n time* periods. The domain is nonnegative integers representing the time periods from 0, 1, 2, ... that will be the inputs to the function. Since the domain is discrete then our function is a discrete function. The range is the values of *A(n)* determined for each value of the domain. Thus, *A(n)* also represents the dependent variable. For each input value of the domain from 0, 1, 2, ... the result is one and only one *A(n)*, thus *A(n)* is a function.

There are three components to dynamical systems: an equation for sequence representing *A(n)*, the time period *n* is well defined, and at least one starting value. This starting value is called an **initial condition**. For example, if we start with no antibiotic in our system then *A(0) = 0* mg is our initial condition, but if we started after we took an initial 200 mg Tablet then *A(0) = 200* mg would be our initial condition. An example of a DDS with its initial condition would be:

$$A_{n+1} = 0.5 A_n, \; A(0) = 500$$

We are interested in modeling discrete *change*. Modeling with discrete dynamical systems employs a method to explain certain discrete behaviors or make long-term predictions. A powerful paradigm that we use to model with DDS is:

$$\textit{future value} = \textit{present value} + \textit{change}$$

The dynamical systems that we will study with this paradigm may differ in appearance and composition, but we will be able to solve a large class of these "seemingly" different dynamical systems with similar methods. In this chapter, we will use iteration and graphical methods to answer questions about the DDS.

We will use flow diagrams to help us see how the dependent variable changes. These flow diagrams help to see the paradigm and put it into mathematical terms. Let's consider financing a new Ford Mustang. The cost is $25,000 and you can put down $2,000, so you need to finance $23,000. The dealership offers you 2% financing over 72 months. Consider the flow diagram for financing the car below that depicts this situation in Figure 3.2.

We use this flow diagram to help build the discrete dynamical model. Let A_n = the amount owed after *n* months. Notice that the arrow pointing into the circle is the interest

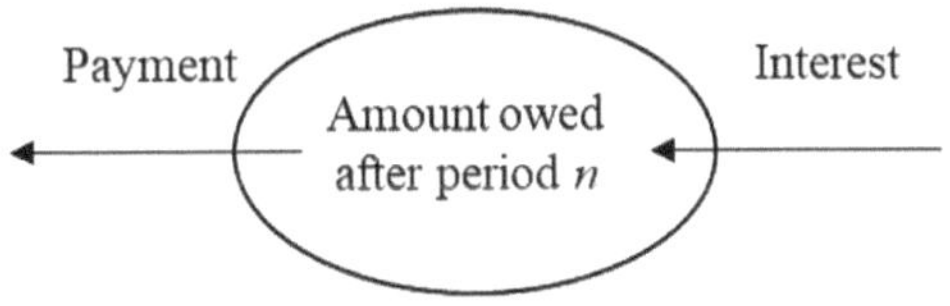

FIGURE 3.2
Flow diagram for financing a car.

to the unpaid balance. This increases your debt. The arrow pointing out of the circle is your monthly payment that decreases your debt. We define the following variables:

$$A_{n+1} = \text{the amount owed in the future}$$
$$A_n = \text{amount currently owed}$$

Change, as depicted in the flow diagram, is $i\,A_n - P$, so the model is

$$A_{n+1} = A_n + iA_n - P$$

where

i = monthly interest rate and
P = monthly payment

We model dynamical systems that have only **constant coefficients**. A dynamical system with constant coefficients may be written in the form

$$A_{n+3} = b_2A_{n+2} + b_1A_{n+1} + b_0A_n$$

where b_0, b_1, and b_2 are arbitrary constants.

Solutions to Discrete Dynamical Systems

Although some DDS have closed form analytical solutions, our emphasis in this chapter will be on iterative and graphical solutions that we will illustrate on Excel. We will show that DDS are relatively easy to model. The solutions may always be obtained by iterative and graphical methods. Not all DDS have closed form analytical solutions.

Let's begin with a simple DDS model with a closed form solution.

Theorem 3.1

The solution to a linear DDS $a(n + 1) = r\,a(n)$ for $r \neq 0$ is

$$a(k) = b^k a(0) \tag{3.1}$$

where $a(0)$ is the initial condition of the system at time period 0 and k is a generic time period.

Technology is an integral part of models with DDS. Every DDS has a numerical and graphical solution, which can easily be attained with technology.

Example 3.1: A Drug Dosage

Suppose that a doctor prescribes that their patient takes a pill containing 100 mg of a certain drug every hour. Assume that the drug is immediately ingested into the bloodstream once taken. Also, assume that every hour the patient's body eliminates 25% of the drug that is in his/her bloodstream. Suppose that the patient had 0 mg of the drug in his/her bloodstream prior to taking the first pill. How much of the drug will be in his/her bloodstream after 72 hours?

Problem Statement: Determine the relationship between the amount of drug in the bloodstream and time.

Assumptions: The system can be modeled by a DDS. The patient is of normal size and health. There are no other drugs being taken that will affect the prescribed drug. We assume that there are no internal or external factors that will affect the drug absorption rate. The patient always takes the prescribed dosage at the correct time. Figure 3.3 shows the change diagram.

Variables: Define $a(n)$ to be the amount of drug in the bloodstream after period n, $n = 0, 1, 2, \ldots$ *hours.*

Change Diagram:

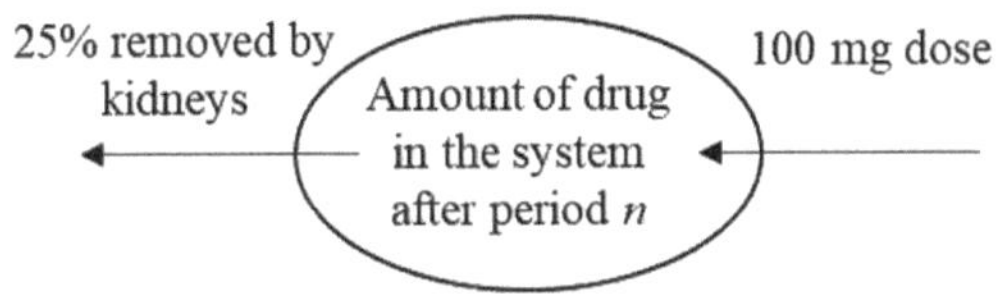

FIGURE 3.3
Change diagram for drugs in system.

Model Construction: Let's define the following variables:

a_{n+1} = amount of drug in the system in the future
a_n = amount currently in system

We define change as follows: change = dose – loss in system

$$\text{change} = 100 - 0.25a_n$$

so, *Future = Present + Change* is

$$a_{n+1} = a_n - 0.25a_n + 100$$

or

$$a_{n+1} = 0.75a_n + 100$$

We note that this is not in the form of Theorem 3.1 so we introduce Theorem 3.2.

Theorem 3.2

The solution to a linear DDS $a(n+1) = r\,a(n) + d$ for $r \neq 0$ and $d \neq 0$ is

$$a(k) = b^k C + (d/(1-r)) \tag{3.2}$$

where $d/(1 - r)$ is the equilibrium value of the system, C is the initial condition of the system at time period 0 and k is a generic time period that provides $a(0) = V$.

We can employ Theorem 3.2 and find

$$A(k) = 0.75kC + d/(1-r)$$

$$\mathrm{D} = 100$$

$$1-7 = 0.25, \text{ so } \mathrm{d}/(1-\mathrm{r}) = 400$$

Now,

$$\mathrm{A}(0) = 0$$

$$\text{So, } 0 = \mathrm{C}(0.75)^{\circ} + 400$$

$$\mathrm{C} = -400$$

The model, in general form, is

$$A(k) = 0.75^{k}(-400) + 400$$

Let's iterate the system numerically. If we let $a(n)$ (we say "a at n" or "a of n") be the number of milligrams of drug in the bloodstream after n hours, and the initial amount in the bloodstream is 0 mg, then

$$\begin{aligned}
&\mathrm{a}(0) = 0\\
&\mathrm{a}(1) = 0.75(0) + 100 = 100\\
&\mathrm{a}(2) = 0.75(100) + 100 = 175\\
&\mathrm{a}(3) = 0.75(175) + 100 = 231.25\,\mathrm{mg}
\end{aligned}$$

We could write these equations where we do not substitute the numerical values that we calculated:

$$\begin{aligned}
&\mathrm{a}(0) = 0\\
&\mathrm{a}(1) = 0.75\,\mathrm{a}(0) + 100\\
&\mathrm{a}(2) = 0.75\,\mathrm{a}(1) + 100\\
&\mathrm{a}(3) = 0.75\mathrm{a}(2) + 100
\end{aligned}$$

Here, we see that the amount of drug in the bloodstream is related to the amount of drug in the bloodstream after the previous time period. Specifically, the amount of drug in the bloodstream after any of the first three time periods is 0.75 times the amount of drug in the bloodstream after the previous time period plus an additional 100 mg that is injected every hour.

We see that a pattern has developed that describes the amount of drug in the bloodstream. We are now prepared to conjecture (make an educated guess) about the amount of drug in the bloodstream after *n* hours. Mathematically, we say that the amount of drug in the bloodstream after *n* hours is

$$a_{n+1} = 0.75a(n) + 100.$$

The relationship above describes the amount of drug in the bloodstream after *n* hours. With this, we can see the change that occurs every hour within this "system" (amount of drug in the bloodstream), and state of the system, after any hour, is dependent on the state of the system after the previous hour. This is a *DDS*.

Analyzing the DDS: We want to find the value of *a(72)*. First, we can apply *Theorems 3.1* and 3.2 as before.

$$a_e = 100/(0.25) = 400$$
$$a(k) = c(0.75)^k + 400$$

$$Since\, a(0) = 0$$
$$0 = c + 400$$
$$c = -400$$

The solution is $a(k) = -400\ (0.75)^k + 400$ as illustrated with Theorem 2.2 before.

$$\text{For } a(72) = -400(0.75)^{72} + 400 = 399.99$$

Interpretation of results: The DDS shows that the drug reaches a value where change stops and eventually the concentration in the bloodstream levels at 400 mg as shown in Figure 3.4. If 400 mg is both a safe and effective dosage level, then this dosage schedule is acceptable. We discuss this concept of change stopping (equilibrium) later in this chapter.

Example 3.2: Simple Mortgage

Five years ago your parents purchased a home by financing \$80,000 for 20 years, paying monthly payments of \$880.87 with a monthly interest of 1%. They have made 60 payments and wish to know what they actually owe on the house at this time. They can use this information to decide whether or not they should refinance their house at a lower interest rate for the next 15 or 20 years. The change in the amount owed each period increases by the amount of the interest and decreases by the amount of the payment. Figure 3.5 shows the flow diagram for Example 3.2.

Problem Identification: Build a model that relates the time with the amount owed on a mortgage for a home.

Assumptions: Initial interest was 12%. Payments are made on time each month.

Variables: Let b_n = amount owed on the home after n months

Flow Diagram:

n	a(n)
0	0
1	100
2	175
3	231.25
4	273.438
5	305.078
6	328.809
7	346.606
8	359.955
9	369.966
10	377.475
11	383.106
12	387.329
13	390.497
14	392.873
15	394.655
16	395.991
17	396.993
18	397.745
19	398.309
20	398.732
21	399.049
22	399.286
23	399.465
24	399.599
25	399.699
26	399.774
27	399.831

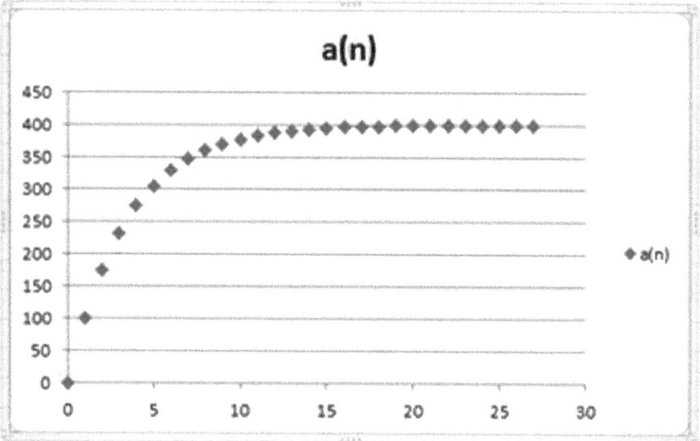

FIGURE 3.4
Behavior of drugs in our systems for Example 3.1.

Model Construction:

$$b_{n+1} = b_n + 0.12/12\, b_n - 880.87, b(0) = 80,000$$
$$b_{n+1} = 1.01\, b_n - 880.87, b(0) = 80,000$$

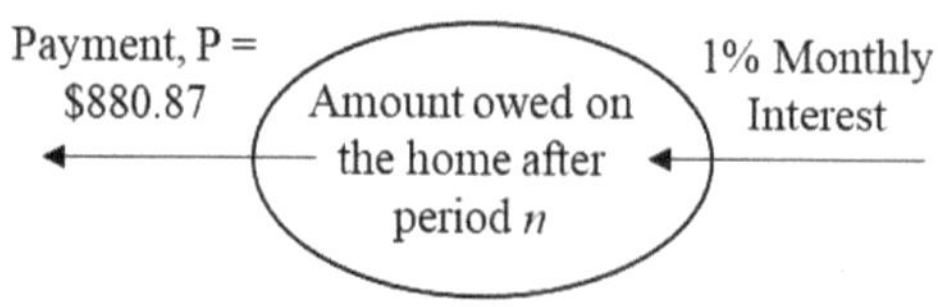

FIGURE 3.5
Flow diagram for Example 3.2.

Model Solution:

$$\text{Mortgage Owed}(n) = -8087(1.01)n + 88087.$$

We can iterate this over the entire 20 years (240 months) in Excel (Figure 3.6).

After paying for 60 months, your parents still owe \$73,395. 37 out of the original \$80,000.

n	b(n)		int
0	80000	37	76400.67
1	79919.13	38	76283.8
2	79837.45	39	76165.77
3	79754.96	40	76046.56
4	79671.64	41	75926.15
5	79587.48	42	75804.55
6	79502.49	43	75681.72
7	79416.64	44	75557.67
8	79329.94	45	75432.38
9	79242.37	46	75305.83
10	79153.92	47	75178.02
11	79064.59	48	75048.93
12	78974.37	49	74918.55
13	78883.24	50	74786.86
14	78791.2	51	74653.86
15	78698.24	52	74519.53
16	78604.36	53	74383.85
17	78509.53	54	74246.82
18	78413.76	55	74108.42
19	78317.02	56	73968.64
20	78219.32	57	73827.45
21	78120.65	58	73684.86
22	78020.98	59	73540.84
23	77920.32	60	73395.37
24	77818.66		
25	77715.97		
26	77612.26		
27	77507.51		
28	77401.72		
29	77294.87		
30	77186.95		
31	77077.95		
32	76967.85		
33	76856.66		
34	76744.36		
35	76630.93		
36	76516.37		

FIGURE 3.6
Screenshot of mortgage iterations (Example 3.2) in Excel.

In addition, they have paid in a total of $52,852.20 and only $6,605 went toward the principal payment of the home. The rest of the money went toward paying only the interest. If the family continues with this loan, then they will make 240 payments of $880.87 or $211,400.80 total in payments. This is $133,400.80 in interest. They have already paid $46,647.20 in interest. They would pay an additional $86,753.60 in interest over the next 15 years. What should they do? Perhaps an alternative scheme is available, such as refinance.

Equilibrium and Stability Values and Long-Term Behavior

We have previously mentioned Theorem 3.1 concerning equilibrium values to linear DDS. Let's further discuss these equilibriums.

Equilibrium Values

Let's go back to our original paradigm,

$$Future = Present + Change$$

When change stops, the change equals zero and future equals the present. The value for which this happens, if any, is the equilibrium value. This gives us a context for the concept of the equilibrium value.

Models of the form $a_{n+1} = r\, a_n + b$, where r and b are constants

Let's return to our drug dosage problem (Example 3.1) and consider adding the constant dosage each time period (time periods might be 4 hours). Our model is $\mathbf{a}_{n+1} = 0.5\, \mathbf{a}_n + 16$ mg. We will also assume that there is an initial dosage applied prior to beginning the regime. We will let these initial values be as follows (Figure 3.7).

Regardless of the starting value (Figure 3.7), the future terms of *a(n)* approach 32. Thus, 32 is the equilibrium value. We could have solved this algebraically as well.

$$\begin{aligned} a_{n+1} &= 0.5a_n + 16 \\ ev &= 0.5ev + 16 \\ 0.5ev &= 16 \\ ev &= 32 \end{aligned}$$

Another method of finding the equilibrium values involves solving the equation $a = ra + b$ and solving for a (where a is ev) we find:

$$A = \frac{b}{1-r}, \text{ if } r \neq 1.$$

Using this formula in our previous example, the equilibrium value is

$$a = 16/(1-0.5) = 32.$$

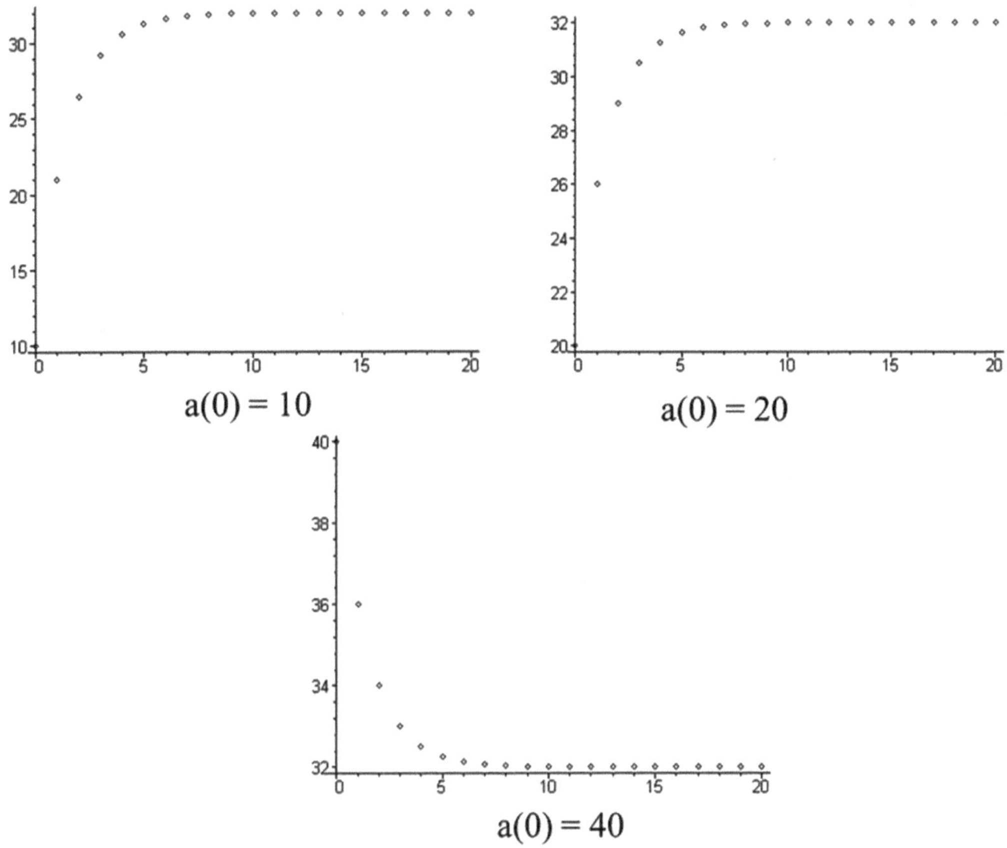

FIGURE 3.7
Plot of drugs in our systems with different initial starting conditions.

Stability and Long-Term Behavior

For a dynamical system, $a(n+1)$ with a specific initial condition, $a(0) = a_0$, we have shown that we can compute $a(1)$, $a(2)$, and so forth. Often these particular values are not as important as the long-term behavior. By long-term behavior, we refer to what will eventually happen to $a(n)$ for larger values of n. There are many types of long-term behavior that can occur with DDS, we will only discuss a few here.

If the $a(n)$ values for a DDS eventually get close to the equilibrium value, ev, no matter the initial condition, then the equilibrium value is called a **stable equilibrium value** or **an attracting fixed point**.

Often, we characterize the long-term behavior of the system in terms of its stability. If a DDS has an equilibrium value and if the DDS tends to the equilibrium value from starting values near the equilibrium value, then the DDS is said to be stable.

Thus, for the dynamical system $a(n+1) = r\,a(n) + b$, where $b \neq 0$:

TABLE 3.1

Stability for Linear Functions

Value of r	DDS Form	Equilibrium	Stability of Solution	Long-term Behavior
$r = 0$	$a(n+1) = b$	b	Stable	Stable equilibrium
$r = 1$	$a(n+1) = a(n) + b$	None	Unstable	
$r < 0$	$a(n+1) = r * a(n) + b$	$b/(1-r)$	Depends on $\|r\|$	Oscillations
$\|r\| < 1$	$a(n+1) = r * a(n) + b$	$b/(1-r)$	Stable	Approaches $b/(1-r)$
$\|r\| > 1$	$a(n+1) = r * a(n) + b$	$b/(1-r)$	Unstable	Unbounded

If $r \neq 1$, an equilibrium exists at $a = b/(1-r)$.
If $r = 1$, no equilibrium value exists.

Let's look at the relationship to analytical solutions.

If a DDS has an equilibrium value, *ev* value, then we can use the *ev* value to find the analytical solution.

Recall the mortgage example (Example 3.2),

$$B_{n+1} = 1.0054B_n - 639.34, B(0) = 73,395$$

The *ev* value is found as 118,031.9274.

The analytical solution may be found using the following form:

$$B(k) = (1.0054^k)C + D \text{ where D is the ev.}$$

$$B(k) = (1.0054^k)C + 118,031.9274, B(0) = 73,395$$

Since $B(0) = 73,395 = 1.0054^0\,(C) + 118,031.9274$

$$C = -44,636.92736$$

Thus,

$$B(k) = -44,636.92737(1.0054^k) + 118,031.9274$$

Let's assume we did not know the payment was 639.34 per month. We could use the analytical solutions to help find the payment.

$$B(k) = (1.0054^k)C + D$$

We build a system of two equations and two unknowns.

$$B(K) = (1.0054^2)C + D$$

$$B(0) = 73,395C + D$$

$$B(180) = 0 = 1.0054^{180}\,C + D$$

$$C = -44,638.70, D = 118,033.7$$
$$B(K) = -44,638.79(1.0054)^k + 118,033.7$$

where D represents the equilibrium value and we accepted some round-off error. From our model form:

$$B_{n+1} = 1.0054B_n - P, \text{ we can find P.}$$

Solving analytically for the equilibrium value,

$$X - 1.0054X = -P$$
$$X = P/0.0054$$
$$X \text{ is } 118,033.70 \text{ so}$$

$$118,033.70 = P/0.00541667$$
$$P = 639.34$$

Example 3.3: Growth of a Bacteria Population

We often model population growth by assuming that the change in population is directly proportional to the current size of the given population. This produces a simple, first-order DDS similar to those seen earlier. It might appear reasonable at first examination, but the long-term behavior of growth without bound is disturbing. Why would growth without bound of a yeast culture in a jar (or controlled space) be alarming?

There are certain factors that affect population growth. Things include resources (food, oxygen, space, etc.) These resources can support some maximum population. As this number is approached, the change (or growth rate) should decrease and the population should never exceed its resource supported amount.

Problem Identification: Predict the growth of yeast in a controlled environment as a function of the resources available and the current population.

Assumptions and Variables: We assume that the population size is best described by the weight of the biomass of the culture. We define $y(n)$ as the population size of the yeast culture after period n. There exists a maximum carrying capacity, M, that is sustainable by the resources available. The yeast culture is growing under the conditions established.

Model:

$$y_{n+1} = y_n + ky_n(M - y_n)$$

where

y_n = population size after period n
n = time period measured in hours
k = constant of proportionality
M = carrying capacity of our system

TABLE 3.2

Bacteria Growth in a Petri Dish

n	0	1	2	3	4	5	6	7	8	9
y	10.3	17.2	27	45.3	80.2	125.3	176.2	256.6	330.8	390.4
n	10	11	12	13	14	15	16	17	18	19
y	440	520.4	560.4	600.5	610.8	614.5	618.3	619.5	620	621

TABLE 3.3

Bacteria Growth in a Petri Dish Solution Values

n	0	1	2	3	4	5	6	7	8
$y(n)$	10.3	17.2	27	45.3	80.2	125.3	176.2	256.6	330.8
$\Delta y(n)$	6.9	9.8	18.3	34.9	45.1	50.9	79.4	74.2	59.6
n	9	10	11	12	13	14	15	16	17
$y(n)$	390.4	440	520.4	560.4	600.5	614.5	618.3	619.5	620
$\Delta y(n)$	46.4	79.4	40	40.1	14	3.9	1.2	0.5	1

We have data shown in Table 3.2 for the growth of bacteria in a Petri dish. The variable, $y(n)$, is the number of bacteria at the end of period n.

It is often convenient to think about the way the variables change between time periods. we compute $\Delta y(n) = y(n + 1) - y(n)$. The values are provided in Table 3.3 and used to find the proportionality constant.

We find the constant slope is 0.0008. We also find the *ev* of 621.

In our experiment, we first plot y(n) versus n and find a stable equilibrium value of approximately 621. Next, we plot $y_{n+1} - y_n$ versus $y(n)$ $(621 - y_n)$ to find the slope, k, is approximately 0.0008, with $k = 0.0008$ and the carrying capacity in biomass is 621. This model is

$$y_{n+1} = y_n + 0.0008 y_n (621 - y_n)$$

Again, this is nonlinear because of the $y^2(n)$ term. There is no closed form analytical solution for this equation, however, we may obtain a solution through iteration and graphing the iterated values in Excel (Figure 3.8).

The model shows stability in that the population (biomass) of the yeast culture approaches 621 as n gets large. Thus, the population is eventually stable at approximately 621 units as shown in Figure 3.8.

Example 3.4: Spread of a Contagious Disease

There are 5,000 students in college dormitories, and some students have been diagnosed with COVID-19, a highly contagious disease. The health center wants to build a model to determine how fast the disease will spread.

Problem Identification: Predict the number of students affected with COVID-19 as a function of time.

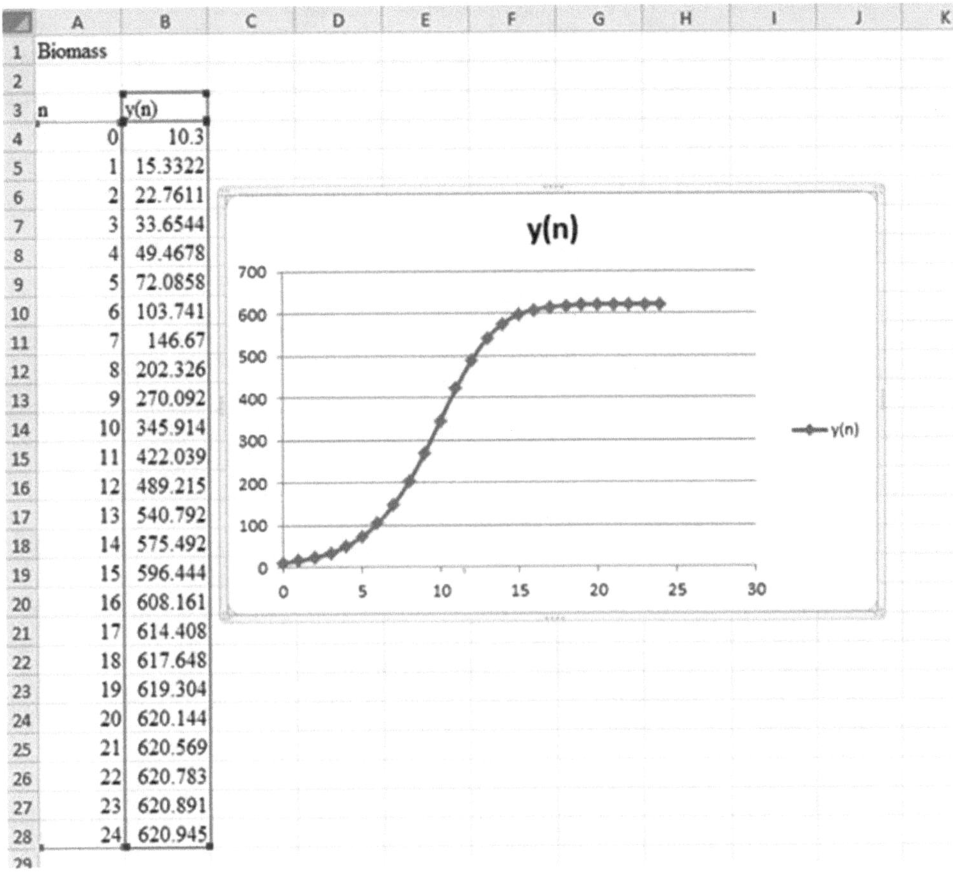

Biomass

n	y(n)
0	10.3
1	15.3322
2	22.7611
3	33.6544
4	49.4678
5	72.0858
6	103.741
7	146.67
8	202.326
9	270.092
10	345.914
11	422.039
12	489.215
13	540.792
14	575.492
15	596.444
16	608.161
17	614.408
18	617.648
19	619.304
20	620.144
21	620.569
22	620.783
23	620.891
24	620.945

FIGURE 3.8
Screenshot of DDS for bacteria growth in a petri dish (Example 3.3).

Assumptions and Variables: Let *m(n)* be the number of students affected with COVID-19 after n days. We assume all students are susceptible to the disease. The possible interactions of infected and susceptible students are proportional to their product (as an interaction term).

The model is,

$$m_{n+1} - m_n = k\,m_n(5{,}000 - m_n)\text{or}$$
$$m_{n+1} = m_n + k\,m(n)(5{,}000 - m_n)$$

Two students returned from spring break with COVID-19, so m*(0)* = 2. The rate of spreading per day is characterized by *k* = *0.00090*. It is assumed that there is no vaccine that can be introduced to slow the spread (Figure 3.9).

Interpretation: The results show that most students will be affected within two weeks. Since only about 10% will be affected within one week, every effort must be made to get the vaccination at the school and get the students vaccinated within one week. This is illustrated graphically in Figure 3.9.

n	M	rate =	0.00009
0	2		
1	2.89964		
2	4.203721		
3	6.093805		
4	8.832676		
5	12.80036		
6	18.54577		
7	26.86042		
8	38.88267		
9	56.2438		
10	81.26881		
11	117.2454		
12	168.7686		
13	242.151		
14	345.8416		
15	490.7058		
16	689.8521		
17	957.4549		
18	1305.805		
19	1739.956		
20	2250.466		
21	2807.361		
22	3361.359		
23	3857.084		
24	4253.833		
25	4539.5		
26	4727.639		
27	4843.525		
28	4911.735		
29	4950.753		
30	4972.696		
31	4984.916		
32	4991.683		
33	4995.42		
34	4997.479		
35	4998.613		
36	4999.237		
37	4999.58		
38	4999.769		

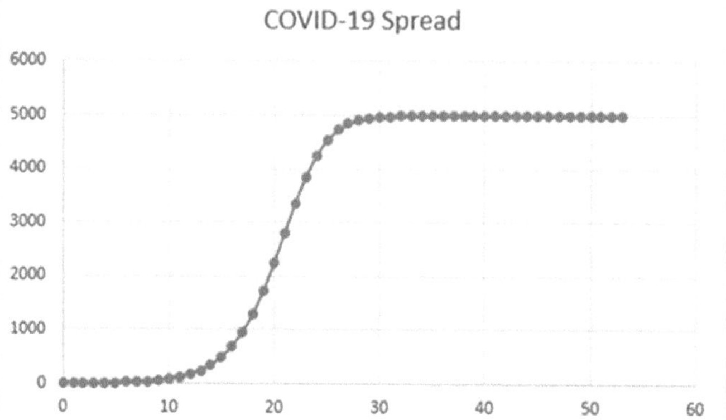

FIGURE 3.9
Screenshot of plot and iteration for the spread of meningitis.

Using Python for a Drug Problem

Example 3.5: Revisiting Drug Dosage

Consider the prescribed drug problem where the body eliminates 50% of the drug every 8 hours and every 8 hours we take 16 mg of the prescribed drug. Let $D(n)$ = amount of the drug in our system after n periods., where n = 0,1,2,.. measured in 8-hour periods. Assume that $D(0) = 0$.

The following is the Python Code:

```
# Program to print the drug DDS up to n_terms

# Recursive function
def recursive_drug(n):

if n <= 1:
      return n

else:
      return(recursive_drug(n - 1) - 0.5* recursive_drug(n - 1) + 16)

n_terms = 20

# check if the number of terms is valid

if n_terms <= 0:
  print("Invalid input ! Please input a positive value")

else:
  print("Drug over time")

for i in range(n_terms):
    print(recursive_drug(i))

import pylab as pl

for i in range(n_terms):

 y1 = [0,1,2,3,4,5,6,7,8,9,10,11,12,13,14,15,16,17,18,19,20]
 y2 = [0,1,16.5,24.25,30.06,31.03,31.52,31.76,31.88, 31.94,
      31.97,31.98, 31.99, 32,32,32,32,32,32,32,32]

 x1 = range(len(y1))
 x2 = range(len(y1))

pl.plot(x2, y2,'bo')
pl.show()
```

```
Output

Drug over time
0
1
16.5
24.25
28.125
30.0625
31.03125
31.515625
31.7578125
31.87890625
31.939453125
31.9697265625
31.98486328125
31.992431640625
31.9962158203125
31.99810791015625
31.999053955078125
```

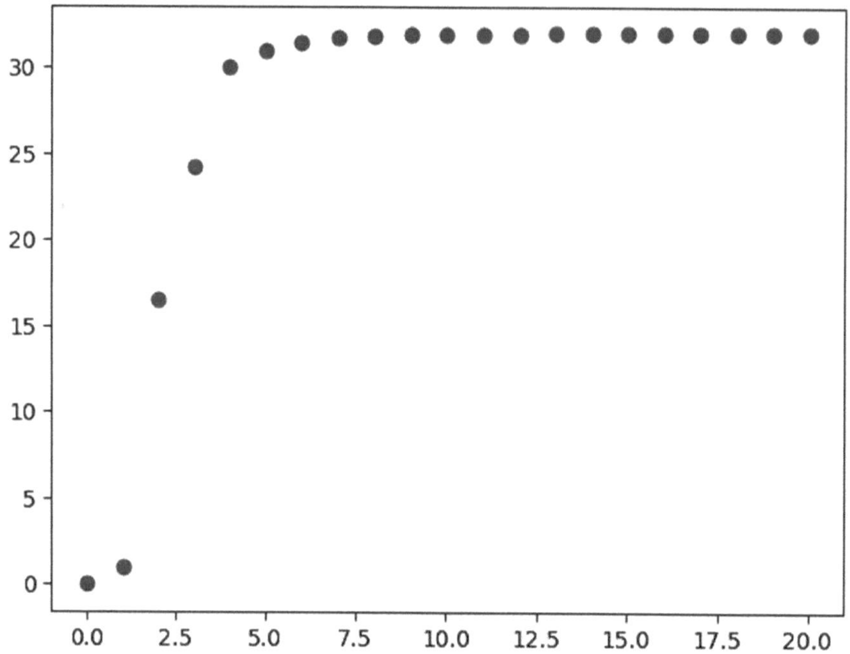

FIGURE 3.10
Python plot of our DDS for Example 3.5.

```
31.999526977539062
31.99976348876953
31.999881744384766
```

The Python plot is given in Figure 3.10.

Technology Examples for Discrete Dynamical Systems

Using DDS is an interesting and productive approach. The use of a computer technology is essential to the methods described in this chapter. It allows for interactions between instructors and students. It provides a means to use technology in a nonstandard way. It provides another way to educate students concerning squares and cubes using discrete mathematics. In this section, we present technology and examples of solving linear and nonlinear DDS.

Excel for Linear and Nonlinear DDS

Example 3.6: Generic DDS

Let's consider a DDS such as $a_{n+1} = 0.5a_n$, with $a(0) = 100$.
The steps to iterate and graph in Excel are as follows:

Step 1. Open a new worksheet and name as DDS or some appropriate name.
Step 2. Label the following columns as n and a_n in cell $a1$ and $b1$.

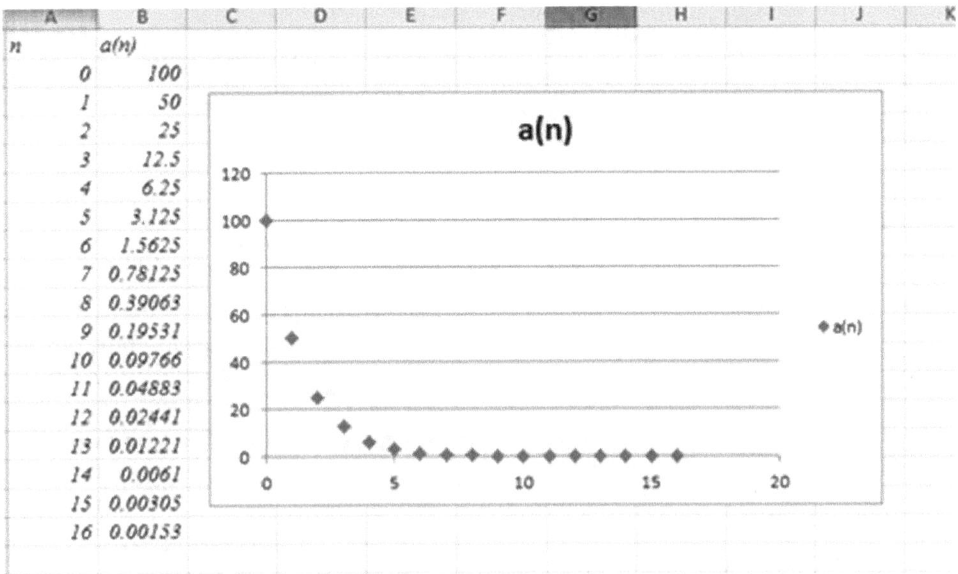

FIGURE 3.11
Screenshot of Excel's solution for Example 3.6.

Step 3. In cells *a2* and *b2* input the initial condition by putting in 0 in cell *a2* and 100 in cell *b2*.
Step 4. In cell *a3* type = *1 + cell a2*
Step 5. In cell *b3* type = *0.5*cell b2*
Step 6. Highlight cells *a3* and *b3* and drag the cursor down to fill in cells as far as desired or needed, in this case to about a4:b16
Step 7. Highlight cells a1:b16, INSERT scatterplot to obtain the graph
Step 8. Interpret the results.

A screenshot of the model and results is provided in Figure 3.11, where we see that our DDS tends to zero over time.

In Figure 3.12, we show the appropriate formulas used to solve for the level of Drugs in the system for Example 3.6.

Maple for Linear and Nonlinear DDS

In Maple, DDS are referred to as recursion equations. One might obtain closed form solutions if they exist. One might also iterate and graph the behavior of the recursion equation. We will use both commands and libraries from Maple such as *with(plots)* and we will add some commands to our Maple toolbox, **rsolve** and **seq**.

The following is the generic Maple code

```
rsolve - recurrence equation solver

Calling Sequence
     rsolve(eqns, fcns)
     rsolve(eqns, fcns, 'genfunc'(z))
     rsolve(eqns, fcns, 'makeproc')
```

n	a(n)
0	100
=1+A2	=0.5*B2
=1+A3	=0.5*B3
=1+A4	=0.5*B4
=1+A5	=0.5*B5
=1+A6	=0.5*B6
=1+A7	=0.5*B7
=1+A8	=0.5*B8
=1+A9	=0.5*B9
=1+A10	=0.5*B10
=1+A11	=0.5*B11
=1+A12	=0.5*B12
=1+A13	=0.5*B13
=1+A14	=0.5*B14
=1+A15	=0.5*B15
=1+A16	=0.5*B16
=1+A17	=0.5*B17

FIGURE 3.12
Screenshot of Excel formulas used in Example 3.6.

```
Parameters
     eqns - single equation or a set of equations
     fcns - function name or set of function names
     z    - name, the generating function variable
```

```
seq - create a sequence

Calling Sequence
     seq(f, i = m..n)
     seq(f, i = x)

Parameters

     f       - any expression
     i       - name
     m, n    - numerical values
     x       - expression
```

Many of the models that we will solve have closed form solutions so that we can use the command **rsolve** to obtain the closed solution, and then we can use the sequence command (**seq**) to obtain the numerical values in the solution. Many dynamical systems do not have closed form analytical solutions so we cannot use the rsolve and seq commands to obtain solutions. When this occurs, we will write a small program using PROC to obtain the numerical solutions. To plot the solution to the dynamical systems, we will use plot commands to plot the sequential data pairs. We will illustrate all these commands in our example.

Example 3.7: Revisiting Drug Levels from Example 3.1

Solve the DDS an + 1 = 0.75an + 100, a(0) = 0.

Determine the value of *a(72)*.

Let's illustrate the iterative technique for analyzing a DDS in Maple. Figure 3.13 shows the graphical representation of the solution.

We use the following Maple commands for Example 3.1:

```
Drug:= rsolve({a(n+1)=0.75*a(n)+100,a(0)=0}, a(n));
```

$$\text{drug} := 400 - 400\left(\frac{3}{4}\right)^n$$

```
L:= limit(drug, n=infinity);
```

```
L:= 400
```

```
with(plots):
pointplot({seq([I, -400*(3/4)^i+400], i=0..48)};
```

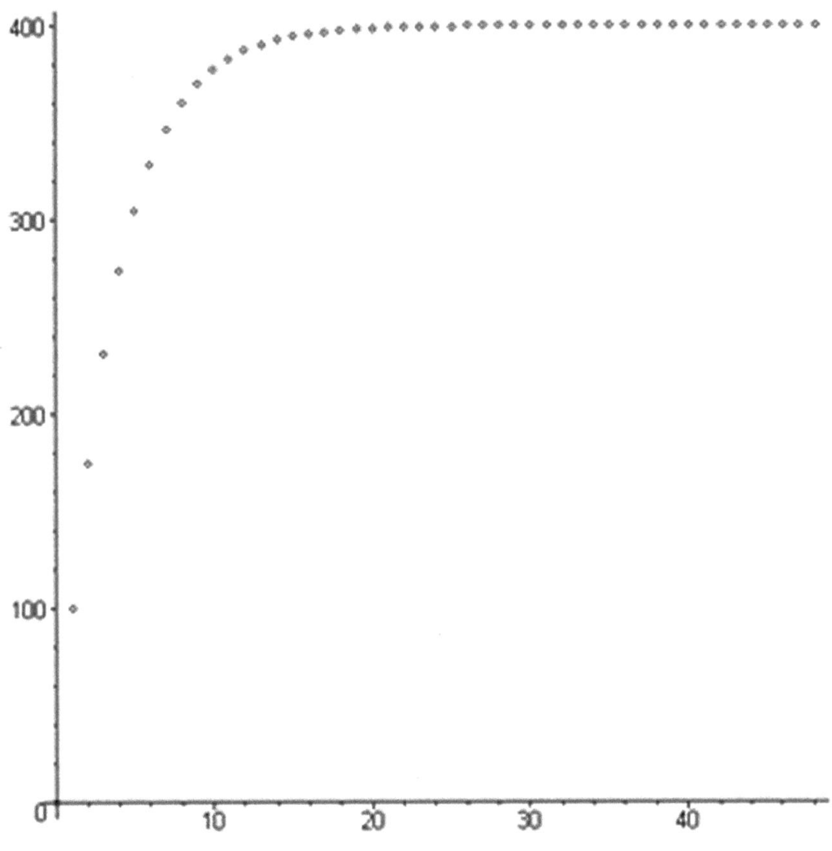

FIGURE 3.13
Behavior of drugs in our systems for Example 3.7.

```
drug_Table:=seq(-400.0*(0.75)^i+400., i=0.48);

      drug_table:= 0., 100.00, 175.00, 231.25, 273.44, 305.08,
      328.81, 346.61, 359.95, 369.97, 377.47, 383.11, 387.33,
      390.50, 392.87, 394.65, 395.99, 396.99, 397.75, 398.31,
      398.73, 399.05, 399.29, …, 399.99
```

Interpretation of Results: The DDS shows that the drug reaches a value where change stops and eventually the concentration in the bloodstream levels at 400 mg. If 400 mg is both a safe and effective dosage level, then this dosage schedule is acceptable.

R for Linear and Nonlinear DDS

Example 3.8: Population Dynamics Using R.

Given the DDS, $N_{t+1} = -\lambda N_t$, where λ is $(1 + r)$

We open R studio and we are going to use the **For Loop to address this problem**. We type the following commands:

```
> generations <- 10

> N <- numeric(generations)

> lambda <- 2.1

> N [1] <- 3

> for (t in 1: (generations-1)) { N [t+1] <- lambda * N [t]}

> N
 [1]     3.0000     6.3000    13.2300    27.7830    58.3443 122.5230
257.2984
 [8]  540.3266 1134.6858 2382.8401


> plot(0:(generations-1),N, type="o",xlab="Time", ylab="Pop Size")
```

We see unbounded growth in the plot shown in Figure 3.14.

Example 3.9: Repeat Example 3.4 where we solved the DDS $a_{n+1} = 0.75a_n + 100$, $a(0) = 0$ but now we will solve it using R.

We type the following R commands:

```
> gener<-20
> D<-numeric(gener)
> lam<-0.75
> D[1]<-0
> for (t in 1:(gener-1)) {D[t+1]<-lam*D[t]+100}

> D
 [1]   0.0000 100.0000 175.0000 231.2500 273.4375 305.0781 328.8086
346.6064
 [9] 359.9548 369.9661 377.4746 383.1059 387.3295 390.4971 392.8728
394.6546
```

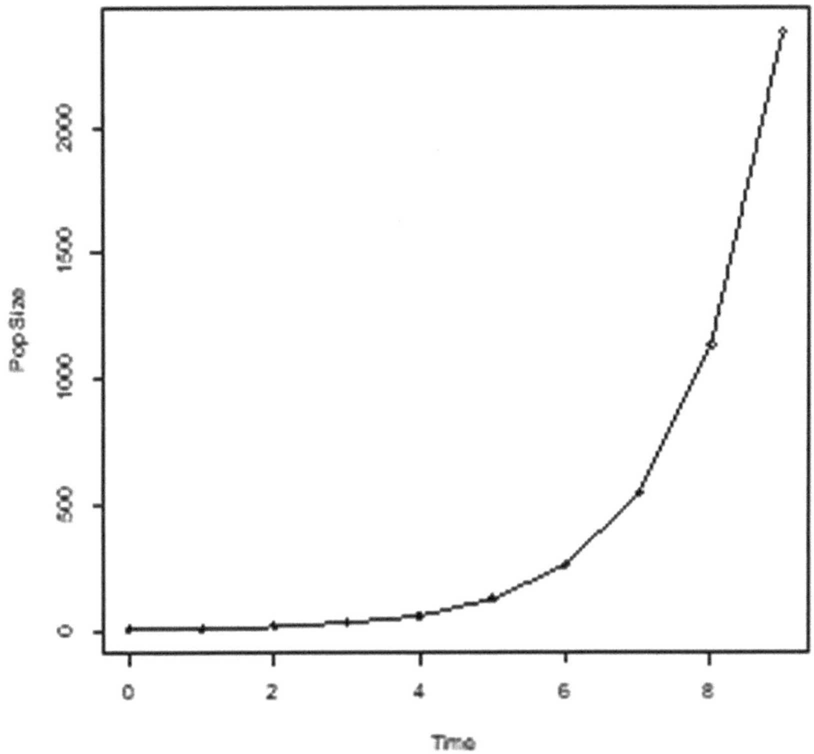

FIGURE 3.14
Screenshot from R for Example 3.8.

```
[17] 395.9910 396.9932 397.7449 398.3087

> plot(0:(gener-1),D,type="o",xlab="Time",ylab="Drug_in_Sys")
```

In R, we can see that the drug becomes stable at approximately 400 units as shown in Figure 3.15.

We now present a drug dosage model analytically.

Example 3.10: Logistics Growth

Let's modify this model to a nonlinear model using a logistics growth DDS in R.

Given the DDS, $N_{t+1} = N_t + r\lambda N_t\left(1 - N_t/K\right)$, we have the following R script using the **function** command.

```
DDSL <- function(K,r,N0,generations)
+ {N <- c(N0,numeric(generations-1))
+ for ( t in 1:(generations-1)) N [t+1] <- {N [t] + r*N [t]*
(1- (N [t]/K))}
+ return(N )}
> Output <-DDSL(K=1000,r=1.5,N0=10,generations=30)
> generations <-30
> plot(0:(generations-1),Output, type='o', xlab="time",ylab="Populat
ion")
```

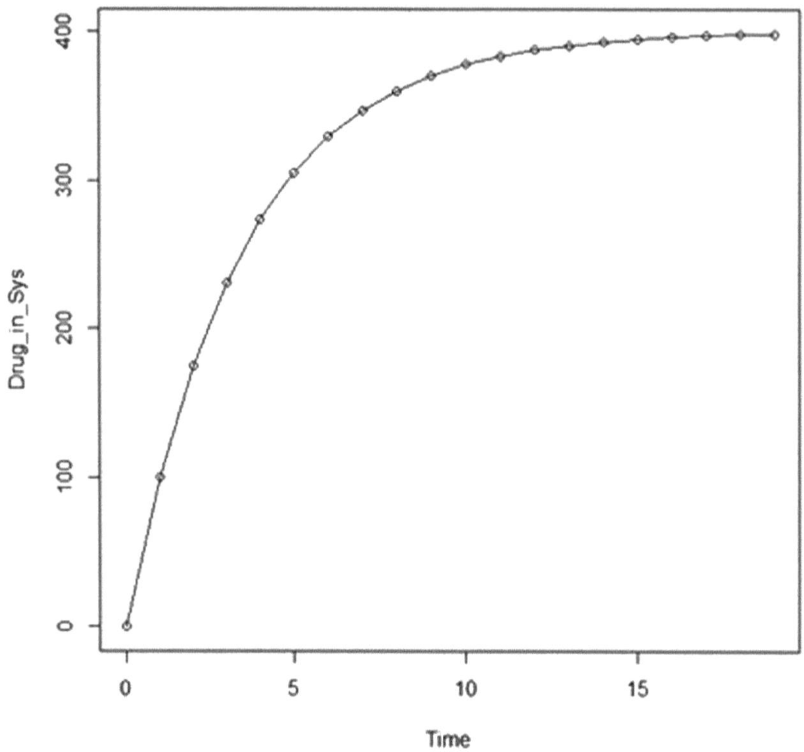

FIGURE 3.15
Example 3.9 R solution graph for DDS $a_{n+1} = 0.75a_n + 100$, $a(0) = 0$.

The plot in Figure 3.16 shows the exponential growth at the beginning and then levels off to the carrying capacity, K = 1000. We can now use the following R code to solve the problem.

```
plot.new()

plot.window(xlim=range(x1),ylim=range(y1))

lines(x1,y1)
axis(1); axis(2); box()

plot.window(xlim=range(x2),ylim=range(y2))

lines(x2,y2)

axis(1); axis(2); box()

## Using the `deSolve` package
library(deSolve)

## Time
t <- seq(0, 100, 1)

## Initial population
N0 <- 10
```

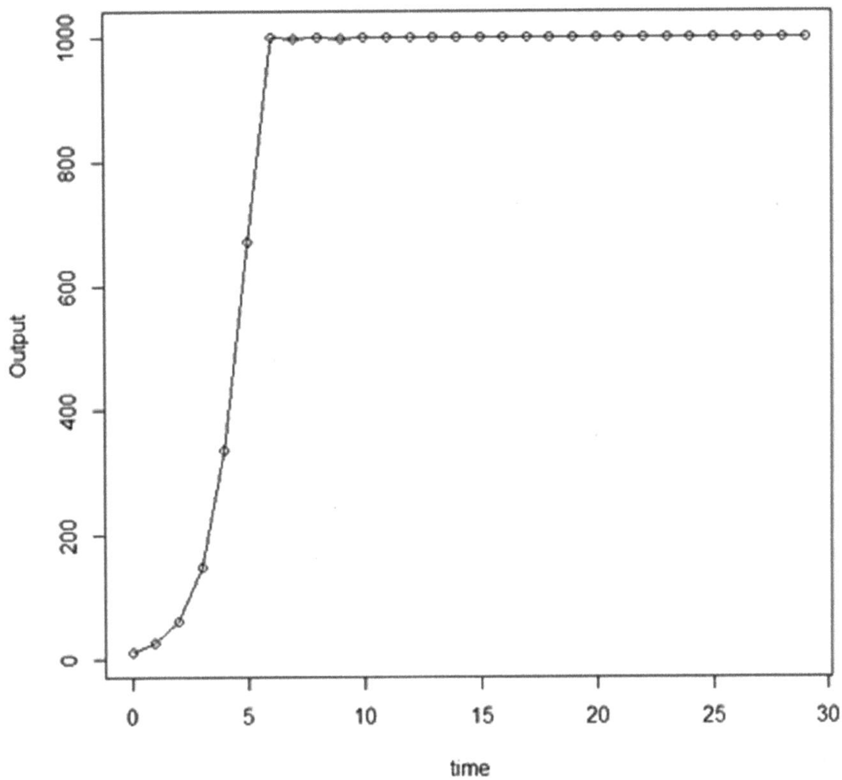

FIGURE 3.16
Example 3.10 R solution graph for DDS $N_{t+1} = N_t + r\,\lambda\, N_t\,(1 - N_t/K)$.

```
## Parameter values
params <- list(r=0.1, K=1000)

## The logistic equation
fn <- function(t, N, params) with(params, list(r * N * (1 - N / K)))

## Solving and plotting the solution numerically
out <- ode(N0, t, fn, params)
plot(out, lwd=2, main="Logistic equation\nr=0.1, K=1000, N0=10")

## Plotting the analytical solution
with(params, lines(t, K * N0 * exp(r * t) / (K + N0 * (exp(r * t) -
1)), col=2, lwd=2))
```

Exercises

3.1 Find the equilibrium value, if any exists, for each of the following dynamical systems,

a. $u_n = 0.5u_{n-1} + 3$
b. $u_n = 5u_{n-1} + 3$
c. $u_n = 0.75u_{n-1} + 16$
d. $u_n = 0.5u_{n-1}$
e. $u_n = -0.5u_{n-1} + 8$

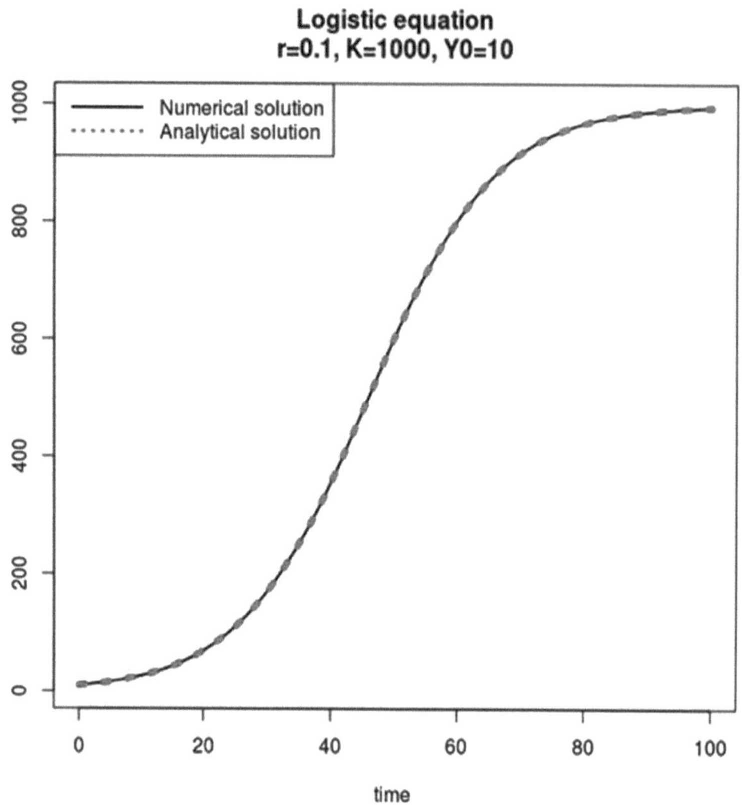

FIGURE 3.17
Example 3.10 R solution graph for DDS $N_{t+1} = N_t + r\,\lambda N_t\,(1 - N_t/K)$.

3.2 Suppose that the equilibrium value for the following dynamical system is 5, $u(n) = 1.3\,u(n-1) + b$. Find b.

3.3 Suppose Susy burns 150 kcal per week and consumes 16500 kcal per week (where 3600 is equivalent to 1 pound). Build a dynamical system and use it to find her equilibrium weight. If Susy weighs 105 lb currently, how many weeks until she is within 1 pound of her equilibrium weight?

3.4 Suppose a patient takes 40 mg of a prescribed drug per day. Through blood tests it is shown the drug has stabilized at 140 mg. What percentage of the drug is being eliminated daily for this to occur?

3.5 Suppose a patient eliminates 40% of a particular drug every 12 hours. The patient takes a mg pill every 12 hours each day. How many mg should the pill be for the drug to stabilize at 45 mg?

3.6 Consider the model $a(n+1) = r\,a(n)\,(1 - a(n))$. Let $a(0) = 0.2$. Determine the numerical and graphical solution for the following values of r. Find the pattern in the solution.

$r = 2$
$r = 3$
$r = 3.6$
$r = 3.7$

For problems 3.7–3.10 find the equilibrium value by iteration and determine if it is stable or unstable.

3.7 $a_{n+1} = 1.7a_n - 0.14a_n^2$

3.8 $a_{n+1} = 0.8a_n + 0.1a_n^2$

3.9 $a_{n+1} = 0.2a_n - 0.2a_n^3$

3.10 $a_{n+1} = 0.1a_n^2 + 0.9a_n - 0.2$

3.11 Consider spreading a rumor through a company of 1,000 employees all working in the same building. We assume that the spread of a rumor is similar to the spread of a contagious disease in that the number of people hearing the rumor each day is proportional to the product of the number hearing the rumor and the number who have not heard the rumor. This is given by the formula:

$$r_{n+1} = r_n + 1000kr_n - kr_n^2,$$

where k is the parameter that depends on how fast the rumor spreads. Assume k = 0.001 and further assume that four people initially know the rumor. How soon will everyone know the rumor?

Iterate and graph the following DDS. Explain their long-term behavior. Try to find realistic scenarios that these DDS might explain.

3.12 $a_{n+1} = 0.5a_n + 0.1$, $a(0) = 0.1$, $a(0) = 0.2$, $a(0) = 0.3$

3.13 $a_{n+1} = 1.01a_n - 1000$, $a(0) = 90{,}000$, $a(0) = 100{,}000$, $a(0) = 110{,}000$

3.14 $a_{n+1} = -4a_n - 50$, $a(0) = 9$

3.15 $a_{n+1} = 0.1a_n + 1{,}000$, $a(0) = 9$

3.16 $a_{n+1} = 0.9987a_n + 0.335$, $a(0) = 72$

3.17 At the beginning of every 24-hour period Oliver takes a 1 mg pill of digoxin that is immediately absorbed into the blood. Every 24 hours the kidneys filter out 30% of the digoxin in the blood. We want to develop a DDS that helps us to model the amount of digoxin in the blood on any given day (24-hour period).

a. Define your variables in words.
b. Make a flow diagram for digoxin in the blood.
c. Write a DDS that models the amount of digoxin in the blood. Also, state the initial condition.
d. How much digoxin is in the blood after 10 days?

3.18 Smoking a pack of cigarettes each day results in absorbing 2.7 mg of cadmium each year from smoking. Assume that 7% of the cadmium is eliminated from the body each year. We want to develop a DDS to model the amount of cadmium in the body after a number of years.

a. Define your variables in words.
b. Make a flow diagram for the amount of cadmium in the body.
c. Write a DDS that models the amount of cadmium in the body.
d. Find the amount of cadmium in this person's body as a result of smoking a pack a day for 30 years.

3.19 Working full-time but fed up with gas prices and excited about their bad press, you decide to buy a Toyota Prius. You get a gift from graduation for a significant down payment. You check with your new bank, and you can get a loan for $16,785 at 4.55% compounded monthly for five years. They tell you that your monthly payments will be $313.31.

a. Define your variables in words.
b. Make a flow diagram for the amount owed to the bank.
c. Write a DDS that models the amount owed to the bank.
d. What will your final monthly payment be? (Month 60)
e. Find the equilibriums for the following DDS.

3.20 $a_{n+1} = 0.3a_n - 10$

3.21 $a_{n+1} = 0.5a_n$

3.22 $a_{n+1} = -0.1a_n + 11$

PROJECTS

3.1 **Sewage Plant**. A sewage-treatment plant processes raw sewage to produce usable fertilizer and clean water by removing all contaminants. The process is such that each hour 12% of the remaining contaminants in the process tank are removed. What percentage is left after one day? How long would it take to lower the amount of sewage by about half? How long until the level of sewage is down to 10% of its original level?

3.2 **Cooling a Hot Soda**. A can of soda has been sitting out at room temperature (~72°) for several days. If that same soda is placed in a refrigerator that is set at 40°, how long will it take the soda to reach 50°? 45°? 40°? Assume a cooling rate of 0.0081 degrees per minute (obtained experimentally).

3.3 You are also a bit worried about the comfort of a smaller car and its ride. You are in luck: *Consumer Reports* did a study and compared the rides of four small cars in terms of their suspension systems. Consider the following alternative models for suspension systems available for a small car. Analyze each alternative, describe the behavior of each suspension system over time, and choose an alternative suspension system. Each car sits at 0.2 initially. It is determined that under 0.6 maintains a smooth ride.
Alternative 1: $S(n + 1) = 2\, S(n)\, (1 - S(n))$
Alternative 2: $S(n + 1) = 3\, S(n)\, (1 - S(n))$
Alternative 3: $S(n + 1) = 3.6\, S(n)\, (1 - S(n))$
Alternative 4: $S(n + 1) = 3.7\, S(n)\, (1 - S(n))$

3.4 Consider the contagious disease as the Ebola virus. Look up on the internet and find out some information about how deadly this virus actually is. Now consider an animal research laboratory in Reston, VA, and a suburb of Washington, DC, with a population of 856,900 people. A monkey with the Ebola virus has escaped its captivity and infected one employee (unknown at the time) during its escape. This employee reports to the University hospital later with Ebola symptoms. The Infectious Disease Center (IDC) in Atlanta gets a call and begins to model the spread of the disease. Build a model for the IDC with the following growth rates to determine the number infected after two weeks:

a. $k = 0.00025$
b. $k = 0.000025$
c. $k = 0.00005$
d. $k = 0.000009$
e. List some ways of controlling the spread of the virus.

3.5 Consider the spread of a rumor concerning termination among 1000 employees of a major company. Assume that the spreading of a rumor is similar to the spread of a contagious disease in that the number hearing the rumor each day is proportional to the product of those who have heard the rumor and those who have not heard the rumor. Build a model for the company with the following rumor growth rates to determine the number having heard the rumor after one week:

a. $k = 0.25$
b. $k = 0.025$
c. $k = 0.0025$
d. $k = 0.00025$
e. List some ways of controlling the spread of the rumor.

3.6 You are a new college graduate on your way back from graduation practice. In your mind, you could not be happier. You are thinking about your plans after graduation. You are going to start a new job in two weeks that will allow you to work full-time during the summer and part-time when college starts in the fall. This job will allow you to save $260 a month. You are considering buying a car with this money. Your mind is a thousand miles away when all of a sudden, you see something in the road ahead of you. Several deer have run out in front of your car. You try everything you can to avoid them, but you still hit one. At this point, you are not too upset, because you know that you have insurance on your car and you assume that everything will be taken care of with no problems. However when you begin talking to your insurance agent, you run into some problems, and things are not as simple as you thought they would be.

You plan on having your car repaired at the local body shop. The owner informs you that it will take two to three weeks to repair your car. You are not able to drive the car, so you have to get a rental car. Further, when the insurance adjuster comes out to talk to you, he informs you that your policy does not cover a rental car and that you have a $300 deductible on your collision coverage. Therefore, you are going to have to pay for the rental car and the deductible. Suddenly you realize you have a problem: you don't have the money to pay for these expenses. Because you are only 17, you cannot borrow money from a financial institution. A friend tells you that he will loan you the money at 10% interest, compounded monthly. As a result, you gather the following information:

- You will need a rental car for 20 days at $25.50 per day plus taxes (8.25%).
- You need to pay the $300 deductible.
    a. How much money do you need to borrow?
    b. What will your monthly payments be if you repay your friend in three monthly payments or four monthly payments? (Set up a DDS for the amount that you owe your friend and figure out the amount for each payment.)
    c. How much interest will you pay if you make three payments? Or four payments?
    d. Is there a difference in the amount of interest that you will pay? Will it be easy to repay the loan in three payments? Or four payments? Which is best and why? (Think about your income and the amount of each payment. Consider the amount of interest you will pay.)

Three or four months later, you are planning on buying a used car with the money you will save from your job. Between the body shop owner and the insurance settlement, you will receive $5,850 for your wrecked car.

The used car you want to purchase will cost a total of $11,000. Your friend agrees to loan you $5,150 for 3 years at 10% annual interest, compounded monthly. He says your payments should be $175 each month.

a. Is this figure $175 a month correct? (Use a DDS to model your new loan and figure the amount of your monthly payment.)
b. What should your monthly payment be? Using iteration what should your last payment be?
c. How much interest will you pay?
d. If you decide to make payments of $260 per month, how many payments should you make? How much will the last payment be?
e. How much interest will you pay?
f. Is there much difference in the amount of interest that you will pay at $175 per month and $260 per month? Which plan is best for you and why?

References and Additional Readings

Albright, B. (2010). *Mathematical Modeling with Excel*. Sudberry, MA: Jones and Bartlett.

Allman, E., and Rhodes, J. (2004). *Mathematical Models in Biology: An Introduction*. Cambridge, UK: Cambridge University Press.

Arney, D., Giordano, F., and Robertson, J. (2002). *Mathematical Modeling with Discrete Dynamical Systems*. Boston, MA: McGraw Hill.

Fox, W. (2010). Discrete Combat Models: Investigating the Solutions to Discrete Forms of Lanchester's Combat Models. *International Journal of Operations Research and Information Systems, (IJORIS). 1*(1): 16–34.

Fox, W. (2012). Mathematical Modeling with Maple. Boston, MA: Cengage Publishing.

Fox, W.P., and Driscoll, P. (2011). Modeling with Dynamical Systems for Decision Making and Analysis. *Computers in Education Journal (COED)*. **2**(1): 19–25.

Fox, W. (2012). Discrete Combat Models: Investigating the Solutions to Discrete Forms of Lanchester's Combat Models. *Innovations in Information Systems for Business Functionality and Operations Management*, IGI Global & SAGE Publishers, pp. 106–122.

Fox, W. (2012). Mathematical Modeling of the Analytical Hierarchy Process Using Discrete Dynamical Systems in Decision Analysis. *Computers in Education Journal*. 3(3): 27–34.

Giordano. F., Fox, W., and Horton, S. (2014). *A First Course in Mathematical Modeling*, 5th edition. Boston, MA: Cengage Publishing, Boston.

Leyendekker, J., and Shannon, A. (2015). The Odd-Number Sequence: Squares and Sums. *International Journal of Mathematical Education in Science and Technology*. 46(8):1222–1228.

Sandefur, J. (1990). *Discrete Dynamical System: Theory and applications*. New York, NY: Oxford University Press.

Sandefur, J. (2002). *Elementary Mathematical Model: A Dynamic Approach*, 1st edition. Belmont, CA: Brooks-Cole Publisher.

Sandefur, J. (2003). *Elementary Mathematical Modeling: A dynamic approach*. Belmont, CA: Thomson Publishing.

4

Systems of First Order Difference Equations

In this chapter, we shall consider systems of equations where each variable has a time index t = 0; 1; 2; ... and variables of different time periods are connected in some manner. These systems are called systems of difference equations and are useful to describe dynamical systems with discrete time.

Modeling Systems of First Order Difference Equations

Suppose two neighboring countries are engaged in a competitive build-up of military armaments, like tanks. During each year, Country A builds 0.3 new tanks for every tank that Country B had the previous year. Meanwhile, Country B builds 0.1 new tanks for every tank that Country A had the previous year. Each year Country A loses 10% of their tank force from the year before due to training accidents, while Country B loses 5% of their tank force from the year before for a similar reason. Assume that if a tank is lost due to an accident it is then used for spare parts to build new tanks. Country A starts with 25,000 tanks and Country B starts with 20,000 tanks. In the long term, Country A wants to maintain numerical superiority so as to deter an attack. What is the long-term behavior of this system? What is the long-term ratio of tanks possessed by Country A and Country B? Can Country A maintain numerical superiority?

To answer these questions, we first write two equations which model this situation. If we let $a(n)$ denote the number of tanks possessed by Country A after n years and $b(n)$ denote the number of tanks possessed by Country B after n years, then we can model the problem using the two difference equations

$$a_{n+1} = 0.9a_n + 0.3b_n$$
$$b_{n+1} = 0.1a_n + 0.95b_n$$

with initial conditions $a(0) = 25{,}000$ and $b(0) = 20{,}000$. Notice that we have a system of homogeneous equations. If we let

$$\boldsymbol{A}(n) = \begin{bmatrix} a(n) \\ a(n) \end{bmatrix}$$

then we can write the system in matrix form as

$$\begin{bmatrix} a(n+1) \\ a(n+1) \end{bmatrix} = \begin{bmatrix} 0.9 & 0.3 \\ 0.1 & 0.95 \end{bmatrix} \begin{bmatrix} a(n) \\ a(n) \end{bmatrix}$$

DOI: 10.1201/9781003582274-4

Recall from our earlier study of linear dynamical systems that we can express the form of the solution to a first order system of two equations $\boldsymbol{A}_{n+1} = \boldsymbol{R}\boldsymbol{A}_n$ as

$$\boldsymbol{A}(k) = c_1(r_1)^k \boldsymbol{A}_1 + c_2(r_2)^k \boldsymbol{A}_2$$

where r_1 and r_2 are the characteristic values (eigenvalues) of $\boldsymbol{R}$ and $\boldsymbol{A}_1$ and $\boldsymbol{A}_2$, their corresponding characteristic vectors (eigenvectors).

For our example, the characteristic values are $r_1 = 1.1$ and $r_2 = 0.75$ and the corresponding characteristic vectors are $\boldsymbol{A}_1 = \begin{bmatrix} 3 \\ 2 \end{bmatrix}$ and $\boldsymbol{A}_2 = \begin{bmatrix} 3 \\ 2 \end{bmatrix}$. Therefore, we can write the general solution to our system as

$$\boldsymbol{A}(k) = c_1(1.1)^k \boldsymbol{A}_1 + c_2(.75)^k \boldsymbol{A}_2.$$

But how do we determine the long-term behavior of a homogeneous system of difference equations? We have seen that for the single difference equation $A_{n+1} = \boldsymbol{r}A_n$, if $|r|$ is less than 1 then as k goes to infinity, $A(k)$ approaches 0. Since this behavior occurs for any initial condition, then 0 is a stable equilibrium value. Also, if $|r|$ is greater than 1 then as k goes to infinity, $A(k)$ goes to infinity.

Similarly, we can analyze the long-term behavior of a system of difference equations by examining the characteristic values. We will only consider systems in which there are no multiple roots of the characteristic equation (in other words $r_1 \neq r_2$).

First, note that the zero vector is an equilibrium vector for the homogeneous system of difference equations just like zero is an equilibrium value for the homogeneous difference equation. In other words, when $\boldsymbol{A}(\boldsymbol{0}) = \begin{bmatrix} 0 \\ 0 \end{bmatrix}$ then $\boldsymbol{A}(\boldsymbol{n}+\boldsymbol{1}) = \boldsymbol{A}(\boldsymbol{n}) = \begin{bmatrix} 0 \\ 0 \end{bmatrix}$ for all values of n.

Also, recall that the norm of $\boldsymbol{R}$, $\|\boldsymbol{R}\|$, is defined as the largest number, in absolute value, of the characteristic values, that is,

$$\|\boldsymbol{R}\| = \text{maximum of } \{|r_1|, |r_2|, |r_3|, \ldots\}.$$

If $|r_1| < 1$ and $|r_2| < 1$, or the $\|\boldsymbol{R}\| < \mathbf{1}$, then as k goes to infinity, $A(k)$ approaches the zero vector, $\begin{bmatrix} 0 \\ 0 \end{bmatrix}$. Since this behavior occurs for any initial conditions, then the zero vector is a stable equilibrium vector.

If $|r_1| > 1$ **or** $|r_2| > 1$, or the $\|\boldsymbol{R}\| > \mathbf{1}$, then as k goes to infinity, $A(k)$ does **not** approach the zero vector, and the zero vector is said to be an unstable equilibrium vector. **However**, even though a system approaches infinity as k gets large, we can still determine the long-term ratio at which the system grows.

In our tank example, since the norm of $\boldsymbol{R}$ is greater than one or $|1.1| > 1$, then as k goes to infinity, $A(k)$ does not approach the zero vector and the zero vector is an unstable equilibrium vector. However, even though the total number of tanks is growing without bound Country A is more concerned with always maintaining numerical superiority over Country B so as to deter an attack. In other words, Country A wants to maintain a ratio of tanks of $a(k)$ to $b(k)$ which is better than 1 to 1.

We will discover that the long-term ratio will just be the ratio of the components of the characteristic vector associated with the characteristic value which dictates the norm.

The trick is to study the long-term behavior of $\left[\frac{a(k)}{r_1 k}\right] A(k)$ where r_1 is the characteristic value which is greatest in absolute value. Notice that the ratio of $\left[\frac{a(k)}{r_1 k}\right]$ to $\left[\frac{b(k)}{r_1 k}\right]$ will be the same as the ratio of $a(k)$ to $b(k)$, that is,

$$\frac{\frac{a(k)}{r_1 k}}{\frac{b(k)}{r_1 k}} = \frac{a(k)}{b(k)}$$

Let's return to our tank problem. We have written the solution as

$$A(k) = c_1(1.1)^k A_1 + c_2(0.75)^k A_2$$

and determined the equilibrium vector to be unstable. However, we are concerned with the long-term ratio of tanks. In our example, the characteristic value which is greatest in absolute value is $r_1 = 1.1$. Therefore, let's examine the long-term behavior of

$$\frac{1}{(1.1)^k} A(k) = c_1 \left(\frac{1.1}{1.1}\right)^k A_1 + c_2 \left(\frac{0.75}{1.1}\right)^k A_2$$

Notice that $\left(\frac{1.1}{1.1}\right)^k = 1$, so that

$$\frac{1}{(1.1)^k} A(k) = c_1 A_1 + c_2 \left(\frac{0.75}{1.1}\right)^k A_2$$

Since $\left(\frac{0.75}{1.1}\right)^k$ approaches zero as k gets large then $\frac{1}{(1.1)^k}$ $A(k)$ approaches $c_1 A_1$ where $A_1 = \begin{bmatrix} 3 \\ 2 \end{bmatrix}$ is the characteristic vector corresponding to $r_1 = 1.1$. This means that for large values of k, $a(k)$ and $b(k)$ are at the same ratio as the components of the vector A_1, that is, $\frac{a(k)}{b(k)} = \frac{3}{2}$. In other words, in the long-term Country A will maintain a numerical superiority of 3 to 2.

Systems of Nonhomogeneous Difference Equations

Suppose a scientist is studying a material that over time transforms from one state to another state and finally to a third state. Let's call the three states *a*, *b*, and *c*. During each time interval 10% (by weight) of the material in state *a* transforms to state *b* and 30% (by weight) of the material in state *b* transforms to state *c* and 10% (by weight) of the material in state *b* transforms to a useless by-product. At the end of each time interval, the scientist removes all of the material that has reached state *c*. To ensure a continuous supply of material in state *c*, he also adds 10 grams of material in state *a* during each time interval. The scientist starts with 20 grams of material in state *a*. After a long period of time, how much material in state *c* will the scientist be able to remove at each time interval? How much material will be in each of the other two states? (Changing states does not affect the weight or the volume of the material.)

We can model the problem using three difference equations.

$$a_{n+1} = 0.9a_n + 10$$
$$b_{n+1} = 0.1a_n + 0.6b_n$$
$$c_{n+1} = 0.3b_n$$

with initial conditions $a(0) = 20$, $b(0) = 0$, and $c(0) = 0$.

To study the long-term behavior of the system we first write it in matrix form. Notice that we have a system of nonhomogeneous equations. If we let

$$A(n) = \begin{bmatrix} a(n) \\ b(n) \\ c(n) \end{bmatrix}$$

then we can write the homogeneous part of the system as $\boldsymbol{A}_{n+1} = \boldsymbol{R}\boldsymbol{A}_n$. We can then include the nonhomogeneous constant terms in a vector $\boldsymbol{B}$, and write the entire system as $\boldsymbol{A}_{n+1} = \boldsymbol{R}\boldsymbol{A}_n + \boldsymbol{B}$.

For this system, we get

$$\begin{bmatrix} a(n+1) \\ b(n+1) \\ c(n+1) \end{bmatrix} = \begin{bmatrix} 0.9 & 0 & 0 \\ 0.1 & 0.6 & 0 \\ 0 & 0.3 & 0 \end{bmatrix} \begin{bmatrix} a(n) \\ b(n) \\ c(n) \end{bmatrix} + \begin{bmatrix} 10 \\ 0 \\ 0 \end{bmatrix}$$

But how do we determine the long-term behavior of a nonhomogeneous system of difference equations? Let's see what we can determine based on what we have already learned.

We have seen that the single difference equation A $\boldsymbol{A}_{n+1} = \boldsymbol{r}\boldsymbol{A}_n$ has equilibrium value 0. We have also seen that for the system of difference equations $\boldsymbol{A}_{n+1} = \boldsymbol{R}\boldsymbol{A}_n$, if the norm of $\boldsymbol{R}$ is less than 1 then as k goes to infinity $\boldsymbol{A}(k)$ approaches the $\mathbf{0}$ vector. Since this behavior occurs for any initial conditions, the $\mathbf{0}$ vector is a stable equilibrium vector.

We have also seen that for the nonhomogeneous difference equation $\boldsymbol{A}_{n+1} = \boldsymbol{r}\boldsymbol{A}_n + \text{b}$, we can find the equilibrium value a by solving the equation $a = ra + b$. We can find the equilibrium vector for a nonhomogeneous system of difference equations in a similar manner. For the system of equations $\boldsymbol{A}_{n+1} = \boldsymbol{R}\boldsymbol{A}_n + \boldsymbol{B}$, we'll let $\boldsymbol{A}$ be the equilibrium vector. Then $\boldsymbol{A}_{n+1} = \boldsymbol{A}_n = \boldsymbol{A}$.

Substituting into the system we get $\boldsymbol{A} = \boldsymbol{R}\boldsymbol{A} + \boldsymbol{B}$. Since this is an equation of matrices, we cannot solve it exactly as we would for a scalar equation. But we can solve it by following these steps:

$$\begin{aligned} &\boldsymbol{A} = \boldsymbol{RA} + \boldsymbol{B} \Rightarrow \boldsymbol{A} - \boldsymbol{RA} = \boldsymbol{RA} + \boldsymbol{B} - \boldsymbol{RA} \\ &\Rightarrow \boldsymbol{A} - \boldsymbol{RA} = \boldsymbol{B} \Rightarrow \boldsymbol{IA} - \boldsymbol{RA} = \boldsymbol{B} \\ &\Rightarrow (\boldsymbol{I} - \boldsymbol{R})\boldsymbol{A} = \boldsymbol{B} \Rightarrow (\boldsymbol{I} - \boldsymbol{R})^{-1}(\boldsymbol{I} - \boldsymbol{R})\boldsymbol{A} = (\boldsymbol{I} - \boldsymbol{R})^{-1}\boldsymbol{B} \\ &\Rightarrow \boldsymbol{A} = (\boldsymbol{I} - \boldsymbol{R})^{-1}\boldsymbol{B} \end{aligned}$$

For the single equation $A_{n+1} = rA_n + b$, the solution is $A(k) = c(r)^k + a$. Similarly, the solution to $\mathbf{A}_{n+1} = \mathbf{R}\mathbf{A}_n + \mathbf{B}$ is $\mathbf{A}(k) = c\mathbf{R}^k + \mathbf{A}$, or by using the eigenvalues r_1 and r_2, and their associated eigenvectors $\mathbf{A}_1$ and $\mathbf{A}_2$, $\mathbf{A}(k) = a(r_1)^k\, \mathbf{A}_1 + b(r_2)^k\, \mathbf{A}_2 + \mathbf{A}$.

We can analyze the stability of $\mathbf{A}$ using what we have learned previously. If $|r_1| < 1$ and $|r_2| < 1$, then A is a stable equilibrium vector. If $|r_1| > 1$ or $|r_2| > 1$, then $\mathbf{A}$ is an unstable equilibrium vector. What are the other possibilities?

Let's return to our scientist's problem. We have written the equations in matrix-vector form

$$\begin{bmatrix} a(n+1) \\ b(n+1) \\ c(n+1) \end{bmatrix} = \begin{bmatrix} 0.9 & 0 & 0 \\ 0.1 & 0.6 & 0 \\ 0 & 0.3 & 0 \end{bmatrix} \begin{bmatrix} a(n) \\ b(n) \\ c(n) \end{bmatrix} + \begin{bmatrix} 10 \\ 0 \\ 0 \end{bmatrix}$$

$$\text{or } \mathbf{A}_{n+1} = \mathbf{R}\mathbf{A}_n + \mathbf{B}$$

Now we find the equilibrium vector $\mathbf{A} = (\mathbf{I} - \mathbf{R})^{-1}\, \mathbf{B}$.

$$I - R = \begin{bmatrix} 1 & 0 & 0 \\ 0 & 1 & 0 \\ 0 & 0 & 1 \end{bmatrix} - \begin{bmatrix} 0.9 & 0 & 0 \\ 0.1 & 0.6 & 0 \\ 0 & 0.3 & 0 \end{bmatrix} = \begin{bmatrix} 0.1 & 0 & 0 \\ -0.1 & 0.4 & 0 \\ 0 & -0.3 & 1 \end{bmatrix}$$

So, the equilibrium vector

$$A = \begin{bmatrix} 0.1 & 0 & 0 \\ -0.1 & 0.4 & 0 \\ 0 & -0.3 & 1 \end{bmatrix}^{-1} x \begin{bmatrix} 10 \\ 0 \\ 0 \end{bmatrix} = \begin{bmatrix} 100 \\ 25 \\ 7.5 \end{bmatrix}$$

We need to find the eigenvalues of $\mathbf{R}$ in order to determine its norm. Using Derive, we find the eigenvalues to be 0, 0.6, and 0.9. Therefore the norm of $\mathbf{R}$ is 0.9. Since the norm is less than 1, the equilibrium vector is stable. So in the long term, $\mathbf{A}(k)$ approaches the equilibrium vector

$$\begin{bmatrix} 100 \\ 25 \\ 7.5 \end{bmatrix}.$$

Thus, after a long period of time, the scientist will be able to remove 7.5 grams of material in state c during each time interval and there will be 100 grams of material in state a and 25 grams in state b.

Introduction to Systems

In Chapter 3, we learned about linear and nonlinear DDS models. Now, we extend the discussion to systems of systems, but we still use our paradigm:

$$future = present + change.$$

Consider wanting to retire on a lake that you stock with bass and trout for endless fishing. Will it be endless? Can the species go exist in your lake? How often do you need to restock the lake? This is an example of a competitive hunter model where both species compete for the same resources.

Let's define a system of DDS.

$$A_n = f\left(A_n, B_n\right)$$
$$B_n = g\left(A_n, B_n\right)$$

As before, simple linear systems of DDS have closed form analytical solution, however, most systems do not. We will analyze all those DDS through iteration and graphs.

For a selected set of initial conditions, we build numerical solutions to get a sense of long-term behavior for the system. For the systems that we will study, we will find their equilibrium values. We then explore starting values near the equilibrium values to see if by starting close to an equilibrium value, the system will:

a. remain close
b. approaches the equilibrium value
c. does not remain close

What happens near these values gives great insight concerning the long-term behavior of the system. We can study the resulting pattern of the numerical solutions and the resulting plots.

Simple Linear Systems and Analytical Solutions

Let's consider school vouchers. There are both students in the public school (PS) and the private magnet (PM) school. Suppose a pre-survey of families used as historical records determined that 75% of the magnet school remains while 25% preferred to transfer. We found that 65% of the public school wanted to remain but 35% preferred to transfer. Let's build a model to determine the long-term behavior of these students based upon this historical data. The change diagram is shown in Figure 4.1.

Problem Identification: Determine the school voucher over time.

Assumptions and variables: Let n represent the number of student months. We define

PS_n = the number of students in public school at the end of n months
PM_n = the number of students in the magnet school at the end of n months

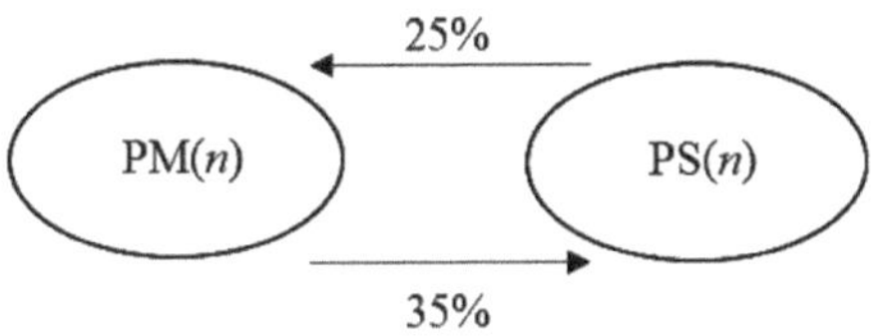

FIGURE 4.1
Change diagram for the school vouchers.

We assume that no other incentives are given to the students for either staying or moving.

The Model: As build the dynamical systems. Mathematically, this is written as:

$$PM_{n+1} = 0.75PM_n + 0.35PS_n$$
$$PS_{n+1} = 0.25PM_n + 0.65PS_n$$

There are initially 1,500 students in the magnet school and 2,000 students in the public school. We seek to find the long-term behavior of this system.

We rewrite the model as a system of DDS:

$$PM_{n+1} = 0.75PM_n + 0.35PS_n$$
$$PS_{n+1} = 0.25PM_n + 0.65PS_n$$

PM(0) = 1,500 and PS(0) = 1,000, respectively.

This is a simple linear system. We may use the initial conditions to iterate this system of discrete dynamical systems as shown in Figure 4.2. We see from the solution plot that there are stable equilibriums at about 2,042 and 1,458 are in the magnet and public schools respectively.

Analytical Solutions

Analytical solutions assume knowledge of linear algebras through eigenvalues and eigenvectors.

We rewrite the DDS in matrix form:

$$X_{n+1} = MX_n, \ X(0) = B$$

$$\begin{bmatrix} PM(n+1) \\ PS(n+1) \end{bmatrix} = \begin{bmatrix} 0.75 & 0.35 \\ 0.25 & 0.65 \end{bmatrix} \begin{bmatrix} PM(n) \\ PS(n) \end{bmatrix}, \begin{bmatrix} PM(0) \\ PS(0) \end{bmatrix} = \begin{bmatrix} 1500 \\ 2000 \end{bmatrix}$$

We define the analytical solution with two distinct real eigenvalues and eigenvectors. The general form of the solution is

$$\mathbf{X}(\boldsymbol{k}) = \lambda_1^k c_1 \mathbf{V}_1 + \lambda_2^k c_2 \mathbf{V}_2$$

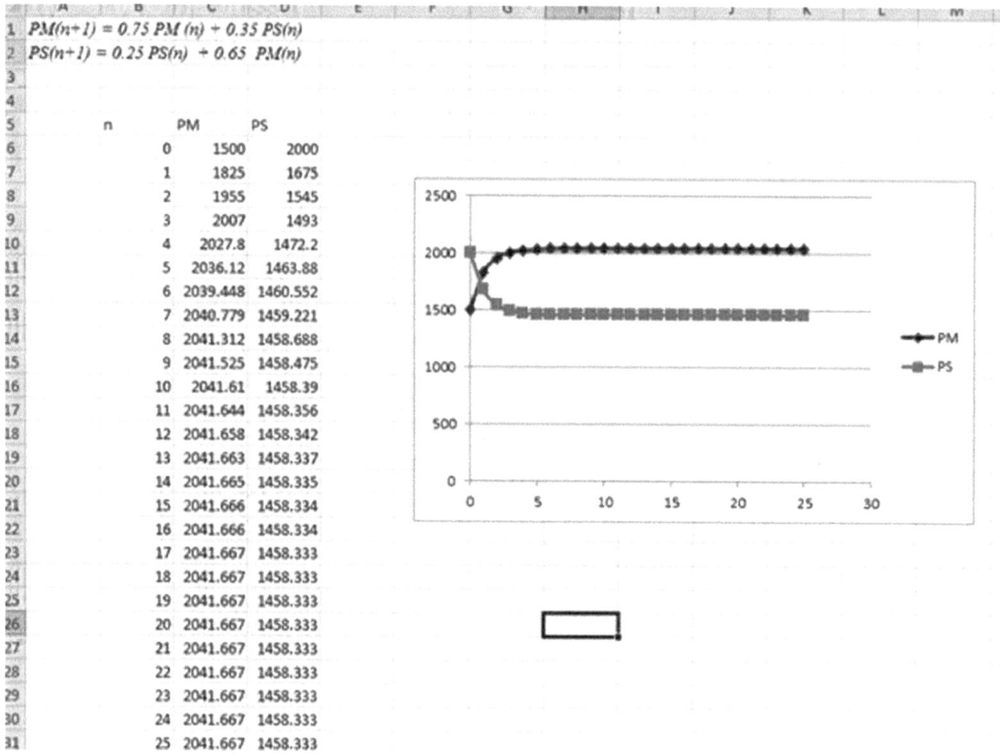

PM(n+1) = 0.75 PM (n) + 0.35 PS(n)
PS(n+1) = 0.25 PS(n) + 0.65 PM(n)

n	PM	PS
0	1500	2000
1	1825	1675
2	1955	1545
3	2007	1493
4	2027.8	1472.2
5	2036.12	1463.88
6	2039.448	1460.552
7	2040.779	1459.221
8	2041.312	1458.688
9	2041.525	1458.475
10	2041.61	1458.39
11	2041.644	1458.356
12	2041.658	1458.342
13	2041.663	1458.337
14	2041.665	1458.335
15	2041.666	1458.334
16	2041.666	1458.334
17	2041.667	1458.333
18	2041.667	1458.333
19	2041.667	1458.333
20	2041.667	1458.333
21	2041.667	1458.333
22	2041.667	1458.333
23	2041.667	1458.333
24	2041.667	1458.333
25	2041.667	1458.333

FIGURE 4.2
Screenshot iterative solution and plot for voucher students.

where

λ_1 & λ_2 = two distinct eigenvalues
V_1 & V_2 = corresponding eigenvectors
c_1 & c_2 = constant.

The characteristic polynomial is $\lambda^2 - 1.4\,\lambda + 0.40 = 0$. This provides two eigenvalues: 1, 0.4. The corresponding eigenvalues can be found easily as vectors for $\lambda_1 = 1$ of [0.35, 0.25] and for $\lambda_2 = 0.40$ as [1, –1].

The general solution is

$$X(k) = c_1\left(1^k\right)\begin{bmatrix}0.35\\0.25\end{bmatrix} + c_2\left(0.4^k\right)\begin{bmatrix}1\\-1\end{bmatrix}$$

With our initial conditions we find c_1 and c_2 to be 416.6666 and –4.16666, so

$$X(k) = 5833.333\left(1^k\right)\begin{bmatrix}0.35\\0.25\end{bmatrix} - 541.6666\left(0.4^k\right)\begin{bmatrix}1\\-1\end{bmatrix}$$

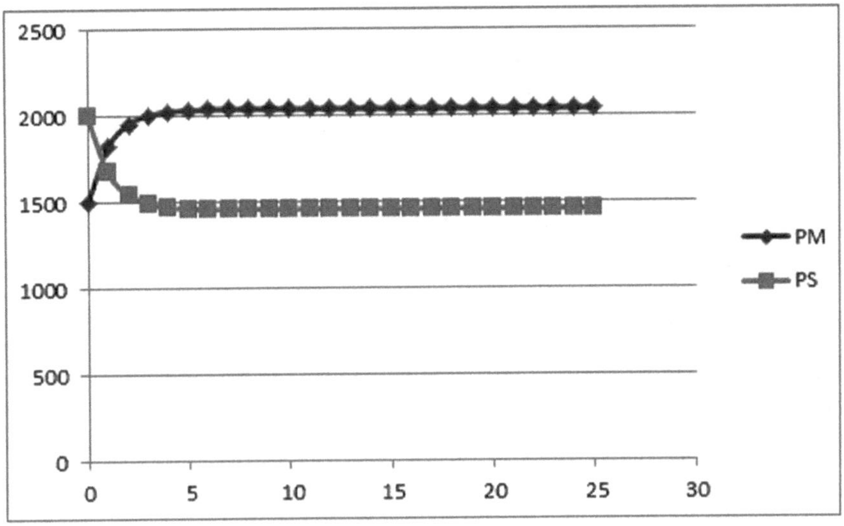

FIGURE 4.3
Plot of DDS for student's relocation example.

When $k = 10$, we find 2041.609 and 1458.31 for students, respectively. We see this again in Figure 4.3.

Analytically, we can solve for the equilibrium values. We let $X = D(n)$ and $Y = M(n)$. From the DDS, we obtain the equations:

$$X = 0.75X + 0.35Y$$
$$Y = 0.25x + 0.65Y$$

and both equations reduce to $X = 0.35/0.25\ Y$. There are 2 unknowns, so we need a second equation.

From the initial conditions, we know that $X + Y = 3500$. We can use the equations

$$X + Y = 3{,}500 \text{ and } X = 0.35/0.25\ Y \text{ to find the equilibrium values:}$$

$$\mathrm{X} = 2041.6667 \text{ and } Y = 1458.3333$$

We previously iterated the solution and now we start with initial conditions near f those equilibrium values and we find the sequences tend toward those values. We conclude that system has *stable* equilibrium values.

You should go back and change the initial conditions and see what behavior follows.

Interpretation: The long-term behavior shows that eventually (without other influences) that of the 3,500 students about 1,458 remain in public school and 2,042 go to the magnet school. We might want to try to attract students with advertising and perhaps add incentives.

Iteration and Graphical Solution

Often we are unable to solve the system of DDS in closed form, so must rely on iteration and graphical models.

Example 4.1: Competitive Hunter Models

Competitive hunter models involve species vying for the same resources (such as food or living space) in the habitat. The effect of the presence of a second species diminishes the growth rate of the first species. We now consider a specific example concerning trout and bass in a small pond. Hugh Ketum owns a small pond that he uses to stock fish and eventually allows fishing. He has decided to stock both bass and trout. The fish and game warden tells Hugh that after inspecting his pond for environmental conditions he has a solid pond for growth of his fish. In isolation, bass grow at a rate of 20% and trout at a rate of 30%. The warden tells Hugh that the interaction for the food affects trout more than bass. They estimate the interaction affecting bass is 0.0010 bass*trout and for trout is 0.0020 bass*trout. Assume no changes in the habitant occur.

Model: Let's define the following variables:

B_n = the number of bass in the pond after period n.
T_n = the number of trout in the pond after period n.
$B_n * T_n$ = interaction of the two species.

$$B_{n+1} = 1.20\,B_n - 0.0010\,B_n * T_n$$
$$T_{n+1} = 1.30\,T_n - 0.0020\,B_n * T_n$$

The equilibrium values can be found by allowing X = B(n) and Y = T(n) and solving for X and Y.

$$X = 1.2X - 0.001X * Y$$
$$Y = 1.3Y - 0.0020X * Y$$

We rewrite these equations as

$$0.2X - 0.001X * Y = 0$$
$$0.3Y - 0.002X * Y = 0$$

We can rewrite the two equations to obtain:

$$X(0.2 - 0.001Y) = 0$$
$$Y(0.3 - 0.002X) = 0$$

Solving we find X = 0 or Y = 2000 and Y = 0 or X = 1500

We want to know the long-term behavior of the system and the stability of the equilibrium points.

Hugh initially considered 151 bass and 199 trout for his pond. The solution is left to the student as an exercise. From Hugh's initial conditions, bass will grow without bound and trout will eventually die out.

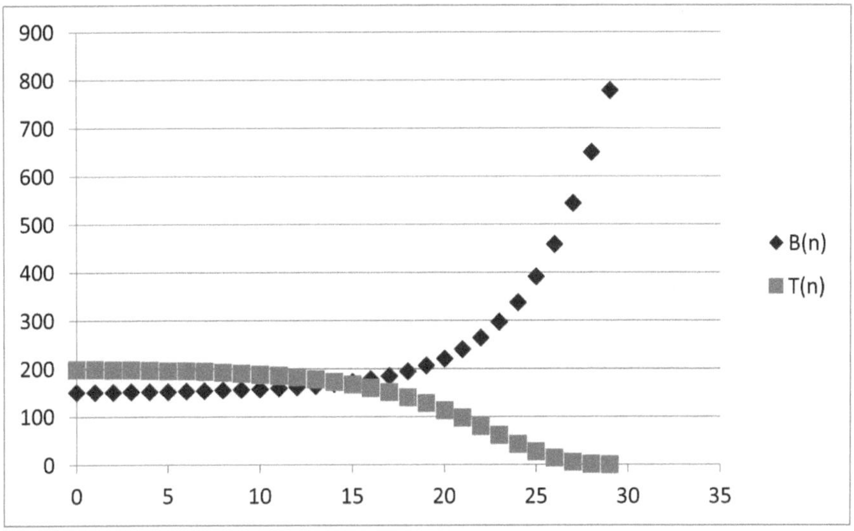

FIGURE 4.4
Bass and trout over time.

We iterated the system and obtained the plot, Figure 4.4, of bass and trout over time. Trout died out at about period 29.

This is certainly not what Hugh had in mind so he must find ways to improve the environment for the fish that alter the parameters from the model.

Example 4.2: Fast Food Tendencies

Consider that your student union center desires to have three fast food chains available to students serving: burgers, tacos, and pizza. These chains run a survey of students and find the following information concerning lunch: 75% that ate burgers will eat burgers again at the next lunch, 5% will eat tacos next, and 20% will eat pizza next. Of those who ate tacos last, 20% will eat burgers next, 60% will stay with tacos, and 35% will eat pizza next. Of those who ate pizza, 40% will eat burgers next, 20% tacos, and 40% pizza again.

We formulate the problem as follows.

Let n represent the nth day's lunch and so we define the following variables:

B_n = the number of burger eaters in the nth lunch.
T_n = the number of taco eaters in the nth lunch.
P_n = the number of pizza eaters in the nth lunch.

Formulating the system, we have the following dynamical system:

$$\begin{aligned} B_{n+1} &= 0.75B_n + 0.20T_n + 0.40P_n \\ T_{n+1} &= 0.05B_n + 0.60T_n + 0.20P_n \\ P_{n+1} &= 0.20B_n + 0.20T_n + 0.40P_n \end{aligned}$$

Analytically, we let $X = B_n$, $Y = T_n$, and $Z = P_n$ so that

$$X = 0.75X + 0.2Y + 0.4Z$$
$$Y = 0.05X + 0.6Y + 0.2Z$$
$$Z = 0.2X + 0.2Y + 0.4Z$$

These equations reduce to

$$X = 20/9Z$$
$$Y = 7/9Z$$
$$Z = Z$$

Since we have 14,000 students, then we assume that $X + Y + Z = 14{,}000$
We substitute and solve for Z first.

$$4Z = 14,000$$
$$Z = 3,500$$
$$X = 20/9Z = 20/9(3,500) = 7,777.77$$
$$Y = 7/9Z = 7/9(3,500) = 2,722.222$$

Suppose the campus has 14,000 students who eat lunch. The graphical results also show that an equilibrium value is reached at a value of about 7,778 burger eaters, 2,722 taco eaters, and 3,500 pizza eaters. This allows the fast food establishments to plan for a projected future. We see this in the iterated Table and Figure 4.5. By varying the initial conditions for 14,000 students we find that these values are stable equilibrium values.

n	B(n)	T(n)	P(n)
0	14000	0	0
1	10500	700	2800
2	9135	1505	3360
3	8496.25	2031.75	3472
4	8167.338	2338.263	3494.4
5	7990.916	2510.204	3498.88
6	7894.78	2605.444	3499.776
7	7842.084	2657.961	3499.955
8	7813.137	2686.872	3499.991
9	7797.224	2702.778	3499.998
10	7788.473	2711.528	3500
11	7783.66	2716.34	3500
12	7781.013	2718.987	3500
13	7779.557	2720.443	3500
14	7778.756	2721.244	3500
15	7778.316	2721.684	3500
16	7778.074	2721.926	3500
17	7777.941	2722.059	3500
18	7777.867	2722.133	3500
19	7777.827	2722.173	3500
20	7777.805	2722.195	3500

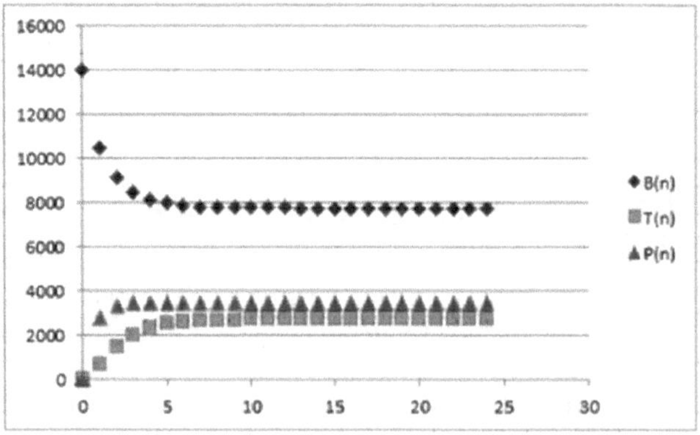

FIGURE 4.5
Screenshot Excel iterated solution and plot for fast food on campus.

Example 4.3: A Predator-Prey Model: Foxes and Rabbits

In the study of the dynamics of a single population, we typically take into consideration such factors as the "natural" growth rate and the "carrying capacity" of the environment. Mathematical ecology requires the study of populations that interact, thereby affecting each other's growth rates. In this module, we study a very special case of such an interaction, in which there are exactly two species, one of which the predators eat the prey. Such pairs exist throughout nature such as lions and gazelles, birds and insects, pandas and eucalyptus trees, and Venus fly traps and flies.

To keep our model simple, we will make some assumptions that would be unrealistic in most of these predator-prey situations. Specifically, we will assume that

- the predator species is totally dependent on a single prey species as its only food supply,
- the prey species has an unlimited food supply, and
- there exist no other threats to the prey other than the specific predator.

In this modeling process, we will use the Lotka-Volterra model for Predator-Prey. Students can read more about the Lotka-Volterra models in the suggested readings. Here we simply present the model that we use.

We repeat our two key assumptions:

- The predator species is totally dependent on the prey species as its only food supply.
- The prey species has an unlimited food supply and no threat to its growth other than the specific predator.

If there were no predators, the second assumption would imply that the prey species grows exponentially without bound, i.e., if $x = x_n$ is the size of the prey population after a discrete time period n, then we would have $x_{n+1} = ax_n$.

But there *are* predators, which must account for a negative component in the prey growth rate. Suppose we write $y = y(n)$ for the size of the predator population at time t. Here are the crucial assumptions for completing the model:

- The rate at which predators encounter prey is jointly proportional to the sizes of the two populations.
- A fixed proportion of encounters lead to the death of the prey.

These assumptions lead to the conclusion that the negative component of the prey growth rate is proportional to the product xy of the population sizes, i.e.,

$$x_{n+1} = x_n + ax_n - bx_n y_n$$

Now we consider the predator population. If there were no food supply, the population would die out at a rate proportional to its size, i.e. we would find $y_{n+1} = -cy_n$.

We assume that is the simple case that the "natural growth rate" is a composite of birth and death rates, both presumably proportional to population size. In the absence of food, there is no energy supply to support the birth rate. But there is a food supply: the prey. And what's bad for hares is good for lynx. That is, the energy to support growth of the predator population is proportional to deaths of prey, so

$$y_{n+1} = -cy_n + px_n y_n$$

This discussion leads to the discrete version of the Lotka-Volterra Predator-Prey Model:

$$x_{n+1} = x_n + (1+a)x_n - bx_n y_n$$
$$y_{n+1} = -cy_n + px_n y_n$$
$$n = 0, 1, 2, \ldots$$

where *a*, *b*, *c*, and *p* are positive constants.

The Lotka-Volterra model consists of a system of linked dynamical systems equations that cannot be separated from each other and that cannot be solved in closed form. Nevertheless, they can be solved numerically and graphed in order to obtain insights about the scenario being studied.

Let us return to our foxes and hares' scenario. Let's assume this discrete model is as explained above. Further, data investigation yields the following estimates for the parameters that we require: {a, b, c, p}={0.04, 0.0004, 0.09, 0.001}. Let's further assume that initially there are 600 rabbits and 125 foxes.

We iterate and plot the results for rabbits and foxes versus time and then plot rabbits versus foxes shown in Figures 4.6 and 4.7.

If we ran this model for many more iterations, we would find the plot of foxes versus rabbits spiral in a similar fashion as above. We conclude that the model appears reasonable. We could find the equilibrium values for the system. There is a set of feasible equilibrium points for rabbits and foxes at (0, 0) and at (2725, 960). The orbits of the spiral indicate that the system is moving away from both (0, 0) and (2725, 960) so we conclude the system is not stable.

In models of predator-preys, it is often that "managers" of the ecological system must intervene in some way to keep both species flourishing.

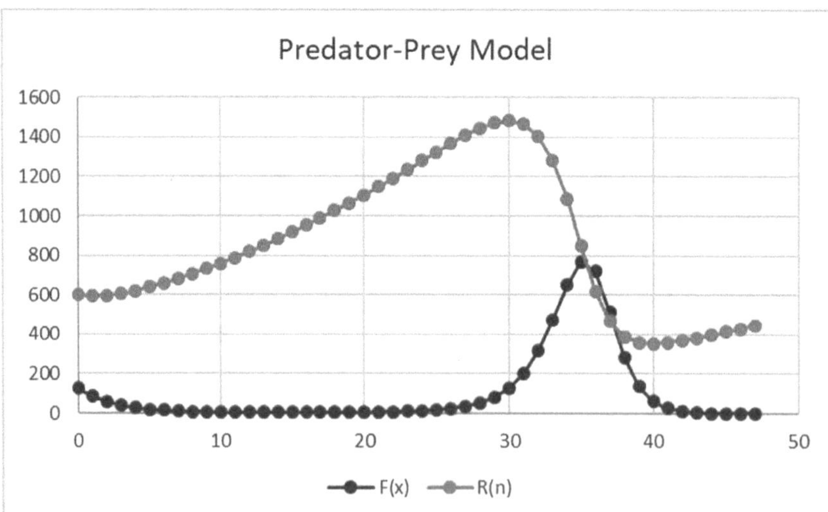

FIGURE 4.6
Foxes and rabbits over time for Example 4.3.

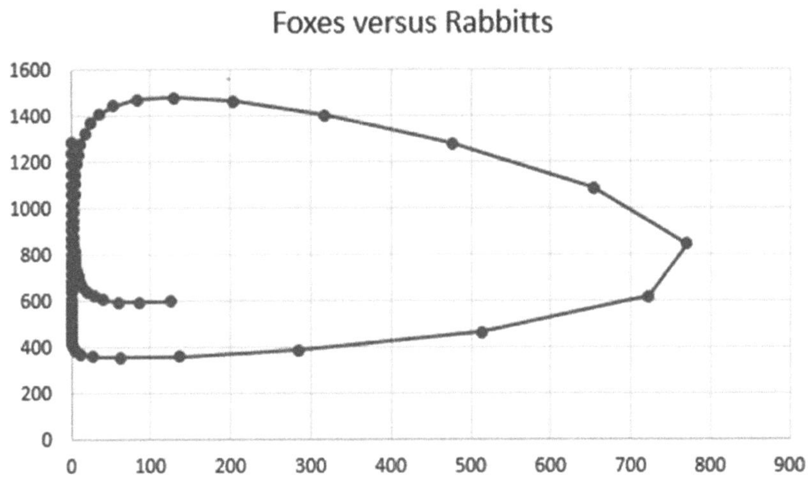

FIGURE 4.7
Foxes versus rabbits in a spiral motion for Example 4.3.

Example 4.4: Discrete SIR Model of Epidemics

Consider a disease that is spreading throughout the United States such as the new flu. The CDC is interested in knowing and experimenting with a model for this new disease prior to it actually becoming a "real" epidemic. Let us consider the population being divided into three categories: susceptible, infected, and removed. We make the following assumptions for our model:

- No one enters or leaves the community and there is no contact outside the community.
- Each person is either susceptible, S (able to catch this new flu); infected, I (currently has the flu and can spread the flu); or removed, R (already had the flu and will not get it again that includes death).
- Initially, every person is either S or I.
- Once someone gets the flu this year they cannot get it again.
- The average length of the disease is two weeks over which the person is deemed infected and can spread the disease.
- Our time period for the model will be per week.

The model we will consider is the SIR model.
Let's assume the following definition for our variables.

S_n = number in the population susceptible after period n.
I_n = number infected after period n.
R_n = number removed after period n.

Let's start our modeling process with R_n. Our assumption for the length of time someone has the flu is two weeks. Thus, half the infected people will be removed each week,

$$R_n = R_n + 0.5 I_n$$

The value, 0.5, is called the removal rate per week. It represents the proportion of the infected persons who are removed from infection each week. If real data is available, then we could do "data analysis" in order to obtain the removal rate.

I_n will have terms that both increase and decrease its amount over time. It is decreased by the number that are removed each week, $0.5*I_n$. It is increased by the number of susceptible that come into contact with an infected person and catch the disease, aS_nI_n. We define the rate, a, as the rate at which the disease is spread or the transmission coefficient. We realize this is a probabilistic coefficient. We will assume, initially, that this rate is a constant value that can be found from initial conditions.

Let's illustrate as follows. Assume we have a population of 1,000 students in the dorms. Our nurse found 3 students reporting to the infirmary initially. The next week, 5 students came into the infirmary with flu-like symptoms. $I(0) = 3$, $S(0) = 997$. In week 1, the number of newly infected is 30.

$$5 = a\,I_nS_n = a(3) * (995)$$
$$a = 0.00167$$

Let's now consider $S(n)$. This number is decreased only by the number that becomes infected. We may use the same rate, a, as before to obtain the model:

$$S_{n+1} = S_n - aS_nI_n$$

Our coupled SIR model is

$$\begin{aligned} R_{n+1} &= R_n + 0.5I_n \\ I_{n+1} &= I_n - 0.5I_n + 0.00167\,S_nI_n \\ S_{n+1} &= S_n - 0.00167S_nI_n \\ I_0 &= 3, S_0 = 997, R_0 = 0 \end{aligned}$$

The SIR Model can be solved iteratively and viewed graphically. Let's iterate the solution and obtain the graph, Figure 4.8, to observe the behavior to obtain some insights.

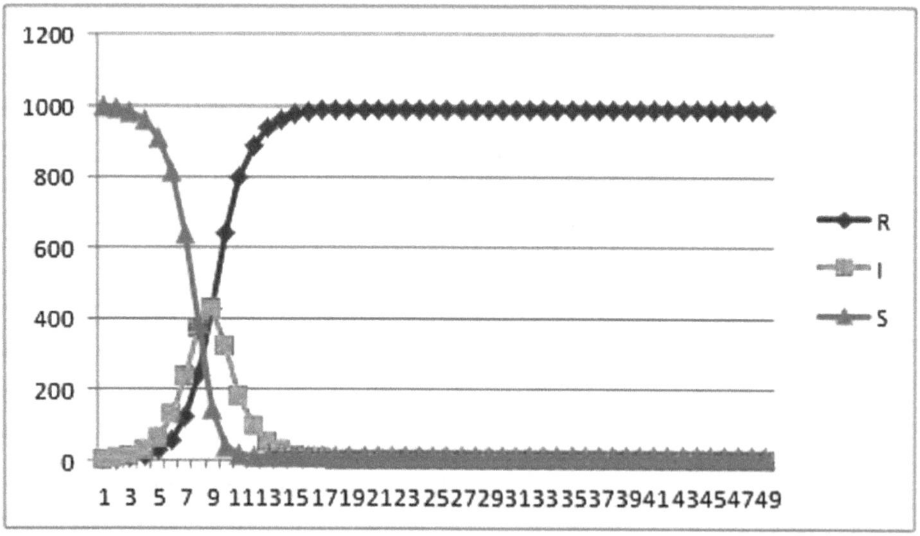

FIGURE 4.8
Plot of SIR model over time for Example 4.4.

The worst of the flu epidemic occurs around week 8, at the maximum of the infected graph. The maximum number is slightly larger than 400, from Figure 4.8 it is 427. After 25 weeks, slightly more than 9 persons never get the flu. You will be asked to check for sensitivity to the coefficient in the exercise set.

Technology Examples for Discrete Dynamical Systems

Using DDS is an interesting and productive approach. The use of a computer technology is essential to the methods described in this chapter. It allows for interactions between instructors and students. It provides a means to use technology in a nonstandard way. It provides another way to educate students concerning squares and cubes using discrete mathematics. In this section, we present technology and examples of solving linear and nonlinear discrete dynamical systems.

Excel for Linear and Nonlinear DDS

Example 4.5: Discrete Dynamical System

Let's consider a discrete dynamical system such as $a(n+1) = 0.5a(n)$, with $a(0) = 100$.

The steps to iterate and graph in Excel are as follows:

Step 1. Open a new worksheet and name it as DDS or some appropriate name.
Step 2. Label the following columns as *n* and *a(n)* in cells *a1* and *b1*.
Step 3. In cells *a2* and *b2* input the initial condition by putting in 0 in cell *a2* and 100 in cell *b2*.
Step 4. In cell *a3* type = *1 + cell a2*.
Step 5. In cell *b3* type = *0.5*cell b2*.
Step 6. Highlight cells *a3* and *b3* and drag the cursor down to fill in cells as far as desired or needed, in this case to about a4:b16.
Step 7. Highlight cells a1:b16, INSERT scatterplot to obtain the graph.
Step 8. Interpret the results.

A screenshot of the model and results have been shown in Figure 4.9, where we see that our DDS tends to zero over time.

In Figure 4.10, we show the appropriate formulas we used in Excel.

Maple for Linear and Nonlinear DDS

In Maple, DDS are referred to as recursion equations. One might obtain closed form solutions if they exist. One might also iterate and graph the behavior of the recursion equation. We will use both commands and libraries from Maple such as *with(plots)* and we will add some commands to our Maple toolbox, **rsolve** and **seq**.

```
rsolve - recurrence equation solver

Calling Sequence
  rsolve(eqns, fcns)
  rsolve(eqns, fcns, 'genfunc'(z))
  rsolve(eqns, fcns, 'makeproc')
```

n	a(n)
0	100
1	50
2	25
3	12.5
4	6.25
5	3.125
6	1.5625
7	0.78125
8	0.39063
9	0.19531
10	0.09766
11	0.04883
12	0.02441
13	0.01221
14	0.0061
15	0.00305
16	0.00153

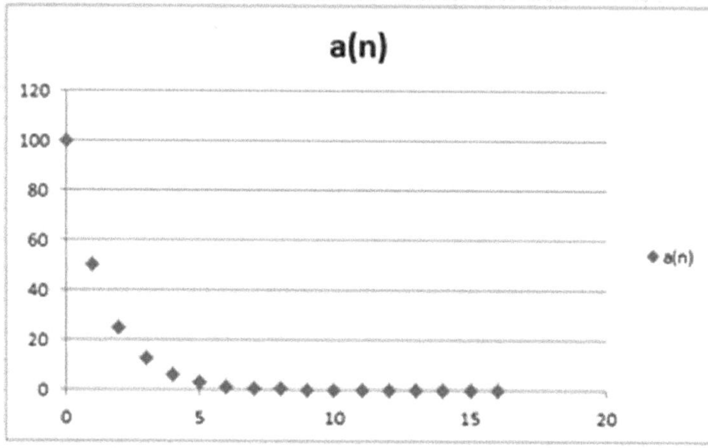

FIGURE 4.9
Screenshot of Excel's solution for Example 4.5.

n	a(n)
0	100
=1+A2	=0.5*B2
=1+A3	=0.5*B3
=1+A4	=0.5*B4
=1+A5	=0.5*B5
=1+A6	=0.5*B6
=1+A7	=0.5*B7
=1+A8	=0.5*B8
=1+A9	=0.5*B9
=1+A10	=0.5*B10
=1+A11	=0.5*B11
=1+A12	=0.5*B12
=1+A13	=0.5*B13
=1+A14	=0.5*B14
=1+A15	=0.5*B15
=1+A16	=0.5*B16
=1+A17	=0.5*B17

FIGURE 4.10
Screenshot of Excel formulas used in Example 4.5.

```
Parameters
  eqns - single equation or a set of equations
  fcns - function name or set of function names
  z - name, the generating function variable
```

`seq` - **create a sequence**

```
Calling Sequence
  seq(f, i=m..n)
  seq(f, i=x)

Parameters
  f - any expression
  i - name
  m, n - numerical values
  x - expression
```

Many of the models that we will solve have closed form solutions so that we can use the command **rsolve** to obtain the closed solution, and then we can use the sequence command (**seq**) to obtain the numerical values in the solution. Many dynamical systems do not have closed form analytical solutions so we cannot use the rsolve and seq commands to obtain solutions. When this occurs, we will write a small program using PROC to obtain the numerical solutions. To plot the solution to the dynamical systems, we will use plot commands to plot the sequential data pairs. We will illustrate all these commands in our next example.

Example 4.6: Revisiting Drugs in Our System

Solve the DDS an + 1 = 0.75an + 100, a(0) = 0.

Determine the value of *a(72)*.

Let's illustrate the iterative technique for analyzing a DDS in Maple. Figure 4.11 shows the graphical representation of the solution.

We use the following Maple commands for Example 4.2:

```
Drug:= rsolve({a(n+1)=0.75*a(n)+100,a(0)=0}, a(n));
```

$$drug := 400 - 400\left(\frac{3}{4}\right)^n$$

```
L:= limit(drug, n=infinity);
```

$$L := 400$$

```
with(plots):
pointplot({seq([I, -400*(3/4)^i+400], i=0..48)};

drug_Table:=seq(-400.0*(0.75)^i+400., i=0.48);
```

```
drug_table:= 0., 100.00, 175.00, 231.25, 273.44, 305.08, 328.81, 346.61,
359.95, 369.97, 377.47, 383.11, 387.33, 390.50, 392.87, 394.65, 395.99,
396.99, 397.75, 398.31, 398.73, 399.05, 399.29, …, 399.99
```

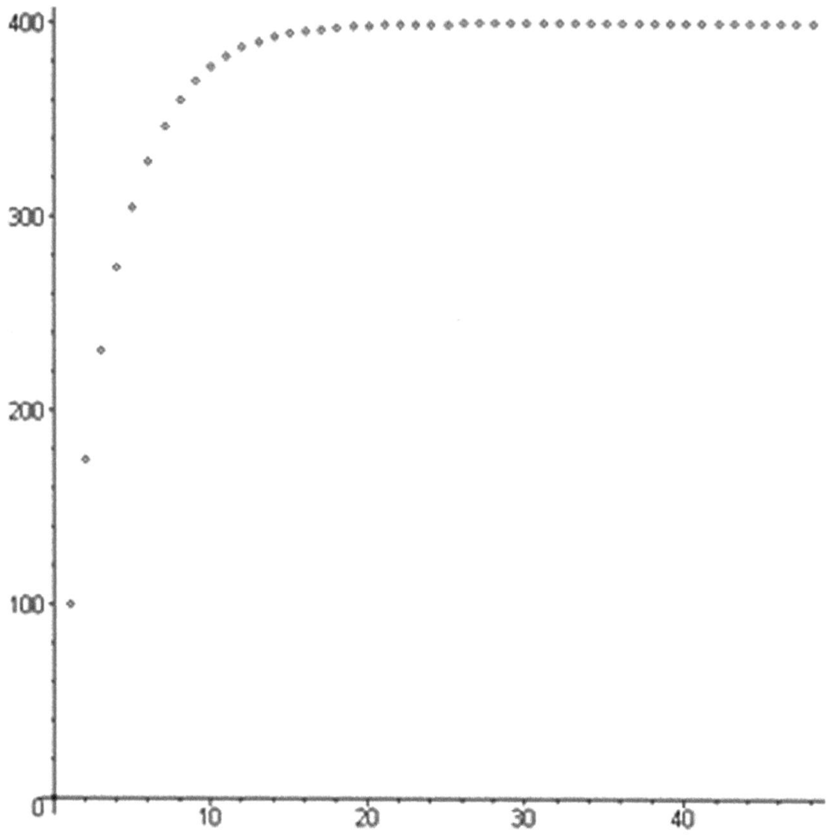

FIGURE 4.11
Behavior of drugs in our systems (Example 4.6).

Interpretation of Results: The DDS shows that the drug reaches a value where change stops and eventually the concentration in the bloodstream levels at 400 mg. If 400 mg is both a safe and effective dosage level, then this dosage schedule is acceptable.

Using Maple for a System of DDS

Again, we use systems of DDS with rsolve, numerical, and plotting for the dynamical systems problem. Mathematically, this is written as:

$$PM_{n+1} = 0.75\,PM_n + 0.35\,PS_n$$
$$PS_{n+1} = 0.25\,PM_n + 0.65\,PS_n$$

There are initially 1,500 students in the magnet school and 2,000 students in the public school. We seek to find the long-term behavior of this system.

$$PM(0) = 1,500 \text{ and } PS(0) = 1,000, \text{ respectively.}$$

The Maple code for solving this problem is:

```
>dds := rsolve({PS(n) = .65*PS(n-1)+.25*PM(n-1), PM(n) = .35*PS (n-1)
+.75*PM(n-1), PS(0) =2000, PM(0) =1500}, {PM, PS});
```

$$\text{dds} := \left\{ \text{PM}(n) = -1\frac{1625}{3}\left(\frac{2}{5}\right)^n + \frac{6125}{3},\ \text{PS}(n) = 1\frac{1625}{3}\left(\frac{2}{5}\right)^n + \frac{4375}{3} \right\}$$

```
>plot({-(1625/3)*(2/5)^n+6125/3, (1625/3)*(2/5)^n+4375/3},n=0..15,thick
ness=3, title=`Student Vouchers`);
```

Figure 4.12 shows the graphical representation of the solution using Python.

```
>public:=n-> if n=0 then 2000 else .65*public(n-1)+.25*magnent(n-1)
end if;
```

```
  public:=n → if n=0 then 2000 else
    0.65public(n - 1) + 0.25magnent(n - 1)
    end if
```

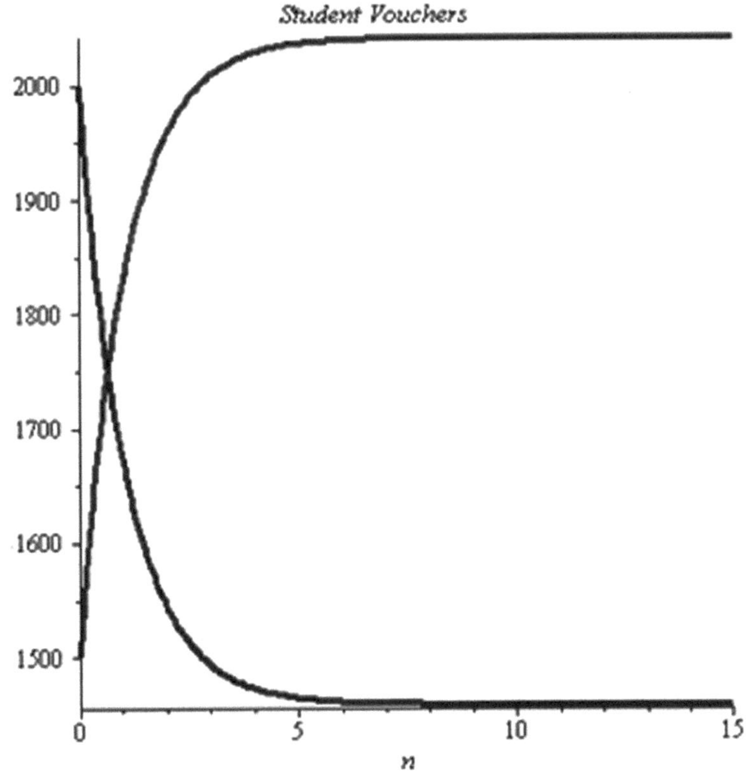

FIGURE 4.12
Python solution to student voucher.

```
>magnent:=n-> if n=0 then 1500 else 0.35*public(n-1)+.75*magnent(n-1)
end if;

  public:=n → if n=0 then 1500 else
    0.35public(n - 1) + 0.75magnent(n - 1)
    end if

>seq([public(n),magnent(n)],n=0..10);

  [2000, 1500], [1675, 1825], [1545, 1955],
  [1493, 2007], [1472.2, 2027.8],
  [1463.88, 2036.12], [1460.552, 2039.448],
  [1459.2208, 2040.7792], [1458.688329, 2041.311680],
  [1458.475328, 2041.524672], [1458.390131, 2041.609869]

>u:=seq(public(n),n=0..10);

  u:= 2000, 1675 1545, 1493, 1472.2, 1463.88, 1460.552, 1459. 2208,
458.688329, 1458.475328, 1458.390131

>w:=seq(magnent(n),n=0..10)

  w:= 1500, 1825, 1955, 2007, 2027.8, 2036.12, 2039.448, 2040.7792,
  2041.311680, 2041.524672, 2041.609869

The following code plots the student voucher as displayed in
Figure 4.13.

>with(plots):
>a:=plot({-(1625/3)*(2/5)^n+6125/3, (1625/3)*(2/5)^n+4375/3},n=0..15,
thickness=3,
>title=`Student Vouchers`):
>b:=pointplot({seq([n,public(n)],n=0..10)}):
>c:=pointplot({seq([n,magnent(n)],n=0..10)}):
>display(a,b,c);

>pointplot({seq([public(n),magnent(n)],n=0..20)});
```

More on Systems of Discrete Dynamical Systems

In this section, we examine models of systems of difference equations (DDS). For selected initial conditions, we build numerical solutions to get a sense of long-term behavior for the system. For the systems that we will study, we will find their equilibrium values. We then explore starting values near the equilibrium values to see if by starting close to an equilibrium value, the system will:

a. remain close
b. approach the equilibrium value
c. Not remain close

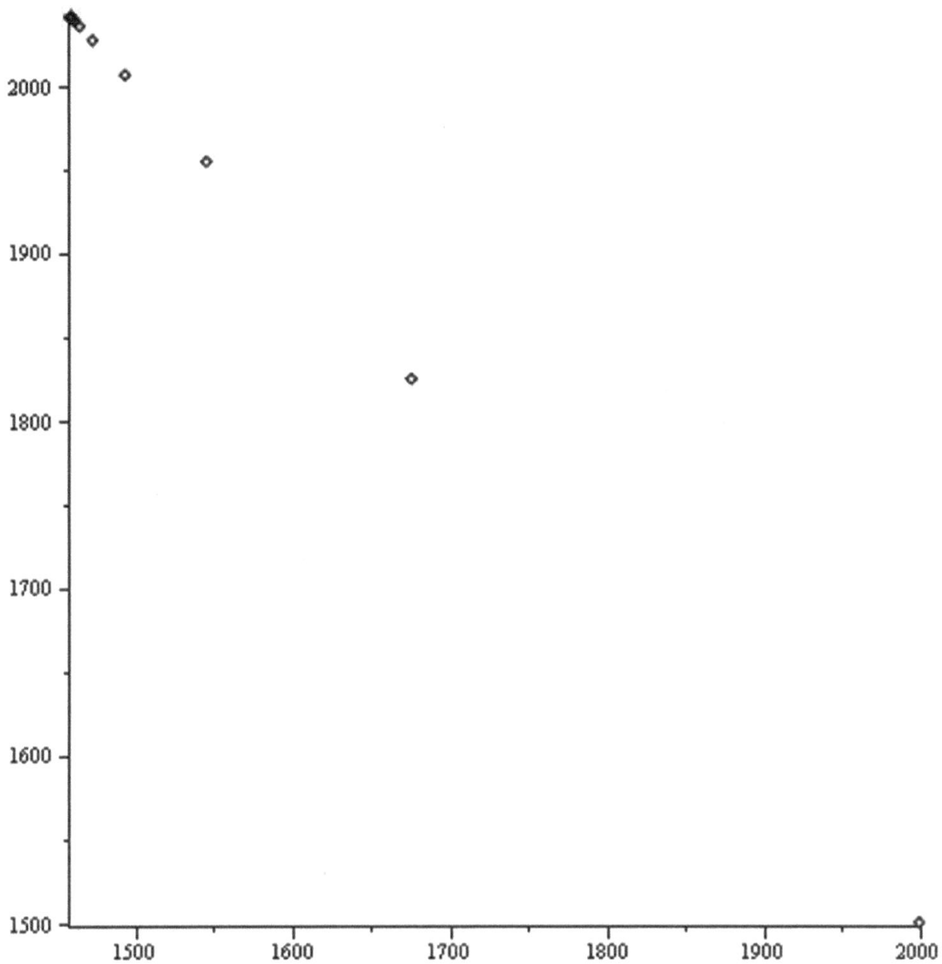

FIGURE 4.13
Python plot for student voucher.

What happens near these values gives great insight concerning the long-term behavior of the system. We can study the resulting pattern of the numerical solutions.

Example 4.7: Merchants Located Downtown and in Malls

Let's consider the attempt to revitalize the downtown section of a small city with merchants. There are merchants downtown and others in the large mall. Suppose historical records determined that 60% of the downtown merchants remain downtown while 40% move to the mall. We find that 70% of the mall merchants want to remain in the mall but 30% want to move to downtown. Figure 4.14 provides a graphical representation of the problem. Build a model to determine the long-term behavior of these merchants based upon this historical data.

Problem Identification: Determine the relationship of the merchants over time.

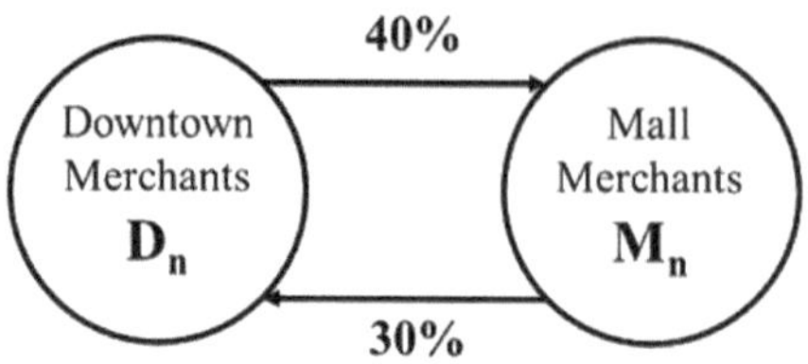

FIGURE 4.14
Flow diagram for Example 4.7.

Assumptions and Variables: Let n represent the number of business months. We define

D(n) = the number of merchants operating downtown at the end of n months

M(n) = the number of merchants operating at the mall at the end of n months

We assume that no other incentives are given to the merchants for either staying or moving.

The Model: The number of merchants downtown in any time period is equal to the number of downtown merchants that stay downtown plus the number of mall merchants that relocate downtown. The same is true for the number of mall merchants in any time period is equal to the number that remain in the mall plus the number of downtown merchants that move to the mall.

Mathematically, this is written as:

$$D_{n+1} = 0.60D_n + 0.30M_n$$
$$M_{n+1} = 0.4D_n + 0.7M_n$$

There are initially 150 merchants in the mall and 100 merchants downtown. We seek to find the long-term behavior of this system.

$$D(0) = 150 \text{ and } M(0) = 100 \text{ merchants, respectively.}$$

Figure 4.15 shows a graphical representation of the solution using Maple.

The following is the Maple code to solve Example 4.7.

```
u:=n->if n=0 then 150 else .6*u(n-1)+.3*v(n-1) fi:v:=n->if n=0 then
100 else .4*u(n-1)+.7*v(n-1) fi;

    v:= proc(n)
    option operator, arrow,
    if n=0 then 100 else 0.4u(n - 1) + 0.7v(n - 1) end if
    end proc

u(0);

    150

v(0);

    100

merchants:=seq(u(n),n=0..10);

    merchants:= 150, 120.0, 111, 108.30, 107.49, 107.247, 107.174,
    107.152, 107.146, 107.144, 107.143
```

```
> dds := rsolve({d(n) = .6*d(n-1)+.3*m(n-1), m(n) = .4*d(n-1)+.7*m(n-1), d(0) = 150, m(0) = 100}, {m, d});
```

$$dds := \left\{ d(n) = \frac{300\left(\frac{3}{10}\right)^n}{7} + \frac{750}{7},\ m(n) = -\frac{300\left(\frac{3}{10}\right)^n}{7} + \frac{1000}{7} \right\}$$

```
> plot([300/7*(3/10)^n+750/7,  -300/7*(3/10)^n+1000/7],n=0..25,thickness=3);
```

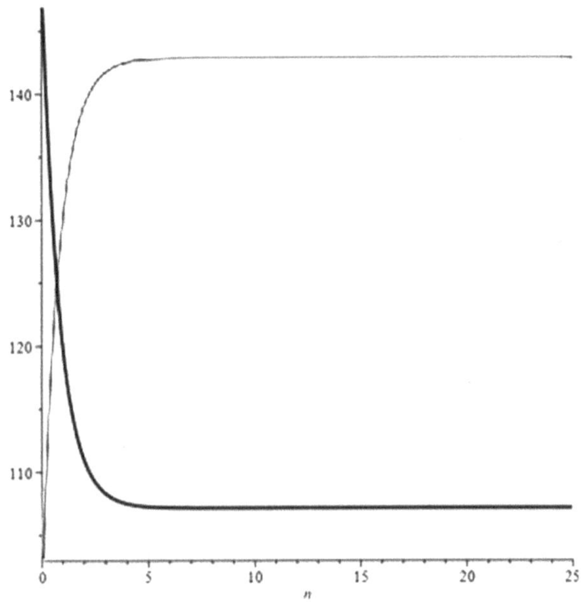

FIGURE 4.15
Maple solution for Example 4.7.

```
downtown:=seq(v(n),n=0..10);

    downtown:= 100, 130.0, 139.00, 141.70, 142.51, 142.75,
    142.826, 142.848, 142.854, 142.856, 142.857

with(plots):
n1:=pointplot({seq([n,u(n)],n=0..10)},title=`Retail`):
n2:=pointplot({seq([n,v(n)],n=0..10)}):
display({n1,n2});
```

The lot is seen in Figure 4.16.

Analytically, we can solve for the equilibrium values. We let $X = D_n$ and $Y = M_n$. From the DDS, we obtain the equations:

$$X = 0.6X + 0.3Y$$
$$Y = 0.4X + 0.7Y$$

and both equations reduce to X = ¾Y. There are two unknowns so we need a second equation. From the initial conditions, we know that X + Y = 250. We can use the equations

$$X + Y = 250 \text{ and } X = ¾\, Y \text{ to find the equilibrium values:}$$

$$X = 107.1428571 \text{ and } Y = 142.85714329$$

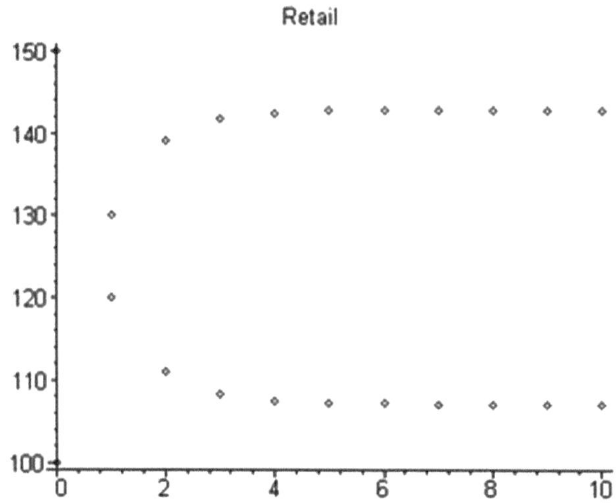

FIGURE 4.16
Plot of DDS for merchant's relocation (Example 4.7).

We iterate from near those equilibrium values and we find the sequences tend toward those values. We conclude that system has *stable* equilibrium values.

You should go back and change the initial conditions and see what behavior follows.

Interpretation: The long-term behavior shows that eventually (without other influences) of the 250 merchants about 107 merchants will be downtown and about 143 will be in the mall. We might want to try to attract new businesses to the community and add incentives for operating the business downtown.

The final plot is seen in Figure 4.17.

Exercises

4.1 For the example above, verify that the long-term ratio of $a(k)$ to $b(k)$ will be 3 to 2 by computing **A** (10) and comparing $a(10)$ to $b(10)$. Remember, we said that c_1 and c_2 values did not matter, therefore, do this for the following different initial conditions as well.

a. $a(0)=10 \quad b(0) = 40$

b. $a(0)=100 \quad b(0)=90$

c. $a(0)=200 \quad b(0)=10$

4.2 Determine the long-term behavior of the following solutions.

a. $\mathbf{A}(k)=c_1(.75)^k\begin{bmatrix}1\\2\end{bmatrix}+c_2(.1)^k\begin{bmatrix}3\\2\end{bmatrix}$

b. $\mathbf{A}(k)=c_1(4)^k\begin{bmatrix}4\\.5\end{bmatrix}+c_2(1/2)^k\begin{bmatrix}1\\1\end{bmatrix}$

c. $\mathbf{A}(k)=c_1(3)^k\begin{bmatrix}4\\.5\end{bmatrix}+c_2(-4)^k\begin{bmatrix}1\\1\end{bmatrix}$

```
> solve({150 = -.707106781186547·c1 + .6·c2, 100 = .707106781186547·c1 + .8·c2}, {c1, c2});
                                          {c1 = -60.60915267, c2 = 178.5714286}
> p1 := plot(-60.609·.3^k + .8·178.57, k = 0..20) :
> p2 := plot(60.609·.3^k + .6·178.57, k = 0..20) :
> with(plots) :
> display(p1, p2);
```

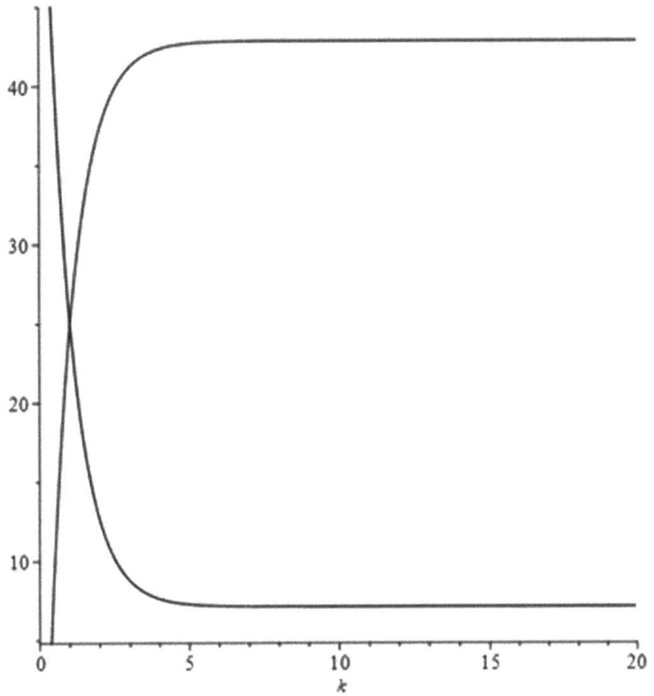

FIGURE 4.17
Final plot.

d. $A(k) = c_1(3/4)^k\begin{bmatrix}1\\2\\2\end{bmatrix} + c_2(4/5)^k\begin{bmatrix}2\\3\\5\end{bmatrix} + c_3(-5/4)^k\begin{bmatrix}3\\2\\1\end{bmatrix}$

4.3 A Move-It-Yourself Trucking Center operates from three major centers in Boston, Los Angeles, and Chicago. Each month, half of the trucks in Boston and Los Angeles go to Chicago while the other half stay where they are. The trucks in Chicago are divided equally between Boston and Los Angeles. Initially, there are 150 trucks in Boston and 75 apiece in Los Angeles and Chicago. They need to determine information about the future numbers of trucks in the three cities.

a. Draw a diagram with arrows for input and output which describes this situation.
b. Model this problem with a system of difference equations.
c. Find the equilibrium vector.
d. Determine the long-term distribution of trucks among the three cities.

4.4 A rental car agency divides its national market into three regions and has regional headquarters in Los Angeles, Chicago, and Miami. The company inventories and records the location of its cars monthly. Historical records show that of the cars rented in the Los Angeles region each month, 60% remain in the Los Angeles region, 30% go to the Chicago region, and 10% go to the Miami region. Of the cars rented in the Chicago region each month, 50% remain in the Chicago region, 10% go to the Los Angeles region, and 40% go to the Miami region. Of the cars rented in the Miami region each month, 50% remain in the Miami region, 20% go to the Los Angeles region, and 30% go to the Chicago region.
   a. Model this problem as a system of difference equations. It may help to sketch a diagram.
   b. Find the general solution in terms of characteristic values and characteristic vectors.
   c. What is the long-term distribution of cars between the three regions?
   d. Suppose there are originally 1000 cars in the Los Angeles region, 4000 cars in the Chicago region, and 3000 cars in the Miami region. How many cars will be in each region after three months?

4.5 Two supermarket chains, called A and B, are in competition. Eighty percent of the people who shop at market A in one month return to market A the next month, while the other 20% go to market B the next month. Seventy percent of the people who shop at market B in one month return to market B the next month, while the other 30% go to market A the next month. Assuming that these are the only two supermarkets available, what is chain A's percentage of the total customers (in the long term)?

4.6 Find the equilibrium values for the Predator-Prey model presented in Example 4.3.

4.7 In the Predator-Prey model, presented in Example 4.3, determine the outcomes with the following sets of parameters.
   a. Initial foxes are 200 and initial rabbits are 400.
   b. Initial foxes are 2000 and initial rabbits are 10,000
   c. Birth rate of rabbits increases to 0.1

4.8 In the SIR model, presented in Example 4.4, determine the outcome with the following parameters changing.
   a. The flu lasts one week.
   b. Initially 5 are sick and 10 the next week.
   c. The flu lasts four weeks.
   d. There are 4,000 students in the dorm and 5 are initially infected and 30 more the next week.

4.9 Consider spreading a rumor through a company of 1,000 employees all working in the same building. We assume that the spread of a rumor is similar to the spread of a contagious disease in that the number of people hearing the rumor each day is proportional to the product of the number hearing the rumor and the number who have not heard the rumor. This is given by the formula:

$$r(n+1) = r(n) + 1000\,k\,r(n) - k\,r(n)^2,$$

where k is the parameter that depends on how fast the rumor spreads. Assume k = 0.001 and further assume that four people initially know the rumor. How soon will everyone know the rumor?

PROJECTS

4.1 Consider the following chemical reaction problem adapted from the chemical reaction example in **Differential Equations, A Modeling Approach** by Giordano and Weir (1991)

Consider a chemical system involving three substances whose concentrations at any discrete time n are $x(n)$, $y(n)$, and $z(n)$, respectively. Over time, a portion of each substance is converted into one of the other two substances by a simple reaction involving only the substances being converted. The rate of change from one substance a to substance b is proportional to the amount of substance a with proportionality constant k_{ab}. This proportionality constant can be thought of as the percentage of substance lost for any given time period. Figure 4.18 shows the situation for the three substances.

To model the amount of substance x present in any time n, we must account for the amount of substance x lost to y and z and the amount of substance x gained from y and z in the given time period. Thus the change in substance x can be modeled by the equation:

$$x_{n+1} - x_n = k_{yx}y_n + k_{zx}z_n - k_{xy}x_n - k_{xz}x_n.$$

This equation can be written as:

$$x_{n+1} = (1 - k_{xy}k_{xz})x_n + k_{yx}y_n + k_{zx}z_n$$

Requirement 1.

a. Write the transfer equations for the other two substances (y and z) and give the first order system of difference equations which describes the reactions between the three chemicals for any time period n.

b. Write the system of equations using matrix notation.

Requirement 2. Given the following proportionality constants:

$$k_{xy} = 0.2 \quad k_{xz} = 0.3 \quad k_{yx} = 0.2 \quad k_{yz} = 0.3 \quad k_{zx} = 0.2 \quad k_{zy} = 0.3$$

a. Use a computer to find the characteristic values (eigenvalues) and the characteristic vectors (eigenvectors) of the coefficient matrix. (Note: Keep all of your work in exact numbers and fractions ... don't use decimal approximation.) Find a general solution to the system of dynamical equations. What happens to the substance after a long period of time? Does your solution make sense?

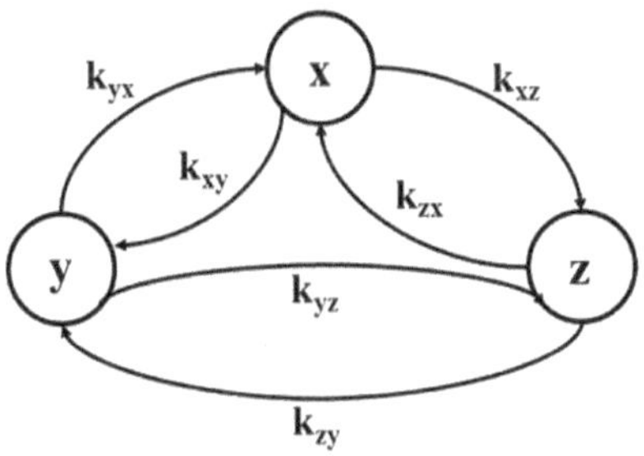

FIGURE 4.18
Chemical system for Project 4.1.

b. Find a particular solution if you initially have 3 units of substance x, 2 units of substance y, and 4 units of substance z. Describe the long-term behavior of the chemical system.
c. Verify your conclusions in requirement 2 by iterating the system on Quattro Pro.

Requirement 3. Each time period an additional 0.8 units of substance x and substance z are introduced into the chemical reaction.
a. Write the transfer equations for the three substances and give the first order system of difference equations which describes the reactions between the three chemicals for any time period n.
b. Write the system of equations using matrix notation.
c. Iterate the new system on Quattro Pro. Describe the long-term behavior of the chemical system.

4.2 Situation: Suppose we have a population of rare birds on a tropical island in the South Pacific. We know that there are numerous factors that affect the size of the population. However, we want to simplify the problem so that we can model the size of the population using a linear homogeneous discrete dynamical system. Therefore, we will consider only the birth and death rates associated with the population. We know that the difference in the size of the population between years is equal to the difference between the number of births and deaths that occur during the year.

Requirement 1:
a. Model the size of the population of birds on the island after n years, where the initial size of the population is 180 birds, the birth rate is 0.55, and the death rate is 0.51.
b. Iterate the model using technology, and analyze the long-term behavior of the population.
c. Suppose the death rate is 0.57. Model this new situation and analyze the long-term behavior of the population.
d. Suppose we use the logistics growth model to make our problem more realistic: $a(n+1) = (1+r)a(n) - ba^2(\text{n})$, where $b = r/\text{L}$. Again, use an initial population of 180 birds. Let the carrying capacity L be 190. Using the following values for the unrestricted growth rate r, iterate the model using technology and analyze the long-term behavior of the population using graphs that you generate: $r = 0.5$, $r = 1$, $r = 1.5$, $r = 2$, $r = 2.5$, $r = 2.75$, $r = 3$, $r = 3.1$.

4.3 Consider the contagious disease as the Ebola virus. Look up on the internet and find out some information about how deadly this virus actually is. Now consider an animal research laboratory in Reston, VA, and a suburb of Washington, DC, with a population of 856,900 people. A monkey with the Ebola virus has escaped its captivity and infected one employee (unknown at the time) during its escape. This employee reports to the University hospital later with Ebola symptoms. The Infectious Disease Center (IDC) in Atlanta gets a call and begins to model the spread of the disease. Build a model for the IDC with the following growth rates to determine the number infected after two weeks:
a. $\text{k} = 0.00025$
b. $\text{k} = 0.000025$
c. $\text{k} = 0.00005$
d. $\text{k} = 0.000009$
e. List some ways of controlling the spread of the virus.

4.4 Consider the spread of a rumor concerning termination among 1000 employees of a major company. Assume that the spreading of a rumor is similar to the spread of a contagious disease in that the number hearing the rumor each day is proportional to the product of those who have heard the rumor and those who have not heard the rumor. Build a model for the company with the following rumor growth rates to determine the number having heard the rumor after one week:

a. $k = 0.25$
b. $k = 0.025$
c. $k = 0.0025$
d. $k = 0.00025$
e. List some ways of controlling the spread of the rumor.

4.5 Lions and Spotted Hyena: Predict the number of lions and spotted hyenas in the same environment at a function of time.

Assumptions:

The variables: $L(n)$ = number of lions at the end of period n

$H(n)$ = number of hyenas at the end of period n

Assume the Model:

$$L(n+1) = 1.2L(n) - 0.001L(n)H(n)$$
$$H(n+1) = 1.3H(n) - 0.002H(n)L(n)$$

i. Find the equilibrium values of the system.
ii. Iterate the system from the following initial conditions and determine what happens to the lions and the spotted hyenas in the long term (Table 4.1).

4.6 It is getting close to election day. The influence of the new Independent Party is of concern to the Republicans and Democrats. Assume that in the next election that 75% of those who vote Republican vote Republican again, 5% vote Democratic, and 20% vote Independent. Of those that voted Democratic before, 20% vote Republican, 60% vote Democratic, and 20% vote Independent. Of those who voted Independent, 40% vote Republican, 20% vote Democratic, and 40% vote Independent.

Formulate and write the system of discrete dynamical systems that models this situation. Assume that there are 399,998 voters initially in the system, how many will vote Republican, Democratic, and Independent in the long run?

TABLE 4.1

Number of Lions and Hyenas for Project 4.5

Lions	Spotted Hyenas
150	200
151	199
149	210
10	10
100	100

(Hint: you can break down the 399,998 voters in any manner that you desire as initial conditions.)

(New Scenario) In addition to the above, the community is growing (18-year-olds + new people – deaths – losses to the community, *etc.*). Republicans predict a gain of 2,000 voters between elections. Democrats estimate a gain of 2,000 voters between elections. The Independents estimate a gain of 1,000 voters between elections. If this rate of growth continues, what will be the long-term distribution of the voters?

4.7 Pollution in the Great Lakes

Scenario: Let $e(n)$ and $o(n)$ be the total amount of pollution in Lake Erie and Lake Ontario, respectively, in year n. It has been determined that each year the percentage of water replaced in Lakes Erie and Ontario is approximately 36% and 11%, respectively. Since most of the water flowing from Lake Ontario is from Lake Erie, this means that each year, 36% of the water in Lake Erie flows into Lake Ontario and is replaced by rain and water flowing from other sources. Also, each year 11% of Lake Ontario's water flows out and is replaced by the water flowing in from Lake Erie. Assuming the concentration of pollution in each lake is constant throughout the lake, then 36% of the pollution in Lake Erie is removed each year. Each year 11% of the pollution in Lake Ontario is removed, but the pollution removed from Lake Erie is added to Lake Ontario. Further, assume that 2 tons of pollutants are added directly to Lake Erie each year and 12 tons of pollutants are added directly to Lake Ontario each year.

a. Make a flow diagram that summarizes the Scenario.
b. Develop a dynamical system for $e(n)$ and $o(n)$.
c. Assume there are now 10 tons and 30 tons of pollution in Lakes Erie and Ontario, respectively. Iterate your system to find the pollution levels twenty years from now.
d. Analytically find the equilibrium point for Lake Erie and Lake Ontario.
e. Suppose EPA determines that an equilibrium level of a total of 5 tons of pollutants in Lake Erie and a total of 15 tons in Lake Ontario would be acceptable. In order to achieve these equilibrium levels, what restrictions should be placed upon the total amount of additional pollutants that are dumped directly into Lake Erie and Lake Ontario?

References and Suggested Future Readings

Albright, B. (2010). *Mathematical Modeling with Excel*. Jones and Bartlett: Sudberry, MA.

Alfred, U. (1967). Sums of Squares of Consecutive Odd Integers. *Mathematics magazine*. 40(4): 194–199.

Allman, E., and Rhodes, J. (2004). *Mathematical Models in Biology: An Introduction*. Cambridge, UK: Cambridge University Press.

Arney, D., Giordano, F., and Robertson, J. (2002). *Mathematical Modeling with Discrete Dynamical Systems*. Boston, MA. McGraw Hill.

Fox, W. (2010). Discrete Combat Models: Investigating the Solutions to Discrete Forms of Lanchester's Combat Models. *International Journal of Operations Research and Information Systems*, (IJORIS). *1*(1): 16–34.

Fox, W. (2012). *Mathematical Modeling with Maple*. Boston, MA. Cengage Publishing.

Fox, W., and Driscoll, P. (2011). Modeling with Dynamical Systems for Decision Making and Analysis. *Computers in Education Journal* (COED). **2**(1): 19–25.

Fox, W. (2012). Discrete Combat Models: Investigating the Solutions to Discrete Forms of Lanchester's Combat Models. *Innovations in Information Systems for Business Functionality and Operations Management*, pp. 106–122. IGI Global & SAGE Publishers.

Fox, W. (2012). Mathematical Modeling of the Analytical Hierarchy Process Using Discrete Dynamical Systems in Decision Analysis. *Computers in Education Journal*. 3(3): 27–34.

Giordano. F., Fox, W., and Horton, S. (2014). *A First Course in Mathematical Modeling*, 5th edition. Boston, MA: Cengage Publishing, Boston.

Leyendekker, J., and Shannon, A. (2015). The Odd-Number Sequence: Squares and Sums. *International Journal of Mathematical Education in Science and Technology*. 46(8):1222–1228.

Sandefur, J. (2002). *Elementary Mathematical Model: A Dynamic Approach*, 1st edition. Belmont, CA: Brooks-Cole Publishers.

Sandefur, J. (2003). *Elementary Mathematical Modeling: A Dynamic Approach*. Belmont, CA: Thomson Publishing.

Sandefur, J. (1990). *Discrete Dynamical System: Theory and Applications*. New York, NY: Oxford University Press.

5

Introduction, Basic Concepts, and Techniques in Modeling First Order Ordinary Differential Equations

Modeling first order ordinary differential equations (ODEs) involve describing systems where the rate of change of a variable depends on the current state of that variable. First order ODEs will normally take the form:

$$\frac{dy}{dt} = f(t, y)$$

where y is the dependent variable, such as population, number of rental cars, and t is the independent variable (often time), and *f*(t, y) is a function defining the relationship between y and t.

Introduction

Example 5.1: Consider a sports parachutist, as shown in Figure 5.1.

We want to build a model on how far the parachutist free falls before they open their parachute. We might initially assume the person is falling from rest and choose to neglect all resistive forces. Or we might choose to neglect a variable because we want to investigate the problem without the influence of that particular variable. For instance, we might want to study the falling-body by neglecting buoyancy but considering the effects of drag within the earth's atmosphere. We can simplify the model for the resistive force by making a reasonable assumption that the drag force equals some constant times the speed of the falling-body or we might decide to try a constant times the speed squared of the falling -body. Next, we want to construct models for propulsion, drag, and buoyancy forces. The propulsion force acting on the falling body from rest is due to gravity. This gravitational attraction in turn depends on the mass of the falling-body and its distance above the surface of the earth. Thus,

$$\begin{aligned}\text{Propulsion force} &= F_p \\ &= \text{gravitational attraction} \\ &= f(\text{mass}, \text{distance})\end{aligned}$$

DOI: 10.1201/9781003582274-5

FIGURE 5.1
Free falling (https://alchetron.com/Free-fall).

Next, consider the resistive force, which is the sum of the drag and buoyancy forces:

$$\begin{aligned}\text{Drag} &= \text{F}_\text{d}\\ &= \text{f}(\text{speed, air density, cross-sectional area of the body, shape of the body})\end{aligned}$$

$$\begin{aligned}\text{Buoyancy} &= \text{F}_\text{b}\\ &= \text{f}(\text{air density, density of the body})\end{aligned}$$

A free-body diagram displays all the forces acting on the body with a coordinate system indicating directions.

From the free body diagram, Figure 5.2, it can be seen that the total force F acting on the body is given by

$$F = F_p + F_d + F_b$$

We assume that Newton's second law applies, $\Sigma F = ma$ (itself a mathematical model). If we assume to neglect all resisting forces we find

$$\sum F = ma = -mg + 0 + 0$$

or my″= −mg, y″ is a second derivative and subject to initial conditions

$$y(0) = h, y'(0) = 0$$

Here m is the mass of the body, g is the acceleration due to gravity, and y is the distance of the body measured from the surface of the earth. The acceleration a is the second derivative of y with respect to time, and it is this idea of change that led to a differential equation. Since the propulsion force is acting down, then it is negative.

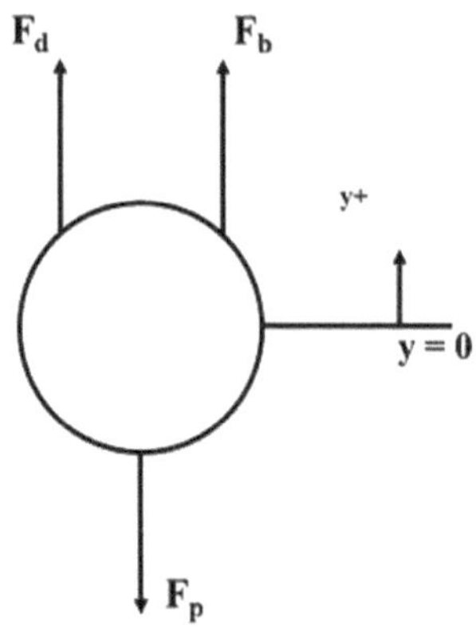

FIGURE 5.2
Free body diagram.

If we were to include the force due to air resistance as a constant times speed then $Fr = kv$. We then would have the model,

$$F = F_p + F_d + F_b$$

$$F = ma = my(t)'' = -mg + kv(t)$$

with the same initial conditions of $y(0) = h$, $y'(0) = 0$.

Bungee Jumping: Consider bungee jumping as shown in Figure 5.3

In bungee jumping from a bridge, a participant attaches one end of a bungee cord to himself (or herself), attaches the other end to a bridge railing, and then drops off the bridge. In this project, the jumper will be dropping off of the Royal Gorge Bridge, a suspension bridge that is 1053 feet above the floor of the Royal Gorge in Colorado.

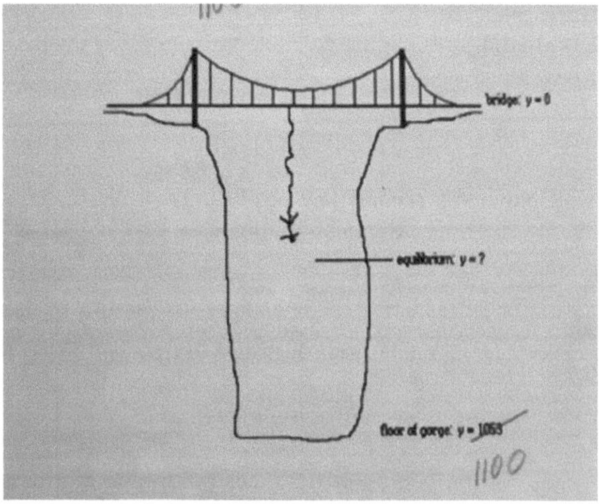

FIGURE 5.3
Bungee jumping diagram for Example 5.1.

The jumper will use a 200-foot-long bungee cord. It would be nice if the jumper has a safe jump, meaning that the jumper does not crash into the floor of the gorge or run into the bridge on the rebound. In this project, you will do some analysis of the fall.

Assume the jumper weighs 160 lbs. The jumper will free fall until the bungee cord begins to exert a force that acts to restore the cord to its natural (equilibrium) position. In order to determine the spring constant of the bungee cord, you found that a MASS weighing 4 lbs stretches the cord 8 feet. Hopefully, this spring force will help slow down the descent sufficiently so that the jumper does not hit the bottom of the gorge. Throughout this project, we will assume that DOWN is the POSITIVE direction.

Before the bungee cord begins to retard the fall of the jumper, the only forces that act on the jumper are his weight and the force due to wind (air) resistance.

If the force due to the wind resistance is 0.9 times the velocity of the jumper, then we use Newton's Second Law (F = ma to write a differential equation that models the fall of the jumper. We are sure to include the initial conditions for the jumper for your differential equation.

What is the velocity of the jumper after the jumper has fallen 200 feet?

What is the terminal velocity of the jumper, if any?

$$m x'' = -mg - kx'$$

Since k = 0.5 since from Hooke's Law ($F = Ks$) 4 = k*8 so k = 4/8 = 0.5

$$\mathrm{g} = 32.00\,\mathrm{ft/s^2}$$

W = mg so W/g = m. This implies m = 160/32.00 = 5

We divide the entire equation by m to get

$$x'' = -g/m - k/mx' - 0.9/mx'$$

$-k/m\ x'$ only after bungee has stretched out. Otherwise, delete during free fall.

$$5x'' = 160 - 0.9x'$$
$$v' + 0.180v = 32.00, \quad v(0) = 0$$

The following is the Maple code for solving the problem.

```
wght:= 160;
mass:= wght/60;
with(DEtools);
eqn:= mass*diff(V(t), t)+0.9*V(t) -160;
```

$$\mathrm{eqn} := 5\left(\frac{d}{dt}V(t)\right) + 0.9V(t) - 160$$

```
dsolve({eqn, V(0) = 0}, V(t));
```

$$V(t) = \frac{1600}{9} - \frac{1600}{9}e^{-\frac{9t}{50}}$$

```
simplify(%);
```

$$V(t) = \frac{1600}{9} - \frac{1600}{9}e^{-\frac{9t}{50}}$$

```
convert(%, float);
```

$$V(t) = 177.78 - 177.78e^{-0.18t}$$

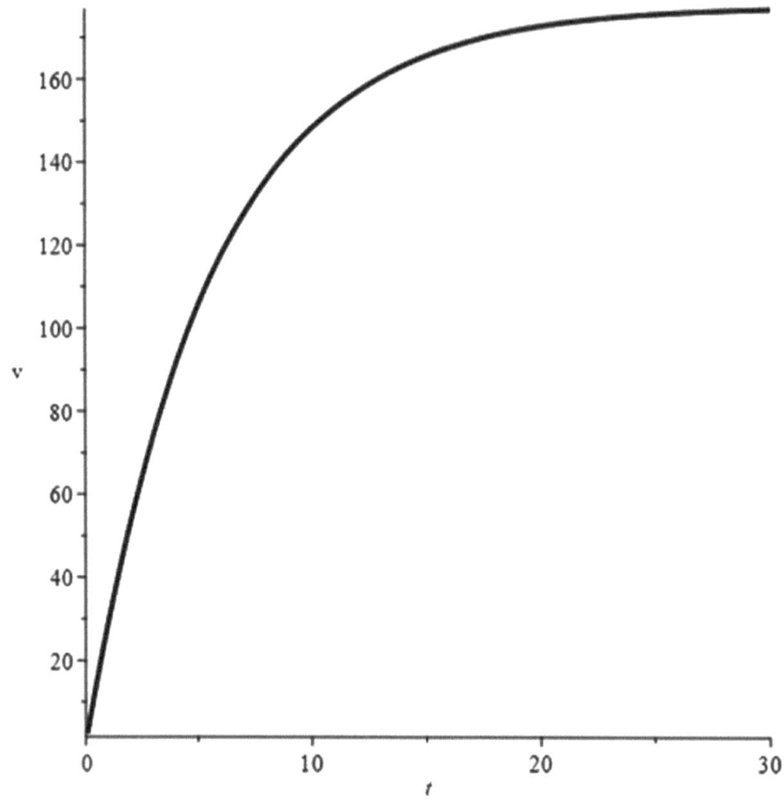

FIGURE 5.4
Bungee jumping velocity plot (Example 5.1).

From the solution and Figure 5.4, we see the terminal velocity is 1777.778.
We will see in more detail the solution methods and solution.

In this chapter, we will consider a variety of models, concepts, and techniques that give the reader some of the basic tools needed in solving and analyzing first order differential equations. Since differential equations are used often in mathematical modeling we devote a more thorough approach in this chapter and provide some models that we will solve and analyze. These models come from a variety of disciplines in science and engineering, including chemistry, physics, biology, fluids mechanics, Newtonian Mechanics, environmental engineering, and financial mathematics. Modeling techniques are discussed that help determine the necessary coefficients in the models and might refer the reader back to Chapter 2.

The **Slope Fields and Qualitative Assessments section** introduces slope fields. Slope fields provided qualitative viewing of solutions of first order differential equations. Recall that we said this book discusses the modeling of change. Slope fields provide information about rates of change since they are derived from derivative information.

We will also cover the analytical method of solution from Maple to help us solve differential equations. Existence and uniqueness are addressed for conceptual and theoretical basis. The focus is on the initial-value problem and the question as to whether or not the initial-value problem has a unique solution and whether the new functions can be defined by the initial-value problem.

We will also discuss various numerical techniques throughout this chapter. Euler, Improved Euler, and Runge-Kutta techniques are introduced as techniques to approximate solutions. Since problems may arise in numerical methods, some errors are discussed as well as methods to compare long-term behavior of the solutions to determine the qualitative accuracy of solutions.

Applied First Order Models

In this section, we introduce many mathematical models from a variety of disciplines. Our emphasis in this section is building the mathematical model and its associated differential equation that will be solved later in the chapter. Recall we have previously discussed the modeling process. In this section, we will confine ourselves to the first three steps: (1) Identifying the problem, (2) Assumptions and Variables, and (3) Building the model.

Example 5.2: Modeling Radioactive Decay (from Chapter 2)

A British physicist Lord Ernest Rutherford (1871–1937) established that an atom consists of a nucleus surrounded by electrons. The nucleus itself consists of protons and neutrons and two nuclei with identical numbers of protons and identical numbers of neutrons belonging to the same nuclear species or nuclide.

The accidental discovery of radioactivity by Henri Becquerel in 1896 was followed quite closely by that of x-rays. Becquerel had left a substance containing uranium on a photographic plate wrapped in black paper. After the plate was developed, it showed images of crystals of uranium compounds. The uranium had emitted penetrating rays, and Becquerel coined the word radioactive to describe this property. Later research by Becquerel, Madam Currie, and Rutherford led to the discovery that there are three different kinds of radioactive emissions: alpha, beta, and gamma radiation. Ordinarily, it is of interest or concern to know how long it will take for a certain radioactive substance to lose some fraction of its total mass through the emission process. Let's construct a model to provide these answers.

Assumptions: Assume we have an object that if exposed to a certain substance will emit radiation. Assume that the emissions are all the same kind and that they all decay by the same process (alpha, beta, gamma). The decaying process is such that once a particular nucleus decays, it cannot repeat the process again. Moreover, the probability of its decaying in any time interval is the same. Suppose that at time t there are N undecayed nuclei present. Then the number of nuclei that decay during the time interval from t to t + Δt must be proportional to the product of N and Δt. That is,

$$\Delta N = -kN\Delta t$$

where $k > 0$ is the proportionality coefficient, called the decay constant. Let us assume that the number of undecayed nuclei is a continuous function of time. Then dividing both sides of the last equation by Δt and passing to the limit as $\Delta t \to 0$ results in the following differential equation.

$$\frac{dN}{dt} = -kN$$

For a given radioactive element, the number k is determined from experimental data.

We desire information concerning the radioactive decay of an element.

Problem Identification: Determine a relationship between time and the amount of radiation emitting from a specific substance after exposure.

Assumptions: We assume the rate at which an item decays is proportional to the amount of that item at any time, t. The model is accurate over a short period of time. Half-life is the measure of stability of a radioactive substance. The half-life is simply the time, ti, takes one-half of the atoms to disintegrate. The longer the half-life, the more stable the item is.

Model Construction: Ordinarily, it is of interest to know how long it takes an object emitting radiation to lose some fraction of its total mass or how much of its mass is remaining after a specified amount of time. Many radioactive materials disintegrate at a rate proportional to the amount present. For example, if x is the radioactive material and $Q(t)$ is the amount of radioactive material present at time, t, then the rate of change of $Q(t)$ with respect to t is given by $\frac{dQ}{dt} = -kQ$, where k represents the proportionality constant and $k > 0$. Let us refer to $Q(0) = Q0$ as the initial quantity of radioactive material at any arbitrary time, $t = 0$.

Finding the proportionality constant k. from the data in Table 5.1:

TABLE 5.1

Nuclei Data for Example 5.2

Time	Number of Nuclei, N
0	100
1	58.5
2	34.2
3	20
4	11.7
5	7
6	4
7	2.3
8	1.4

Estimation using the law of radioactive decay. The law of radioactive decay leads to solution

$$Q(t) = Qo\,e^{-kt}.$$

From the data, we know initially that Q(0) = 100 grams. After 8 hours there are 1.4 grams remaining. So

$$
\begin{aligned}
Q(8) &= 1.4 = 100e^{-t(8)} \\
0.014 &= e^{(-8k)} \\
ln(.014)/-8 &= k \\
k &= 0.5335
\end{aligned}
$$

If we use this method then we might use the model, $Q(t) = Qo\, e^{-.5335kt}$.

If we use all the data, then we must create the change in the number of nuclei and then plot versus N. The slope (using linear regression without an intercept) will be our value for k.

Data (Table 5.2):

TABLE 5.2

Nuclei Data Difference for Example 5.2

Time	Number of Nuclei, N	Difference
0	100	–41.5
1	58.5	–24.3
2	34.2	–14.2
3	20	–8.3
4	11.7	–4.7
5	7	-3
6	4	–1.7
7	2.3	–0.9
8	1.4	

We plot the difference versus N and find the slope as –0.4152. If we use this approach our model is

$$Q(t) = Q0\,e^{(-0.4152\, t)}.$$

From a modeling perspective, if you have all the data, or a substantial amount of data, build the proportionality model to find the slope. If you only have a few data points, then use the model form to estimate the value of *k*.

Example 5.3: Newton's Law of Cooling

You are working as an assistant for the Crime Scene Investigators at a potential homicide. A body was found in a trash dumpster and you were called to the site immediately. The ambient temperature is approximately 68°F. The body temperature is 77°F. After one hour, the body temperature is 74°F.

Problem Identification: Build a mathematical modeling relating a dead body's cooling temperature to time in order to find the time of death.

Assumptions: A simplified list could include normal body temperature, average seasonal temperature outdoors, constant indoor temperature, and no other outside sources of heat or cooling.

Modeling Construction: Newton's Law of Cooling states that the surface temperature of an object changes at a rate proportional to its relative temperature. That is, the difference between its temperature and the surrounding environment's temperature. Let $T(t)$ = the temperature of the object at time t.

According to Newton's Law: $\frac{dT}{dt} = k\left(T - T_s\right)$ where k is the proportionality constant.

Example 5.4: Mixtures

Consider an initial amount of a substance, say barley measured in pounds, dissolved in a large vat in order to make beer. The solution is pumped in at one rate, well-mixed, and pumped out at the same rate. We want to know the concentration of the substance in the vat over time.

Problem Identification: Build a mathematical modeling relating the concentration of barley in the vat to time, t.

Assumptions: We know the size of the vat, 300 gallons. The initial amount dissolved is 50 pounds. The pumping rate is constant both into and out of the vat. No other substances are interacting with the vat and the pumps.

Model Construction: The rate of change of concentration is measured as the difference between the rate pumped in and the mixture being pumped out of the vat. Let $C(t)$ be the concentration or amount of barley in the vat after time, t, $\frac{C(t)}{dt} = R_{in} - R_{out}$.

Example 5.5: Population Models

Consider an experiment measuring the growth of a yeast culture. An analysis yields the growth rate as approximately 0.6. We might try $\frac{dP}{dt} = 0.6P$ as our ODE. Note that this model will predict a population that increases forever. We might think that is not possible.

Model Refinement: Modeling Births, Deaths, and Resources

Both births and deaths during a period are proportional to the population, then the change in population itself should be proportional to the population, as was illustrated above. However, certain resources (food, for instance) can support only a maximum population level rather than one that increases indefinitely. As these maximum levels are approached, growth should slow moving toward the carrying capacity.

Suppose we estimate the carrying capacity to be 665. Nevertheless, as pn approaches 665, the change does slow considerably. Since $665 - pn$ does get smaller as pn approaches 665, consider the following model:

$$\frac{dP}{dt} = k\left(665 - P\right)P$$

If we accept the proportionality argument, we can estimate the slope of the line approximating the data to be about $k \approx 0.00082$, which gives the model:

$$p_{n+1} - p_n = 0.00082(665 - p_n)p_n.$$ Converting to an ODE yields

$$\frac{dP}{dt} = 0.008271 \cdot P \cdot (665 - P)$$

Example 5.6: The Spread of a Contagious Disease

Suppose there are 400 students in a college dormitory and that one or more of the students has a severe case of the flu. Let *in* represent the number of infected students after *n* time periods. Assume some interaction between those infected and those not infected is required to pass the disease. If all are susceptible to the disease, (400 – *in*) represents those susceptible but not yet infected. If those infected remain contagious, we may model the change in infected as a proportionality to the product of those infected by those susceptible but not yet infected or,

$$\Delta i_n = i_{n+1} - i_n = k i_n (400 - i_n).$$

In this model, the product *in*(400 – *in*) represents possible interactions between those infected and those not infected. A fraction k would now become infected. There are many refinements to the above model. For example, we can consider that a segment of the population is not susceptible to the disease, that the infection period is limited, that infected students are removed from the dorm to prevent interaction with uninfected students, and so forth.

$$\frac{di}{dt} = ki(400 - i), \ k > 0$$

Some Classical First Order Models in Differential Equations

Decay:

$$\frac{dQ}{dt} = -kQ, \ k > 0$$

Growth:

$$\frac{dQ}{dt} = kQ, \ k > 0$$

Newton's Law of Cooling:

$$\frac{dT}{dt} = k(T - T_0), \ k > 0$$

Chemical Mixtures, Drug Dissemination, Dialysis:

$$\frac{dA}{dt} = R_{in} - R_{out}$$

L-R Series Circuit:

$$L\frac{di}{dt}+Ri=E(t)$$

Falling Bodies, Newton's Law of Σ Forces = 0

$$m\frac{dv}{dt}=F_{weight}+F_{resistance}=mg-kv, k>0$$

Chemical Reactions:

$$\frac{dX}{dt}=k(\alpha-X)(\beta-X)...$$

Spread of a Disease or Rumor:

$$\frac{dN}{dt}=kN(L-N),\ k>0,\ L>0$$

Tank with cross-sectional area A, filled from the top at rate *K*, while draining from the bottom through an orifice of area α, with height of the tank *h*:

$$\frac{dh}{dt}=\frac{\left(k-\alpha a\sqrt{2gh}\right)}{A}$$

Slope Fields and Qualitative Assessments of Autonomous First Order ODE

We have used the modeling process to create a number of models that are important to differential equations. However, sometimes we can predict the outcomes of models graphically without actually solving them. This is true when we are looking for behavior rather than specific values. For example, a specific population survive is a question that can be answered graphically.

In calculus, we used derivative information to analyze a function or just to gain more information. Knowing where a function is increasing or decreasing, locations of extrema, concavity, and so forth can be important information.

Interpreting the derivative as the slope of the line tangent to the graph of the function is useful in gaining information about the solution to the differential equation. From this information, it is possible to sketch qualitatively the solution curves to the equations from any real initial condition. These graphs provide considerable information regarding the solution and these graphs provide piratical benefits as well. Since most real differential equations cannot be solved analytically, other techniques, such as graphical analysis are needed to analyze the behavior of solutions.

Slope Field

Suppose the first order differential equation $\frac{dy}{dt} = f(t,y)$ is given. Each time an initial value $y(t_0) = y_0$ is specified, the solution curve is required to pass through the coordinate point (t_0, y_0) and we also know the value of slope at that point, $\frac{dy}{dt} = f\left(t_0, y_0\right)$. Therefore, it is possible to draw a short line segment with the correct slope through each point (t, y) in the $t - y$ plane. The result is known as the slope field of the first order differential equation. We can say that the slope field of a differential equation is the set of all line segments with the correct slope or mini-tangent lines.

Let's sketch the possible solution curves of the differential equation, $\frac{dy}{dt} = \frac{(t-y)}{2}$ without solving. Consider the information in Table 5.3 that provides specific points in the plane and the slope at each point.

In Maple, we must first call the library with(DEtools). Then we enter the differential equation and enter the commands to get a specific plot. The following Maple code will provide a complete slope field over the region $-2 \le t \le 2$ as shown in Figure 5.5.

```
with(DEtools):
deq:=diff(y(t),t)=(t-y(t))/2,y(t));

example1:= diff(y(t), t) = ((t-y(t)))/2

DEplot(deq,y(t), t=-2..2, y=0..2);
```

We use Maple to produce the slope field plot shown in Figure 5.5. Although knowledge of how to graph slope fields is important, you will not want to graph complete slope fields without technology. A solution curve is then tangent to the slope line at each point through which the curve passes. Thus, the slope field gives a visual representation of what a family of possible solution curves of the differential equation looks like. Several solution curves are now depicted in Figure 5.5 for our population model dP/dt = 0.008271·P·(665-P) with starting points (0, 2) and (0, 900). The possible solution curves are shown in Figure 5.6.

TABLE 5.3

Specific Points for Modeling

Point	Slope
(0,1)	–1
(0,2)	½
(1,0)	1
(2,0)	–1/2
(–1,0)	–1
(–2,0)	0
(1,1)	–1
(–1,1)	0
(2,2)	–1/2
(1,2)	–1/2

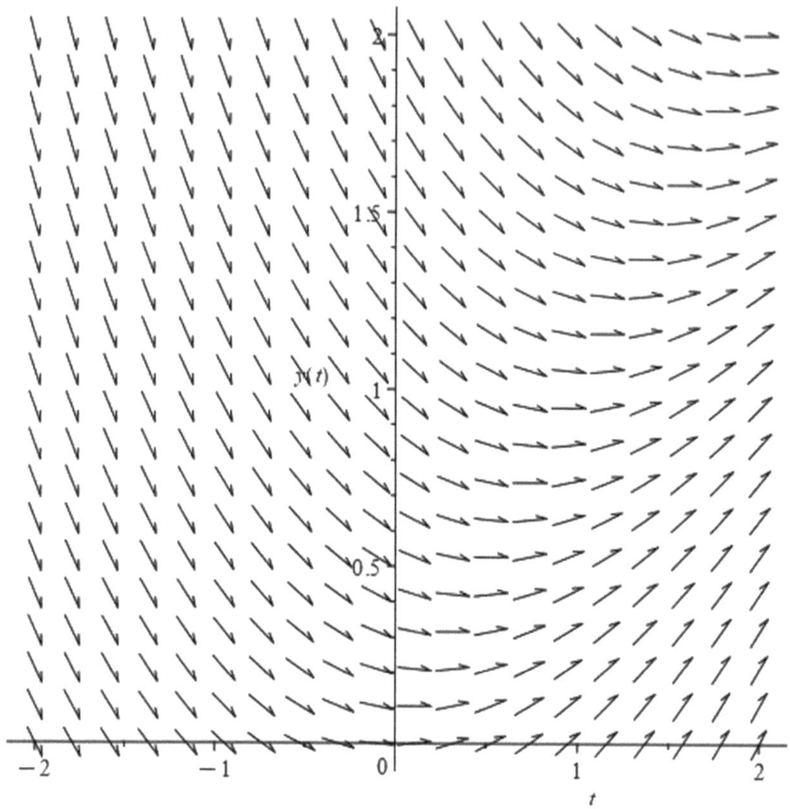

FIGURE 5.5
Slope filed for $\frac{dy}{dt} = \frac{(t-y)}{2}$.

```
>pop:=diff(P(t),t)=0.00871P(t)(665-P(t));

pop:=d/dt P(t)=0.00871P(t)(665-P(t))

>DEplot(pop,P(t),t=0…P=0…1000,color=black,thckness=3,{P(0)=900});
```

We examine the slope field and we notice that the solution curves do not cross tangent lines. We also see that the solution curves from the different starting points appear to converge to about 665.

Below are the Maple commands to obtain the slope fields and possible solution curves for the following first order differential equations:

```
y^'=y-t^2
> example1:=diff(y(t),t)=y(t)-t^2;
example1:=d/dt y(t)=y(t)-t^2

> DEplot(example1,y(t),t=0..5,y=0..10,{[0,2],[3,3]});
```

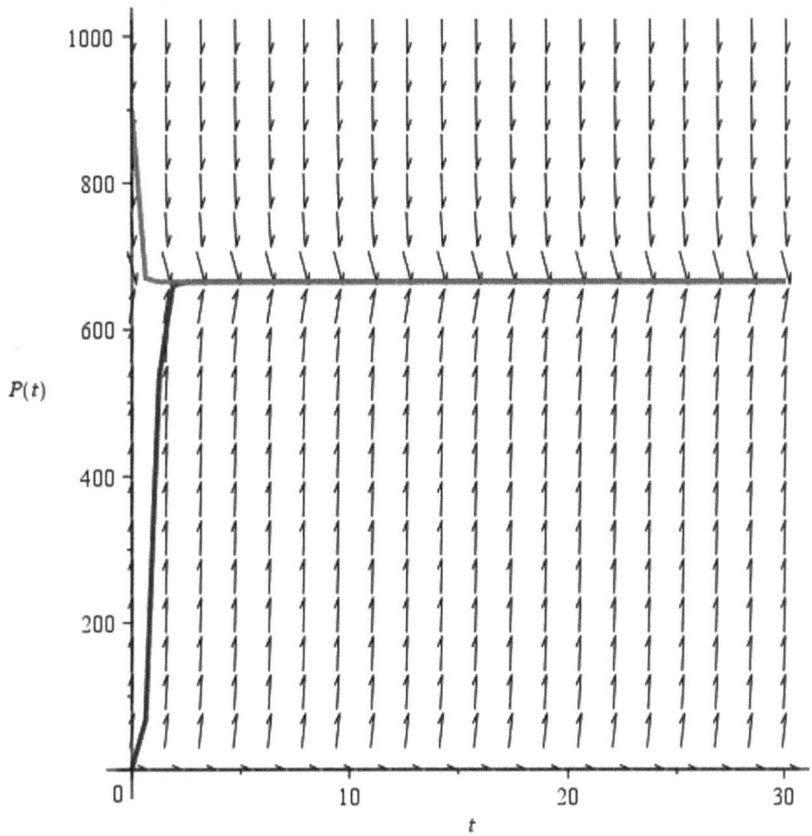

FIGURE 5.6
Slope field for $\frac{dP}{dt} = 0.008271 \cdot P \cdot (665 - P)$ with starting points (0,2) and (0,900).

```
y^'=y· (2-y)
> example2:=diff(y(t),t)=y(t)*(2-y(t));

example2:=d/dt y(t)=y(t)(2-y(t))

> DEplot(example2,y(t),t=0..5,y=0..5,{[0,.5],[0,2],[0,4]});

y^'=sin(t+y)

> example3:=diff(y(t),t)=sin(t+y(t));

example2:=d/dt y(t)=sin(t+y(t))

> DEplot(example3,y(t),t=0..5,y=0..7,{[0,.5],[0,4],[0,6]});
```

Slope fields provide us with possible solution curves for a differential equation. These slope fields are shown in Figures 5.7–5.9.

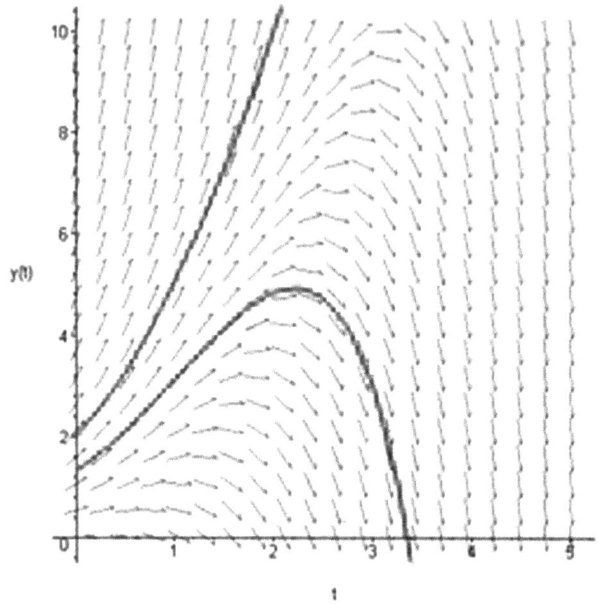

FIGURE 5.7
Slope field with IC plotted $y' = y - t^2$.

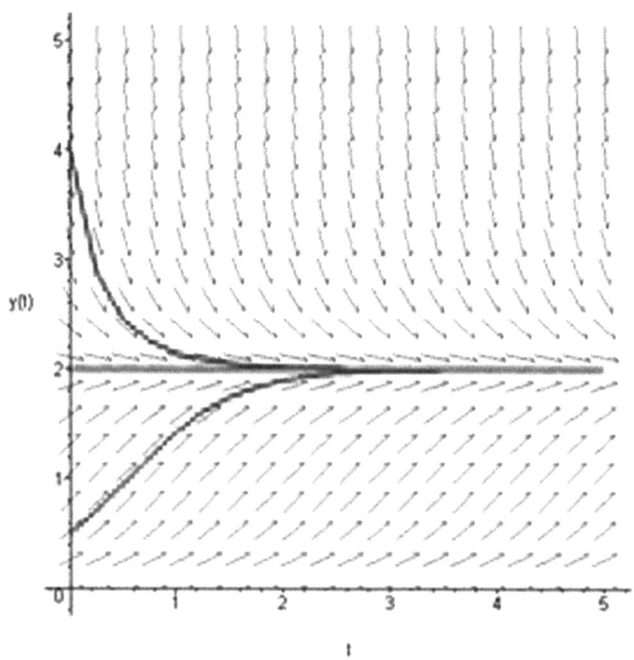

FIGURE 5.8
Slope field $y' = y(2 - y)$.

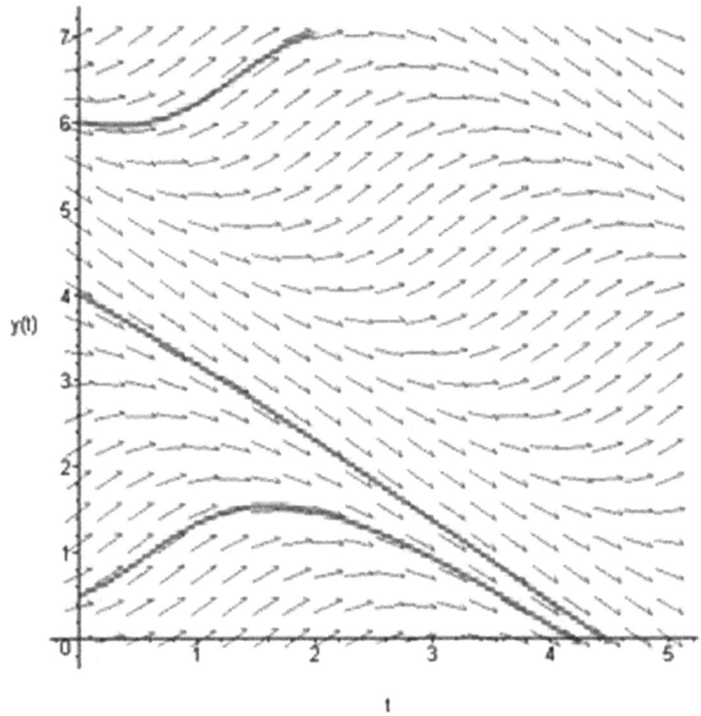

FIGURE 5.9
Slope field for $sin(t + y)$ with a few ICs.

Autonomous Qualitative Assessment

In many modeling applications the instantaneous rate of change $\frac{dy}{dt}$ is assumed to be proportional to some function of the dependent variable y alone. These are called autonomous differential equations, $\frac{dy}{dt} = f(y)$. In these cases, it is possible to analyze the nature of the solution curve by analyzing $\frac{dy}{dt}$ versus y. Let's explore this qualitative assessment.

Rest Points

Qualitative information concerning the solutions can often be obtained from the differential equation. For example, consider the differential equation describing the Malthusian model of population growth discussed earlier: $\frac{dP}{dt} = kP, k > 0$. Thus, the Malthusian model assumes a simple proportionality between growth rate and population (Figure 5.10). Since P > 0 and k > 0, you can see that $\frac{dP}{dt} > 0$, which means that the population curve P(t) is everywhere increasing. Moreover, the smaller the value of k, the less rapid the growth in the population over time. For a fixed value of k, as P increases so does its rate of change dP/dt. Thus, the solution curves must appear qualitatively as depicted in Figure 5.11. Assuming k is constant, at each population level P the derivative

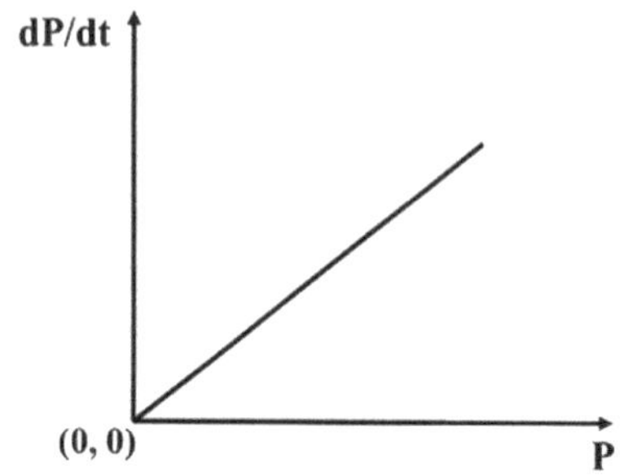

FIGURE 5.10
Graph of dP/dt versus P.

dP/dt is constant. Therefore, all the solution curves are horizontal translations of one another. The initial population P(0) = P0 distinguishes these various curves.

Next, recall the refined population model studied also from Chapter 4: $\frac{dP}{dt} = r(M - P)P$. The graph of $\frac{dP}{dt}$ versus P is the parabola shown in Figure 5.11. Since the second derivative $P'' = rP'(M - 2P)$ is zero when $P = \frac{M}{2}$, positive for $P < \frac{M}{2}$, and negative for $P > \frac{M}{2}$, we conclude that dP/dt is at a maximum at the population level. Note that for 0 < P < M, the derivative dP/dt is positive and the population P(t) is increasing; for M < P, dP/dt is negative and P(t) is decreasing. Moreover, if $P > \frac{M}{2}$, the second derivative P″ is positive and the population curve is concave upward for $P > \frac{M}{2}$, P″ is negative and the population curve is concave downward.

The parabola in Figure 5.11 has zeros at the population levels P = 0 and P = M. At those levels, dP/dt=0, so no change in the population P can occur. That is, if the population level is at 0, it will remain there for all time; if it is at level M, it will remain there for all time. Points for which the derivative dP/dt at 0 are called rest points, equilibrium points, stationary points, or critical points of the differential equation. The behavior of solutions near rest points is of significant interest to us. Let's examine what happens to the population P is near the rest points P = M and P = 0.

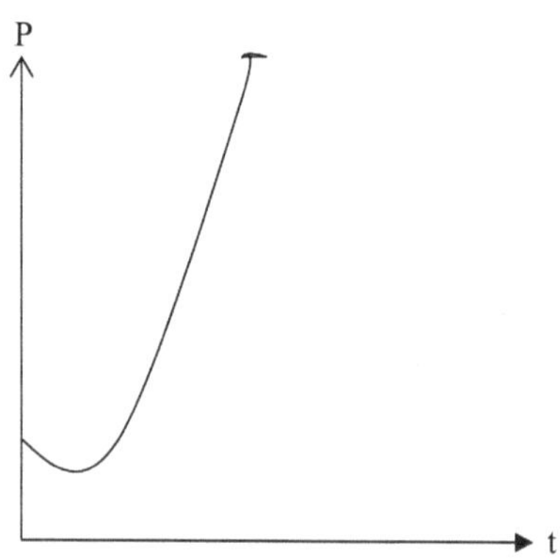

FIGURE 5.11
Plot of P versus t from Figure 5.8.

Suppose the population P is slightly less than M. Then dP/dt is positive, so the population increases and gets closer to M. On the other hand, if P > M, dP/dt < 0, so the population will decrease toward M. Thus no matter what positive value is assigned to the starting population, it will tend to the limiting value M as time tends to infinity. We say that M is an asymptotically stable rest point because whenever the population level is perturbed away from that level it tends to return there again. However, if the starting population is not at level M the population P(t) cannot reach M in a finite amount of time. This fact follows from the property that P = M is a solution to the limited growth equation and the two solutions cannot cross.

Next, consider the rest point P = 0 in Figure 5.9. If P is perturbed slightly away from 0 so that M > P > 0, then dP/dt is positive and the population increases toward M. In this situation, we say that P = 0 is an unstable rest point because any population not starting at that level tends to move away from it.

In general, equilibrium solutions to differential equations are classified as stable or unstable according to whether, graphically, nearby solutions stay close to or converge to the equilibrium, or diverge away from the equilibrium, respectively, as the independent variable t tends to infinity.

At $P = \frac{M}{2}$, P″ is zero and the derivative $\frac{dP}{dt}$ is at a maximum (see Figure 5.12). Thus the population P is increasing most rapidly when $P = \frac{M}{2}$ and a point of inflection occurs in the graph there. These features give each solution curve its characteristic "S" or sigmoid shape. From the information we have obtained, the family of solutions to the limited growth model must appear (approximately) as shown in Figure 5.12. Notice again that at each population level P, the derivative $\frac{dP}{dt}$ is constant, so that the solution curves are horizontal translations of one another. The curves are distinguished by the initial population level $P(0) = P0$. In particular, note the solution curve when

$$\frac{M}{2} < P0 < M \text{ in Figures 5.12 and 5.13.}$$

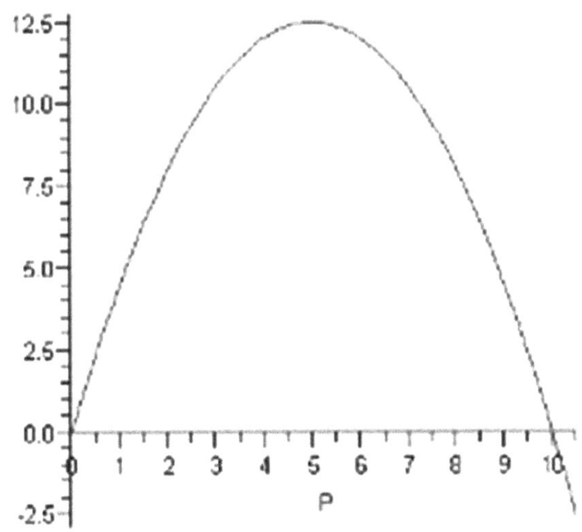

FIGURE 5.12
Plot of $\frac{dP}{dt}$ versus P assuming that $M = 10$.

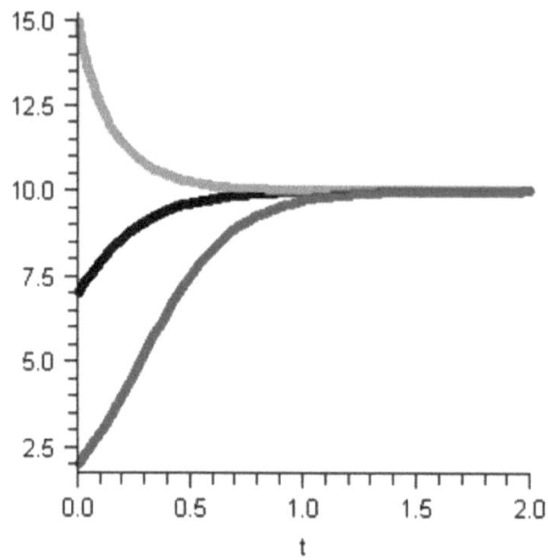

FIGURE 5.13
Possible plots of P versus t with M = 10.

Analytical Solution to 1st Order ODEs

In this section, we present analytical techniques for separable and linear differential equations.

Separable ODEs

The differential equation of the form $\frac{dy}{dt} = f(t,y)$ is called separable, if $f(t,y) = h(t) \cdot g(y)$. That is, $\frac{dy}{dt} = h(t) \cdot g(y)$. You need to be able to write the function $f(t,y)$ as $h(t) \cdot g(y)$.

Example 5.7

$$\frac{dy}{dt} = t \cdot y$$

This is separable because the function $f(t,y) = h(t) \cdot g(y) = t \cdot y$.

Example 5.8

$$\frac{dy}{dt} = t + 9y$$

This is not separable because $f(t,y) = t + 9y$ cannot be written as the product of two functions, $h(t) \cdot g(y)$.

Sometimes this is more complicated to see as in the following example.

Example 5.9

$$\frac{dy}{dt} = 8ty - 12y - 2t + 3$$

This can be written as $\frac{dy}{dt} = (2t-3)\cdot(4y-1)$ and is separable in this form.
In order to solve separable ODEs, we perform the following steps:

1. Separate the differential equation into a clear form of

$$\frac{dy}{dt} = h(t)\cdot g(y).$$

2. Rewrite the differential equation as

$$\frac{dy}{g(y)} = h(t)dt$$

3. Integrate both sides of the expression and include the constant of integration, C,

$$\int \frac{dy}{g(y)} = \int h(t)\cdot dt$$

4. to obtain the general solution:

$$G(y) = H(t) + C$$

5. If you are given an IVP, use the initial conditions to find the particular solution.

Example 5.10: Solve $\frac{dy}{dt} = t\cdot y$

$$\frac{dy}{dt} = tdt$$

$$\int \frac{dy}{y} = \int t\,dt$$

$$ln|y| + C = \frac{t^2}{2} + C$$

$$e^{ln|y|} = e^{\frac{t^2}{2}+C}$$

$$y = ce^{\frac{t^2}{2}}$$

If the ODE $\frac{dy}{dt} = t \cdot y$ passes through the point (1, 1), then we use (1, 1) as the initial condition.

$$1 = ce^{\frac{1}{2}}$$

$$c = \frac{1}{e^{\frac{1}{2}}} = e^{-.5} = 0.60653$$

$$y = 0.60653e^{\frac{t^2}{2}}$$

We plot the solution in Figure 5.14:

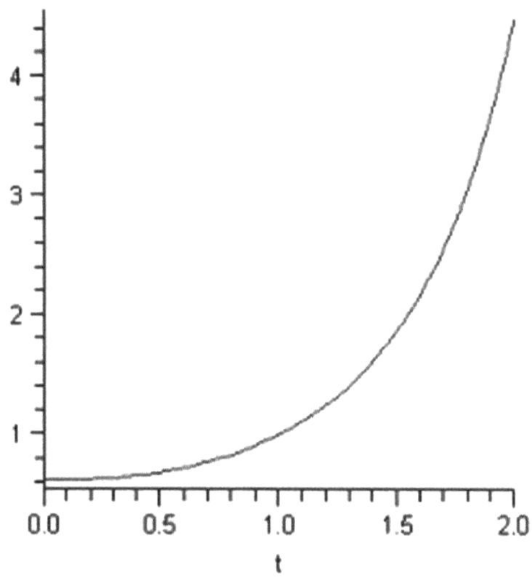

FIGURE 5.14
Plot of $y = 0.60653e^{\frac{t^2}{2}}$.

Example 5.11: Radioactive Decay

Consider the following radioactive decay model, $\frac{dQ}{dt} = -kQ$, $k > 0$ with a radioactive isotope with a half-life of 16 days. You wish to have 30 grams at the end of 25 days.

$$\frac{dQ}{dt} = -kdt$$

$$ln|Q| = -kt + C$$

$$Q = ce^{-kt}$$

Since the half-life is 16 days, then

$$Q = ce^{-kt}$$

$$\frac{Q}{2} = ce^{-16k}$$

$$k = \frac{ln|2|}{16} = 0.043322$$

$$Q = ce^{-.043322t}$$

We want 30 grams after 25 days, so

$$30 = ce^{(-.043322 \cdot 25)}$$

$$c = \frac{30}{e^{-.043322 \cdot 25}} = 88.61\,g$$

Initially, we should start with 88.61 grams of the radioactive substance.
The model is $Q = 88.61 \cdot e^{-.043322t}$.

Example 5.12: Malthusian Population Model

The population of the USA has been given in Table 5.4.
The model that we will initially use is our simple growth model, $\frac{dP}{dt} = kP, k > 0$.

SOLUTION

$$\frac{dP}{p} = kt$$

$$ln|P| = kt + C$$

$$P = ce^{kt}$$

Our initial population in 1990 was 249.4 so $P(0) = 249.4$.

$$P = 249.4e^{kt}$$

TABLE 5.4

US Population Data for Example 5.12

Year	Population in Millions
1990	249.4
1991	252.2
1992	254.9
1993	257.7
1994	260.2
1995	262.7
1996	265.1
1997	267.7
1998	270.2
1999	272.6
2000	280.5

We need to find a good estimate for the growth parameter k.
We find the estimate for k is 0.00493.

$$P = 249.4e^{0.00493t}$$

Predicting the population in 2020 yields

$$P = 249.4e^{0.00493 \cdot 20} = 275.24 \text{ million people}$$

There are two problems with this model. First, it is not realistic to assume that populations tend to infinity. Second, the predictions of this model always underestimate the real populations. We could use a refinement to this model which will be suggested later that does not yield slow exponential growth toward infinity.

Example 5.13: Newton's Law of Cooling - Revisted

Suppose a cold can of soda is taken from a refrigerator and placed in a warm classroom and the temperature is measured periodically. The temperature of soda is initially 40°F and the room temperature is a cozy 72°F.
Newton's Law of Cooling gives

$$\frac{dT}{dt} = -k(T - T), T(0) = 40$$

$$\frac{dT}{dt} = -k(72 - T), T(0) = 40$$

$$\frac{dT}{(72 - T)} = -kdt$$

$$ln|72 - T| = -kt + C$$

Assume $k = 0.05$.

$$72 - T = Ce^{-.5t}$$

When $T(0) = 40$, $C = 32$

$$T = 72 - 32e^{-.05t}$$

After 10 minutes the soda's temperature is

$$T(10) = 72 - 32 \cdot e^{-.05 \cdot 10} = 52.591$$

After 30 minutes the soda's temperature is

$$T(10) = 72 - 32 \cdot e^{-.05 \cdot 30} = 64.860$$

Note the temperature of the soda warms and eventually will reach the temperature of the room, a cozy 72°F (but not cozy for drinking a soda).

Example 5.14: Chemical Mixtures

Initially 50 pounds of salt is dissolved in a 300-gallon tank of water. A brine solution is pumped into the tank at a rate of three gallons per minute, and a well-stirred mixed solution is then pumped out at a rate of three gallons per minute. We want to determine the amount of salt present at any time t if the concentration of the entering solution is two pounds per gallon.

$$\frac{dA}{dt} = R_{in} - R_{out}$$

$$R_{in} = \frac{3\ gallons}{1\ minute} \cdot \frac{2\ pounds}{1\ gallon} = 6\ pounds\ per\ minute$$

$$R_{out} = \frac{3\ gallons}{1\ minute} \cdot \frac{\frac{A}{300\ pounds}}{1\ gallon} = \frac{A}{100}\ pounds\ per\ minute$$

$$\frac{dA}{dt} = 6 - \frac{A}{100},\ A(0) = 50$$

We will rewrite the problem as:

$$\frac{dA}{dt} = 0.01(600 - A), A(0) = 50$$

This is in separable form.

$$\frac{dA}{600 - A} = 0.01dt$$

$$-ln(600 - A) = 0.01t + C$$

$$ln(600 - A) = -0.01t + C$$

$$\mathrm{A} = 600 + \mathrm{ce}^{-.01\mathrm{t}}$$

Using the initial condition, $A(0) = 50$

$$C = -550$$
$$A = 600 - 550e^{-.01t}$$

Example 5.15: Refined Populations Model

Consider the logistic growth model defined for blue crabs in the Chesapeake Bay,

$$\frac{dB}{dt} = 0.25B(10 - B), B(0) = 4. \quad (B \text{ is in millions of bushels.})$$

$$\frac{dB}{B(10 - B)} = .25\,dt$$

$$\int \frac{dB}{B(10 - B)} = \int .25\,dt$$

$$\frac{A}{B}+\frac{D}{10-B}=1$$

$$A\cdot(10-B)+B\cdot D=1+0B$$

$$10A=1,\ A=\frac{1}{10}$$

$$-A+D=0, A=D, D=\frac{1}{10}$$

So, $\int\frac{dB}{B(10-B)}=\int B\frac{1}{10B}dB+\frac{1}{10}\int B\frac{1}{(10-B)}dB=\frac{1}{10}\left[ln(B)-ln(10-B)\right]$

$$\frac{1}{10}\left[ln(B)-ln(10-B)\right]=.25t+C$$

Multiply both sides by 10,

$$\left[ln(B)-ln(10-B)\right]=2.5t+C$$

Solve for C, using $B(0)=4$

$$ln(4)=ln(6)=C$$

$$C=-.405465$$

Using the laws of logarithms and exponents, we obtain

$$\frac{B}{10-B}=0.666666\cdot e^{2.5t}$$

We simplify to

$$B=\frac{6.66666\cdot e^{2.5t}}{1+.66666\cdot e^{2.5t}}\quad\text{or}\quad B=\frac{6.6666666}{.666666+e^{-2.5t}}$$

The result is known as a logistics curve. It approaches the maximum sustainable amount of 6.666 million bushels.

Linear Differential Equations

Let's consider differential equations of the form, $a_1(t)\frac{dy}{dt}+a_0(t,y)=g(t)$, which can be simplified by dividing every term by $a1(t)$ to obtain $\frac{dy}{dt}+P(t)y=f(t)$. We seek a solution to this differential equation. Using the exact method, $dy + P(t)y - f(t))dt = 0$, it is probably not true that

$$\frac{\partial M}{\partial y}=\frac{N}{\partial t}\cdot\partial$$

If there exists a function $u(t)$ such that $\frac{\partial}{ty}u(t) = \frac{\partial}{\partial y}\left[u(t)\cdot(P(t)y - f(t))\right]$, then we will have an exact equation. $\frac{\partial}{ty}u(t) = \frac{\partial}{\partial y}\left[u(t)\cdot\left(P(t)y - f(t)\right)\right]$ reduces to $\frac{du}{dt} = u \cdot P(t)$. This is now separable so

$$\frac{du}{u} = P(t)dt$$

$$ln|u| = \int P(t)dt$$

$$u = e^{\int P(t)dt}$$

Now, this function $u = e^{\int P(t)dt}$ is called the **integrating factor** for the linear equation. We will use this concept of the integrating factor in the linear solution method, also called **variation of parameter.**

Step 1: Put the differential equation in the form: $\frac{dy}{dt} + P(t)y = f(t)$.

Step 2: Find the integrating factor: $e^{\int P(t)dt}$.

Step 3: Multiply through the differential equation by the integrating factor

$$e^{\int P(t)dt\frac{dy}{dt}} + e^{\int P(t)dt} \cdot P(t)y = e^{\int P(t)dt} \cdot f(t).$$

Step 4: This simplifies to $\frac{d}{dt}\left[y \cdot e^{\int P(t)dt}\right] = e^{\int P(t)dt} \cdot f(t)dt$.

Step 5: Integrate both sides, include constant of integration:

$$\int \frac{d}{dt} \cdot \left[y \cdot e^{\int P(t)dt}\right] = \int e^{\int P(t)dt} \cdot f(t)dt$$

$$y \cdot e^{\int P(t)dt} = \int e^{\int P(t)dt} \cdot f(t)dt + C$$

Step 6: Solve for y.

Example 5.16: Let's take a look at our example from slope fields, $\frac{dy}{dt} = \frac{(t-y)}{2}$.

Step 1: $\frac{dy}{dt} + \frac{y}{2} = \frac{t}{2}, P(t) = \frac{1}{2}$

Step 2: The integrating factor is $e^{\int \frac{dt}{2}} \cdot u = e^{\frac{t}{2}}$

Steps 3 & 4: $e^{\frac{t}{2}} \cdot \frac{dy}{dt} + e^{\frac{t}{2}} \frac{y}{2} = e^{\frac{t}{2}} \frac{t}{2}$ simplifies to $\frac{d}{dt}\left[y \cdot e^{\frac{t}{2}}\right] = \frac{t}{2} \cdot e^{\frac{t}{2}}$.

Step 5: $\int \frac{d}{dt}\left[y \cdot e^{\frac{t}{2}}\right] = \int \frac{t}{2} \cdot e^{\frac{t}{2}} dt$ (use integration by parts)

$$y \cdot e^{\frac{t}{2}} = t \cdot e^{\frac{t}{2}} - 2 \cdot e^{\frac{t}{2}} + C$$

Step 6: $y = t - 2 + C \cdot e^{\frac{-t}{2}}$

Example 5.17: Newton's Law of Cooling – One More Time

$$\frac{dT}{dt} = -k\left(T_m - T_s\right),$$

where

k > *0, T_s* = constant temperature of the surroundings,
Tm = temperature of the mass being cooled, and
T(0) = *α* (the initial temperature of the mass).

Let's assume that $k = .190$, $Ts = 70$, $T(0) = 300$. Now, $\frac{dT}{dt} = -0.190\left(T_m - 70\right), T(0) = 300.$ We notice that this equation is both separable and linear. We will use the linear method.

Step 1: Standard form is $\frac{dT}{dt} + .190\left(T_m\right) = 0.190 \cdot (70) = 13.3, T(0) = 300.$
Step 2: $P(t) = .190$

$$u = e^{\int P(t)dt} = e^{.19t}$$

Steps 3 & 4: $\frac{d}{dt}\left[T_m \cdot e^{.19t}\right] = 13.3 \cdot e^{.19t}$

Step 5: $\frac{d}{dt}\left[T_m \cdot e^{.19t}\right] = 13.3 \cdot e^{.19t}$

$$T_m \cdot e^{.19t} = 70e^{.19t} + C$$

Step 6: $T_m = 70 + Ce^{-.19t}$

Use $T(0) = 300$

$300 + 70 + C$

$C = 230,$ so $T_m = 70 + 230 + e^{-.19t}$

We note that the $\lim_{t \to \infty} 70 + 230 \cdot e^{-.19t} = 70$ (the room temperature) as expected.

First Order ODE Models

We will start with the "The Canon Problem." I first used the Cannon problem in 1983 at the United States Military Academy in our differential equations course. The version is an adaptation of it and a model by Arney et al. For a historical background see Arney et al.

Let's start with projectile motion. Let's assume the projectile behaves like a particle moving in a vertical plane (two-dimensions) and that the only force acting on it is gravity, which points down. Initially, we are ignoring the rotation of the earth, the force of air resistance, and the changes in the force of gravity as altitude changes. Let us also assume the projectile is launched from the origin at time, $t = 0$ with an initial velocity, $\overrightarrow{v_0}$.

As a vector we will consider both x and y components. If θ is the angle that $\overrightarrow{v_0}$ makes with the horizontal axis, then

$$\overrightarrow{v_0} = |v_0|\cos\theta\vec{i} + |v_0|\sin\theta\vec{j}.$$

From our assumptions the projectile's initial position,

$$\overrightarrow{r_0} = 0\mathrm{i} + 0\mathrm{j}$$

The only force acting on our projectile is gravity so $r'' = -g\,j$.
As a first order ode this becomes

$$r' = (-gt)j + \overrightarrow{v_0}.$$

Let's further assume the initial |v0| = 500 m/s and $\theta = 50(\pi/180)$ radians, and $g = 9.8$ m/sec^2.

$$r' = (-9.8t) + 500\left[\cos 50(\pi/180)\mathrm{i} + \sin 50(\pi/180)\right)\mathrm{j}.$$

We easily find that

$$x(t) = 500\cos(50(\pi/180)\mathrm{t}$$
$$y(t) = 500\sin\left(50(\pi/180)\right)t - 4.9t^2$$

If we are interested in the position after $t = 20$ sec we find by substitution the coordinates are

$$(x, y) = (6428, 5700).$$

To determine where the projectile lands, we set the vertical position to 0 and solve for t. We find the roots are $t = 0$ and $t = 2$ v0 sinθ/g = 78.16 sec. The root that makes the most sense to use is 78.16 sec.

The distance is using $x(78.16) = 500\cos((50(\pi/180)) * 7816) = 25{,}120$ m.

Now, let's introduce air resistance and see the effect this force has on our projectile. As in previous models we have examined, let's assume the force due to air resistance is proportional to the velocity, $Fa \propto v$, or $Fa = -kv$, where k is the proportionality constant and v is the velocity of the projectile.

Our modeling assumptions:

The weight of the projectile is negligible in relation to the weight of the cannon

There is no friction inside the gun tube.

At t = 0, gas pressure in the tube is 4,500 lbs.

The forces are only in horizontal direction.

F_1 = gas pressure and F_2 = damping force.

The cannon weight 4800 lbs

We want to use 1st order ODEs.

First, we find the mass of the cannon from: $w = mg$. So, $m + w/g = 4800/32.174$ ft/sec $lbs = 149.1887$ slugs

Newton's second law is $\sum F = ma$.

We separate the model into two parts, $0 \le t \le (1/5)$ and $t \ge 1/5$

First, using $0 \le t < (1/5)$

$$\sum F = ma = F_1 - F_2 = ma$$

We substitute v' for a, $F_1 = 4500$, and $F_2 = 600\sqrt{2}v$. This simplifies to

$$30.1621 - 4.02175\sqrt{2}v = \frac{dv}{dt}$$

Now, using $t \ge 1/5$:

$$\sum F = ma = -F_2 = ma$$

We substitute v' for a, and $F_2 = 600\sqrt{2}v$. This simplifies to

$$-600\sqrt{2}\,v = 149.1887\frac{dv}{dt} \text{ or } -4.02175\sqrt{2}v = \frac{dv}{dt}$$

Therefore our differential equation is:

$$\frac{dv}{dt} + 4.02175\sqrt{2}v = \begin{Bmatrix} 30 & 0 \le t < 1/5 \\ 0 & t \ge 1/5 \end{Bmatrix}.$$

Our initial conditions at $t = 0$ are velocity and displacement are both 0.

There is a safety wall behind the cannon at 5 feet. Will firing the cannon cause it to hit the safety wall during its recoil?

We need to solve $\frac{dv}{dt} + 4.02175\sqrt{2}v = 30.1621$ on the interval $0 \le t < 1/5$.

We can use separation of variables here to solve.

$$\int \frac{dv}{30.1621 - 4.02175\sqrt{2}t} = \int dt.$$

We find after we apply the initial condition of $v(0) = 0$ and solve for C.

Our solution is:

$$v(t) = 5.301242287 - 5.301242287 * exp(-5.689628647 * t).$$

We find the displacement at $t = 1/5$ sec.

The involves integration of the velocity function from $t = 0$ to $t = 1/5$.

$$x\left(\frac{1}{5}\right) = \int_0^{\frac{1}{5}} (5.301242287 - 5.301242287 * exp(-5.689628647 * t)dt$$

$x(1/5) = 0.42711$ *feet* and the velocity at 1/5 sec is 3.671 ft/sec.
Now, we use the second interval $t \geq 1/5$.

$$\frac{dv}{dt} + 4.02175\sqrt{2}v = 0$$

with initial condition $v(1/5)$= 3.671 *ft/sec.*
Again we use separation of variables to solve,

$$\int \frac{dv}{v} = -\int 4.02175\sqrt{2}\,dt,\ v\left(\frac{1}{5}\right) = 3.671$$

We find the general solution is:

$$v(t) = Ce^{-5.687613394*t}.$$

We apply the initial condition $v(1/5) = 3.671$ and find $C = 11.4504$.
Thus our particular solution is:

$$v(t) = 11.45048e^{-5.687613394t}.$$

To determine the total displacement we need to find the displacement in each time interval and add them together. In the first interval, the displacement was 0.42711 feet. The second interval requires the solution of the integral,

$$\int_{1/5}^{\infty} 11.4504e^{-5.687613394t}\,dt = 0.6449735006.$$

We find the total displacement as $0.42711 + 0.6449735006 = 1.072$ feet.
Since the wall is 5 feet away, we will not strike the wall.

First Order ODE and Maple

In Maple, we begin with the command with(DETools) to open Maple's differential equation library. We also introduce the dsolve command that solves many first order ordinary differential equations both without and with initial conditions. In dsolve, the parameters are deqns that list the ordinary differential equation in its variables and vars a variable or set of variables.

```
Maple Commands
With(DETools):
Dsolve(deqns, vars);
```

We will use Maple to resolve many of our previous examples.

Example 5.18: Solve in Maple $\frac{dy}{dt} = t \cdot y$

First, we will solve using Maple without initial conditions.

```
>restart;
>with(DETools):
>eqn:=diff(y(t),t)=y(t)*t;
```

$$eqn := \frac{d}{dt}y(t) = y(t)\,t$$

```
dsolve(eqn,y(t));
```

$$y(t) := C1e^{\left(\frac{t^2}{2}\right)}$$

Now, we solve again after we introduce the initial condition $y(1) = 1$.

```
>with(DETools):
>eqn:=diff(y(t),t)=y(t)*t;
eqn:=d/dt y(t)=y(t)t
>inits:=y(1)=1;
inits:=y(1)=1
>dsolve({eqn,inits},y(t));
y(t):=e^((r^2/2))/e^((1/2))
```

We plot the solution in Figure 5.15 from [–1, 1].

```
>plot(exp(t^2/2)/exp(1/2),t=-1..1);
```

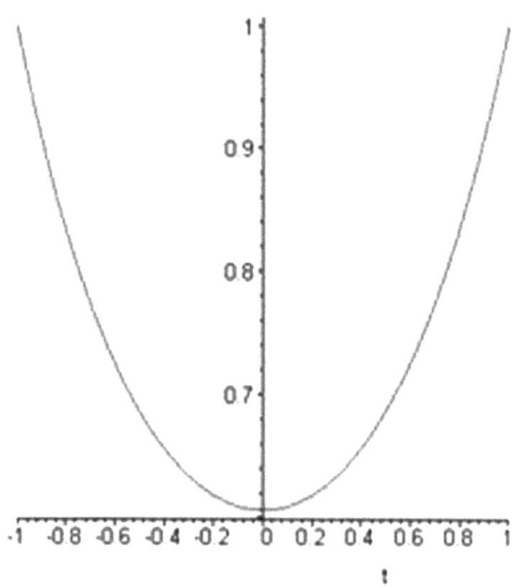

FIGURE 5.15
Example 5.17 Model plotted in Maple.

Example 5.19: Refined Population Model in Maple

Consider the logistic growth model defined for blue crabs in Venezuela,

$$\frac{dB}{dt} = .25B(10-B), B(0)=4. \quad (B \text{ is in millions of bushels.})$$

We can solve this using Maple:

```
>eqn:=diff(B(t),t)=0.25*B(t)*(10-B(t));
eqn:=d/dt B(t)=0.25B(t)(10-B(t))
>inits:=B(0)=4;
inits:=B(0)=4

>dsolve({eqn,inits},B(t));
B(t):=20/2+3e((-5t/2))
```

The following code provides a plot (Figure 5.16) of Example 5.18.

```
>plot(20/(2+3*exp(-5/2*t)), t=0.5);
```

Not all ODES are in fact solvable in closed form. If Maple does not return a solution then there might not be a closed form analytical solution. We might try a numerical method as we will discuss later in this chapter.

```
eqn:=diff(c(t),t)=(c(t)+2*t)/(t^2+c(t)^3);
eqn:=d/dt c(t)=(c(t)+2t)/(t^2 3(t)^3 )
dsolve(eqn,c(t));
```

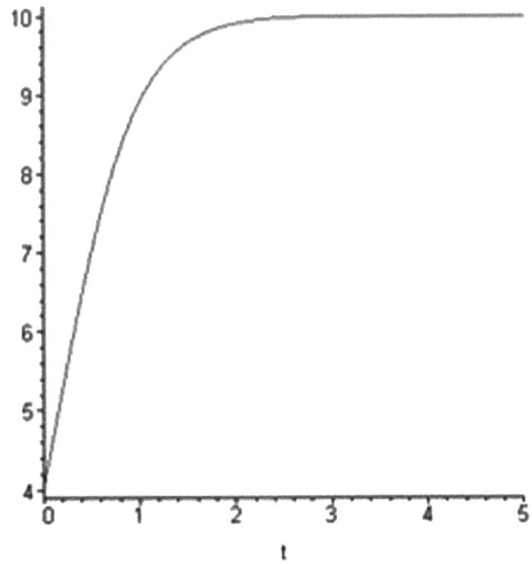

FIGURE 5.16
Plot of Example 5.18 solution in Maple.

Exercises

5.1 Construct a direction field for the following differential equations and sketch in a solution curve.

a. $\frac{dy}{dx} = y$

b. $\frac{dy}{dx} = x$

c. $\frac{dy}{dx} = x + y$

d. $\frac{dy}{dx} = x - y$

e. $\frac{dy}{dx} = xy$

f. $\frac{dy}{dx} = \frac{1}{y}$

5.2 Sketch a number of solutions to the following equations showing the correct slope, concavity, and any points of inflection.

a. $\frac{dy}{dx} = (y+2)(y-3)$

b. $\frac{dy}{dx} = y^2 - 4$

c. $\frac{dy}{dx} = y^3 - 4$

d. $\frac{dy}{dx} = x - 2y$

5.3 Analyze graphically the equation $\frac{dy}{dt} = ry$, when $r < 0$. What happens to any solution curve as t becomes large?

5.4 Develop graphically the following models. First graph $\frac{dP}{dt}$ versus P, then obtain various graphs of P versus t by selecting different initial values $P(0)$ (as in our population example in the text). Identify and discuss the nature of the equilibrium points in each model.

$\frac{dP}{dt} = a - bP, a, b > 0$

$\frac{dP}{dt} = P(a - bP), a, b > 0$

$\frac{dP}{dt} = k(M - P)(P - m), k, M, m > 0$

$\frac{dP}{dt} = kP(m - P)(P - m), k, M, m > 0$

5.5 The Department of Fish and Game in a certain state is planning to issue deer hunting permits. It is known that if the deer population falls below a certain level, m, then the deer will become extinct. It is also known that if the deer population goes above the maximum carrying capacity, M, the population will decrease to M.

Discuss the reasonableness of the following model for the growth rate of the deer population as a function of time: $\frac{dP}{dt} = kP(m - P)(P - m)$ where P is the population of the deer and k is a constant of proportionality. Include a graph of $\frac{dP}{dt}$ versus P as part of your discussion.

Explain how this growth rate model differs from the logistic model $\frac{dP}{dt} = kP(M - P)$. Is it better or worse than the logistic model?

Show that if $P > M$ for all t, then the limit $P(t)$ as $t \to \infty$ is M.

Discuss what happens if $P < m$ for all t.

Assuming that $m < P < M$ for all t, explain briefly the steps you would use to solve the differential equation. Do not attempt to solve the differential equation.

Graphically discuss the solutions to the differential equation. What are the equilibrium points of the model? Explain the dependence of the equilibrium level of P on the initial conditions. How many deer hunting permits should be issued?

5.6 Solve the following ODEs using any analytical method.

$$y' = y + yt, y(0) = 2$$
$$y' = y + 1, y(0) = 2$$

5.7 Consider $v' = 32 - 1.6\,v$, $v(0) = 0$. Find the terminal velocity.
5.8 Consider $P'.P*(15 - 3*P)$, $P(0) = 2$. Find the analytical solution and predict $P(10)$.

Solve the following initial value differential equations using the analytical methods and obtain a graph.

5.9 Suppose the differential equation is $y'(t) + 0.1\,y(t) = c(t)$ with $y(0) = 5$ and $t \geq 0$. Initially, let concentration if pollutant, $c(t) = 10$.
5.10 Suppose the differential equation is $y'(t) + 0.1\,y(t) = c(t)$ with $y(0) = 5$ and $t \geq 0$. Initially, let concentration if pollutant, $c(t) = 2\,e - t$.

PROJECTS

5.1 The Spread of a Contagious Disease

Given the following ODE model for the spread of a communicable disease as:

$$\frac{dN}{dt} = 0.25N(10 - N), N(0) = 2$$

where N is in 100's

Let's completely analyze the behavior of this ODE.

a. Since this is an autonomous ODE perform a complete graphical analysis:
   1. Plot dN/dt versus N. Find and label all rest points (equilibrium points).
   2. Find the value where the rate of change of the disease is the fastest. Why is it this value that you provided?
   3. Plot N versus t from the following initial conditions:

$$N(0) = 2, N(0) = 7, \text{ and } N(0) = 14.$$

   4. Describe the *stability* of each rest point (equilibrium point).

b. Obtain a slope field plot of this ODE from Maple. Briefly discuss how you would analyze the slope field plot. Compare it to the qualitative plot in part (a).
c. Solve this ODE using separable variables. (Hint: You will need partial fraction decomposition as well.) Ensure you find the value of arbitrary constant C using the initial condition $N(0) = 2$.
d. Simplify your solution in (b) and plot N versus t.
e. Compare your actual plot with the graphical solution in (a.3) and the slope field plot in (b).
f. From your graph of the analytical solution in (d), estimate the solution values of $N(.5)$ and $N(5)$.
g. Find the actual values for these two solutions $N(.5)$ and $N(5)$ using your analytical solution from part (c) and substitution.

h. Compute the time, t, when N is changing the fastest using the initial condition $N(0) = 2$. Recall, you have found the value of N already.
i. Use Euler's Method with step sizes of $h = 0.1$ and then $h = 1$ to approximate the solutions to the ODE for $N(.5)$ and $N(5)$. Find the relative error or absolute differences.
j. Plot Euler's approximations for $h = 0.1$ and $h = 1.0$ and compare these graphs to your graphical analysis in part (a) and your other plots. Briefly discuss these plots – you may compare and contrast these plots. What happened with Euler's Method? (Any opinions?)

5.2 **Chemical Reactions**

Purpose: To model the changing amounts of salt dissolved in brine as an initial value problem, and to use the computer to obtain graphical and numerical solutions.

Background: A brine solution is a solution of salt in water. If the brine is in a tank equipped with fill and drain pipes, then the total amount of dissolved salt in the tank varies as the concentration in the inflow stream changes and the inflow and outflow rates are adjusted. The amount of salt in the tank can be modeled by appealing to the

$$\textbf{\textit{Balance law}}: \textit{New rate of change} = \textit{rate in} - \textit{rate out}$$

The term "rate in"refers to the rate at which salt is added to the brine by means of the inflow stream and the term "rate out" is the rate at which salt leaves the tank through the outflow pipe. It is assumed throughout that the inflow stream is instantaneously mixed with the brine in the tank so that at any time the concentration of salt in the tank is uniform. Each of these terms (rate in and rate out) is a product of the appropriate brine flow rate with corresponding salt concentrations.

Consider a tank with a capacity of 4000 liters holds 2000 liters of brine that contains 50 kg of dissolved salt. Brine with a salt concentration of 0.2 kg/liter is piped into the tank at a rate of 40 liters/min. Well-mixed brine is drawn off at a rate of a liter/min.

$$\frac{dx}{dt} = (0.2)(40) + \frac{x}{2000 + (40 - a)t} a$$

Requirement 1. Show that the above is the general ODE; find the appropriate initial condition for this Brine problem.

Requirement 2. Solve the ODE leaving a as a parameter.

Requirement 3. Solve the ODE and obtain plots when the "rate out" $a = 30, 40$, and 50.

Requirement 4. (BULK SALT) Suppose that 200 kg of bulk salt is placed in a tanking holding 2000 liters of brine already containing $x0$ kg of dissolved salt. Suppose that the saturation level of brine is 300 kg of salt. The model representing this is:

$$\frac{dx}{dt} = 0.01(200 - x)(300 - x), x(0) = x_o$$

Solve this ODE, and obtain a plot for various choices of x_0, ($x_0 = 25, 50, 100, 200$).

Requirement 5. Bimolecular Chemical Reactions.

Background: In a bimolecular chemical reaction, two species interact and create one or more products. Consider the reaction,

$$A + B -> C + D$$

The Law of Conservation: $\frac{dC}{dt} = \frac{dD}{dt} = -\frac{dA}{dt} = -\frac{DB}{dt}$

The Law of Mass Action: The rate of an elementary reaction is proportional to the product of the concentrations of the reactants.

The ODE,

$$\frac{dx}{dt} = k_1 (a + c - x)(b + c - x), x(0) = c$$

Consider the following reversible chemical reaction,

$$\frac{dx}{dt} = (0.7 - x)(0.4 - x) - 0.1x^2, x(0) = c$$

Solve qualitatively (autonomous). Solve this ODE (at least numerically) and show the curve and discuss the equilibrium values. Let c = 0.2. This reaction is only valid for $0 \le c \le 0.4$.

5.3 **Harvesting a Species**

Consider harvesting the blue crab in South Carolina. There have been many newspaper reports about the declining populations of blue crabs and the difficulty in harvesting these crabs. Let's model the situation and analysis some "what ifs."

The basic Balance law for Harvesting is:

$$P'(t) = r\left(1 - \frac{P(t)}{k}\right)P(t) - H(t)$$

where r is the intrinsic rate coefficient (growth rate if > 0) and k represents the carrying capacity (or saturation level).

Requirement 1. In the absence of harvesting (H(t) = 0), the ODE is autonomous. Let r = 0.3, k = 12, and P(0) = 5 and

a. Perform a qualitative assessment of this situation since it is autonomous. List and classify each equilibrium value.
b. Solve this ODE.
c. Plot the solution to this ODE and compare to your qualitative solution. Briefly discuss similarities and/or differences.
d. Solve numerically using Euler's Method and obtain a plot of the numerical solution. Compare to the analytical solution.

Requirement 2. Now consider light harvesting where H(t)=0.4 (Assume r and k are the same values as before.)

Obtain a slope field plot and discuss any equilibrium values observed.

Find all equilibrium values of this ODE.

Solve this ODE. Assume initial conditions of $P(0) = 19$, $P(0) = 8$, and $P(0) = 0.8$.

Solve numerically using Euler's Method with $P(0) = 5$.

Requirement 3. Now consider a heavier harvesting where H(t) = 1.5.

Obtain a slope field plot and discuss any equilibrium values observed.

Find all equilibrium values of this ODE.
Solve this ODE. Assume initial conditions of P(0) = 19, P(0) = 8, and P(0) = .8.
Solve numerically using Euler's Method with P(0) = 5.

Requirement 4. Determine which of the models is more likely to represent the current situation for blue crabs in the Chesapeake Bay.

5.4 **Rural Water Supplies**

Consider a cylindrical water tank discharging into a pipe located in the middle of its base. The potential energy given by the pressure of the column of liquid is converted into kinetic energy of the stream of water, as it exists in the tank. If the level of the water in the tank is h ft, then it follows that the exit velocity in ft/sec is $v = \sqrt{2gh}$ where g = 32.2 ft/sec2 is the acceleration of gravity. Since a convergent flow is set up from all sides to the orifice, inertia from the particles in the jet forces them to overshoot the edge and to converge to a smaller cross-section termed the *vena contracta*. The ratio of the diameter of the vena contracta to the diameter of the orifice is called the contraction coefficient, α; and it ranges from 0.5 to 1.0 depending on the smoothness of the edge of the orifice. The diameter of the vena cross-section and the exit velocity of the stream determine the volume rate of the discharge of the stream. Thus, if the tank has cross-section A and is being filled from the top at a rate K, while draining for the bottom through an orifice of area a, the height in the tank is governed by the differential equation:

$$\frac{dh}{dt} = \frac{\left(K - a\alpha\sqrt{2gh}\right)}{A}$$

Now, consider we are in the mountains outside of Stowe, VT. The water for the cabin is supplied from an aquifer some 125 feet below the surface at a maximum rate of 7 gallons per minute. Since the flow rate is insufficient to meet daily needs, you must design a more complicated system. In your design, this water is pumped to a cylindrical tank with a diameter of 10 feet and height 35 feet which is located on a nearby rise about 75 feet from the cabin, with a base 25 feet above the level of the cabin floor. The water is stored in this gravity tank until needed (water is turned on by opening the spicket).

Assume that the coefficient of contraction at the exit is $\alpha = 0.63$ and that the water main consists of 6 inch diameter PVC pipe, so that the friction loses are less than 2 psi.

a. Determine the minimum flow rate in the system and the minimum water pressure during the first two hours.
b. The local fire code demands that water supplies maintain a pressure of at least 20 psi at a flow rate of 30 gallons per minute for a minimum of two hours. Will your design pass inspection?
c. At continuous maximum flow, how long would it take for the level of the top of the water in the tank to fall to 10 feet above the base?

References and Additional Readings

Arney, D., and Clark, C. (1997). The King of Battle: The Cannon.

Fox, W. (2013). *Mathematical Modeling with Maple*. Boston, MA: Cengage Publishing.

Fox, W., and Bauldry, W. (2019). *Problem Solving with Maple. Volume 1*. Boca Raton, FL: Taylor and Francis, CRC Press.

Fox, W., and Bauldry, W. (2020). *Problem Solving with Maple. Volume II*, Boca Raton Fl: Taylor and Francis, CRC Press.

Fox, W., and Burks, R. (2021). *Advanced Mathematical Modeling*. Boca Raton, FL: Taylor and Francis, CRC Press.

Fox, W., and Burks, R. (2022). *Mathematical Modeling Under Change, Uncertainty, and Machine Learning*. Boca Raton, FL: Taylor and Francis, CRC Press.

Giordano, F., Fox, and Horton, S. (2014). *A First Course in Mathematical Modeling*. Boston, MA: Cengage Publishing.

6

Modeling with Numerical Solutions to Differential Equations: IVP for ODEs with Technology

Introduction and Scenarios

The study of differential equations is a key course for applied mathematics, science, and engineering. It is important to motivate the ideas and results of ordinary differential equations through real-world applications that a student may encounter in their discipline. Rapidly changing technologies put more computing power in the hands of the students. Much more difficult problems can now be solved through the use of numerical techniques. In this chapter, we present several modeling scenarios representative of modeling change. These ordinary differential equations with initial values are the type of problems that permeate across disciplines.

In real-world modeling with differential equation, the equation developed may not be solvable in closed form. Numerical methods might be the only choice available. We will discuss Euler's methods, Improved Euler's method, and Runge-Kutta 4 (RK4) methods.

Bridge Bungee Jumping

Consider that in bridge jumping, a participant attaches one end of a bungee cord to himself (or herself), attaches the other end to a bridge railing, and then drops off the bridge. In this problem, the jumper will be dropping off of the Royal Gorge Bridge, a suspension bridge that is 1053 feet above the floor of the Royal Gorge in Colorado. The jumper will use a 200-foot-long bungee cord. It would be nice if the jumper has a safe jump, meaning that the jumper does not crash into the floor of the gorge or run into the bridge on the rebound. In this project, you will do some analysis of the fall.

Assume the jumper weighs 160 lbs. The jumper will free fall until the bungee cord begins to exert a force that acts to restore the cord to its natural (equilibrium) position. In order to determine the spring constant of the bungee cord, you found that a MASS weighing 4 lbs stretches the cord 8 feet. Hopefully, this spring force will help slow down the descent sufficiently so that the jumper does not hit the bottom of the gorge.

DOI: 10.1201/9781003582274-6

Throughout this problem, we will assume that DOWN is the POSITIVE direction. Before the bungee cord begins to retard the fall of the jumper, the only forces that act on the jumper are his weight and the force due to wind (air) resistance.

If the force due to the wind resistance is 0.9 times the velocity of the jumper, then use Newton's second law ($\Sigma F = MA$) to write a differential equation that models the fall of the jumper. Be sure to include the initial conditions for the jumper for your differential equation. (Hint: This problem can be formulated as a second order ODE in position, or as a first order DE in velocity – use the first order ODE model in this problem.)

We will return to this problem later in this chapter.

Spread of a Contagious Disease

Consider modeling the spread of a contagious disease.

Given the following ODE model for the spread of a communicable disease as:

$$\frac{dN}{dt} = 0.25N(10 - N), \; N(0) = 2$$

where N is in 100's.

We want to completely analyze the behavior of this ODE. We return to this problem later as well. First, we are going to discuss numerical methods to solve differential equations.

Numerical Methods

Euler's Method

From the point of view of a mathematician, the *ideal* form of the solution to an initial value problem would be a **formula** for the solution function. After all, if this formula is known, it is usually relatively easy to produce any other form of the solution you may desire, such as a **graphical solution**, or a **numerical solution** in the form of a Table of values. You might say that a formulaic solution contains the recipes for these other types of solutions within it.

Unfortunately, as we have seen in our studies already, obtaining a formulaic solution is not always easy, and in many cases is absolutely impossible. So we often have to "make do" with a numerical solution, i.e., a Table of values consisting of points which lie along the solution's curve. This can be a perfectly usable form of the answer in many applied problems, but before we go too much further, let's make sure that we are aware of the shortcomings of this form of solution.

By its very nature, a numerical solution to an initial value problem consists of a **Table of values** which is **finite** in length. On the other hand, the true solution of the initial value problem is most likely a whole continuum of values, i.e., it consists of an *infinite* number of points. Obviously, the numerical solution is actually leaving out an infinite number of points. The question might arise, "With this many holes in it, is the solution good for anything at all?" To make this comparison a little clearer, let's look at a very simple specific example starting with Euler's Method.

Given the differential equation:

$$\frac{dy}{dt} = g(t,y), y(t_0) = y_0, \; t_0 \le t \le b$$

Euler's method

Step 1. Pick step size, h so that the interval, [b-t0]/n = h, is divided evenly.
Step 2. Start at y(t0) = y0
Step 3. Compute $y_n + 1 = y_n + h * g(t_n, y_n)$
Step 4. Let $t_n + 1 = t_n + h, \; n = 0,1,2,\ldots$
Step 5. Continue until $t_n = b$.

Example 6.1: ***dy/dt = 0.25*t*y(t),* y(0) = 2**

We want to estimate the solution to *y(3)* using a numerical approach. The interval is [0,3].

A few steps by hand:

Let's start with a step size of 1.0 point (*t, y*)

When t = 0, y = 2 This is the initial condition: (0, 2)
When t = 0 + h = 1, y = 2 + h*0.25 t y = 2+ (1)*(0.25)(0)(2) = 2 (1, 2)
When t = 1 + h = 2, y = 2 + h*0.25*t*y = 2+(1)*(0.25)(1)(2.5) = 2.5 (2, 2.5)
When t = 3, y = 2.5 + h*0.25*t*y = 2.5+(1)*(0.25)(2)(2.5) = 3.75 (3, 3.75)

Geometrically,

Note that, we have under-estimated every value between t = 0 and t = 5, as shown in **Figure 6.1.**

The exact solution is $y(t) = 2 \cdot e^{\frac{t^2}{8}}$

Let's make a **Table 6.1** and see how we did:

Notice that as we move farther away from the Initial Condition y(0) = 2, the worse our estimate becomes.

Can we do better?

Maybe, we might improve by changing the step size. In this example, let us make the step size 0.5 instead of 1.

We may use technology, in this case, Maple, EXCEL, or R to obtain our output. Here is the Maple Code:

```
> ans2 := dsolve({eqn, y(0)=2}, numeric,
         method=classical[foreuler],
```

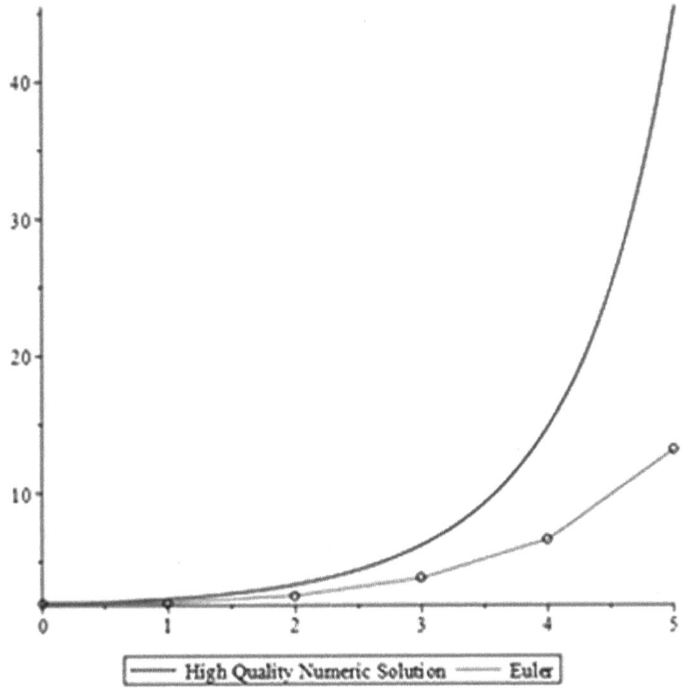

FIGURE 6.1
Euler's method geometrically for Example 6.1.

```
output=array([0,0.5,1,1.5,2,2.5,3]),
stepsize=0.5);
```

$$ans2: = \begin{bmatrix} [t, \ y(t)] \\ \begin{bmatrix} 0 & 2 \\ 0.5 & 2 \\ 1 & 2.13 \\ 1.5 & 2.39 \\ 2 & 2.84 \\ 2.5 & 3.55 \\ 3 & 4.66 \end{bmatrix} \end{bmatrix}$$

TABLE 6.1
Comparison of Euler's Method to Exact Solution for Example 6.1

T	Y (approximate, Ya)	Y (exact, Ye)	Error=|Ye−Ya|	Percent Error = $\frac{100 \cdot \lvert Ye - Ya \rvert}{Ye}$
0	2	2	0	0
1	2.	2.266	0.266	11,73 %
2	2.5	3.297	0.797	24.17 %
3	3.75	6.16	2.41	39.12 %

Let's make a Table 6.2 and see how we did:

TABLE 6.2
Comparison of Euler's Method to Exact Solution for Example 6.1

t	Ya	Ye	Error=|Ye-Ya|	Percent Error
0	2	2	0	0
.5	2	2.063	0.063	3.05 %
1	2.125	2.2633	0.1383	6.11 %
1.5	2.390625	2.6496	0.258975	9.77 %
2	2.838867	3.2974	0.458533	13.91%
2.5	3.54858	4.3684	0.81982	18.76 %
3	4.65752	6.1604	1.503	24.84 %

Figure 6.2 shows the plot of our solution.

We are still underestimating at all points in Figure 6.2. We might be able to do better by making this step size small. (Note: it is not always true that making the step size smaller improves the estimates.)

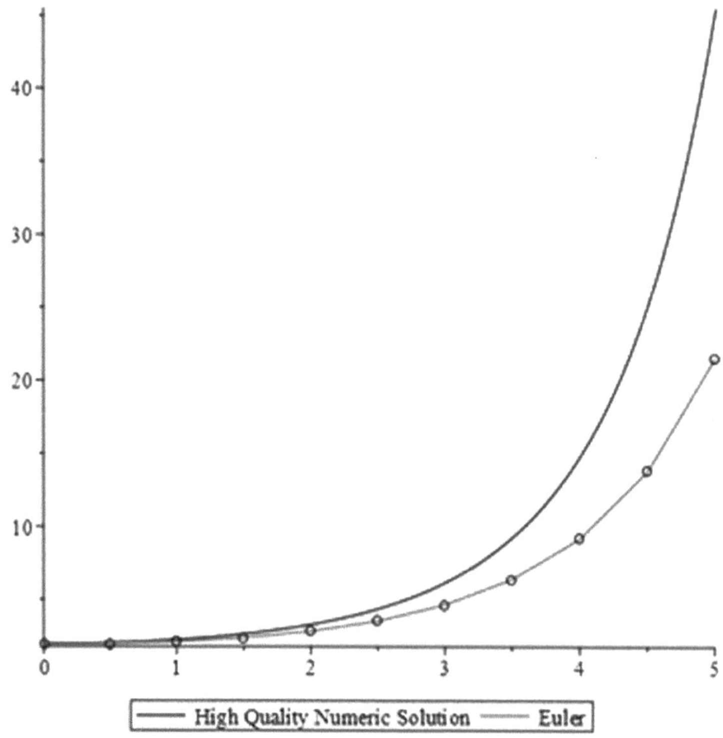

FIGURE 6.2
Smaller step size to Example 6.1.

Example 6.2: Repeat Example 6.1 but with step size = 0.5

We will use EXCEL and the Euler method this time as our technology to solve the problem.

We still did not do well with our estimate (Figure 6.3), so we will turn to Maple

```
> Digits := 20:
ans2 := dsolve({eqn, y(0)=2}, numeric,
             method=classical[foreuler],
             output=array([0,0.5,1,1.5,2,2.5,3]),
             stepsize=0.5);
```

```
                 [t, y(t)]
          0              2
          0.5            2
ans2: =   1              2.125
          1.5            2.390625
          2              2.83886718750
          2.5            3.54858398437500
          3              4.657516479492187500
```

The second method that we will discuss is Improved Euler's method also known as Heun's method.

Euler's Method

*dy/dt=0.25*y(t)*t, y(0)=2*

h 0.5

t	y	dy/dt	new y	exact	error	% Rel error
0	2	0	2	2	0	0
0.5	2	0.25	2.125	2.063487	0.063487	3.076676552
1	2.125	0.53125	2.390625	2.266297	0.141297	6.2347041
1.5	2.390625	0.896484	2.838867	2.64957	0.258945	9.773078825
2	2.838867	1.419434	3.548584	3.297443	0.458575	13.9070006
2.5	3.548584	2.217865	4.657516	4.368402	0.819818	18.76699325
3	4.657516	3.493137	6.404085	6.160434	1.502917	24.39628916

FIGURE 6.3
Excel solution to Example 6.2.

Improved Euler's Method (Heun's Method)

The accuracy of Euler's method is improved by using an average of two slopes in the tangent line approximation. Use the slope obtained at the beginning of the step and the slope obtained at the end of the step in order to improve our accuracy.

Given

$$\frac{dy}{dt} = g(t,y), y(t_0) = y_0, \ t_0 \le t \le b$$

Improved Euler's method algorithm

Step 1. Pick step size, h so that the interval, $[b\text{-}t0]/n = h$, is divided evenly.

Step 2. Start at $y(t0) = y0$.

Step 3. Let $tn + 1 = tn + h$

Step 4. Compute $yn + 1 = yn + (h/2) * [g\ (tn, yn) + g\ (tn + 1, yn + 1)]$

Step 5. Continue until $tn + 1 = b$.

STOP

Step 4 uses the formula:

$$y_{n+1} = y_n + \frac{h}{2}(k_1 + k_2)$$
$$k_1 = f(t_n, y_n)$$
$$k_2 = f(t_{n+1}, y_n + h * k_1)$$

Example 6.3: Improved Euler for y′ = 0.25 t y(t), y(0) = 2

We want to estimate the solution to y(3) using a numerical approach.

$$k_1 = .25 t_n y_n$$
$$k_2 = .25 t_{n+1}(y_n + h * k_1)$$

We start with a few steps of Improved Euler's Method by hand:

Step size of $h = 1$. Point (t, y)

When $t_0 = 0, y_0 = 2$ This is the initial condition:

$$(0, 2)$$

When $t_1 = t_0 + h = 1$, $y_1 = 2 + (1/2)*[(0.25*(0)*(2) + 0.25*(1)*(2 + 0.25*(0)*2) = 2.25$

$$(1, 2.25)$$

When $t_2 = t_1 + h = 2$, $y_2 = 2.25 + (1/2)*(0.25(1)(2.25) + 0.25*(2)(2.25 + (1)0.25*(1)(2.25)) =$ 3.23437

$$(2, 3.234375)$$

TABLE 6.3
Improved Euler's Estimates for Example 6.3

t	Y (approximate, Ya)	Y (exact, Ye)	Error=|Ye-Ya|	Percent Error= $\frac{100 \cdot \lvert Ye - Ya \rvert}{Ye}$
0	2	2	0	0
1	2.25	2.266	0.016	0.706 %
2	3.234375	3.297	0.062625	1.899 %
3	5.862305	6.16	0.2977	4.83 %

When $t_3=2+h=3, y_3=3.234375+(1/2)*((0.25)*(2)*(3.234375)+(0.25)*(3)*(3.234375+0.25*(2)*(3.234375)=5.862305$

$$(3, 5.862305)$$

The exact solution is $y(t)=2\cdot e^{\frac{t^2}{8}}$

Table 6.3 shows how we did in our estimates.

Notice that as we move farther away from the Initial Condition y(0)=2, the worse our estimate becomes. Also, note that these estimates are "better" estimates than Euler's Method. Think about why?

Figure 6.4 shows the improvement in our estimation.

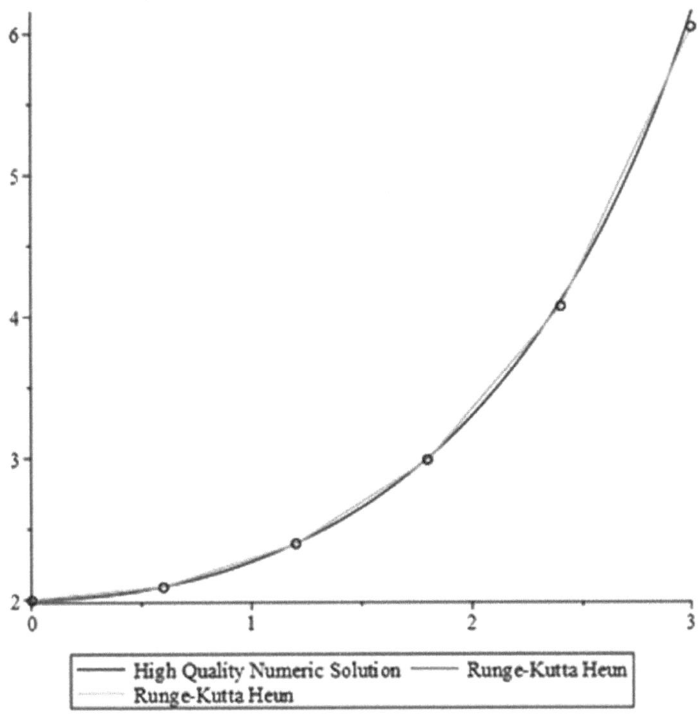

FIGURE 6.4
Improved Euler's (Heun's) estimates of Example 6.1.

Example in Maple: We note that the only command change is our call within

```
dsolve({eqn, y(0)=2}, numeric,method=classical[heunform], output=
array([0,0.5,1,1.5,2,2.5,3]), stepsize=0.5);
```

that calls heunform the improved Euler's algorithm

```
> restart;
> with(DEtools):with(plots):with(linalg):with(LinearAlgebra):
```

Enter your ODE

```
> eqn:=diff(y(t),t)=.25*t*y(t);
```

$$eqb := \frac{d}{dt} y(t) = 0.25 t y(t)$$

Solve the ODE analytically and plot (Figure 6.5):

```
>>dsolve({eqn,y(0)=2},y(t));
```

$$y(t) = 2\ e^{\left(\frac{t^2}{8}\right)}$$

```
> plot(2*exp(1/8*t^2),t=0..3, thickness=3);
```

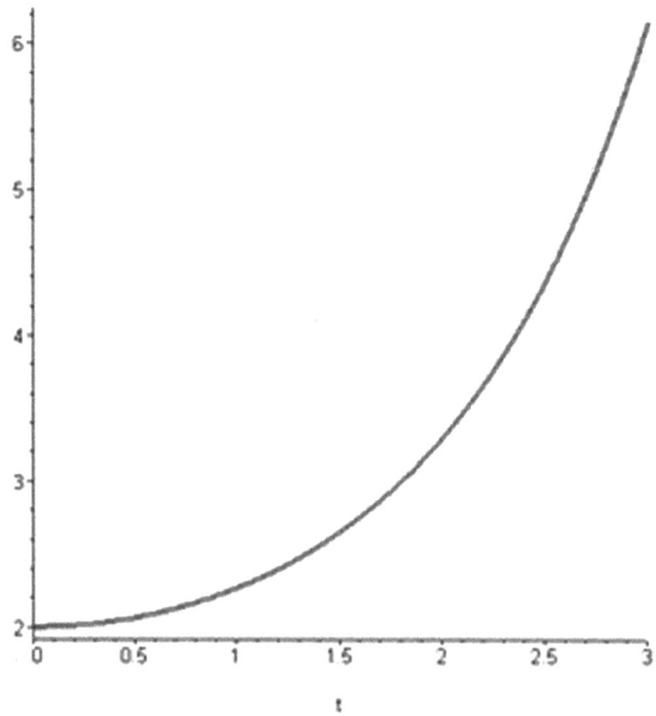

FIGURE 6.5
Plot of Example 6.3.

Note you have to copy and paste the output equation from **dsolve** into the plot command.

Numerical Method–Internal and Classic-OUTPUT only

You will need to specify both the step size you want and the output array you want to see in Maple.

```
> Digits := 20:
ans2 := dsolve({eqn, y(0)=2}, numeric,
            method=classical[heunform],
            output=array([0,0.5,1,1.5,2,2.5,3]),
            stepsize=0.5);
```

ans2: =

t	y(t)
0	2.
0.5	2.0625000000000000000
1	2.2639160156250000000
1.5	2.6441831588745117188
2	3.2845712676644325257
2.5	4.3366605018381960691
3	6.0814887506246577688

First, we plot the actual solution (if we have one) then the numerical solution (Figure 6.6). Second, we will overlay them on one plot.

```
> plot(2*exp(1/8*t^2),t=0..3);
```

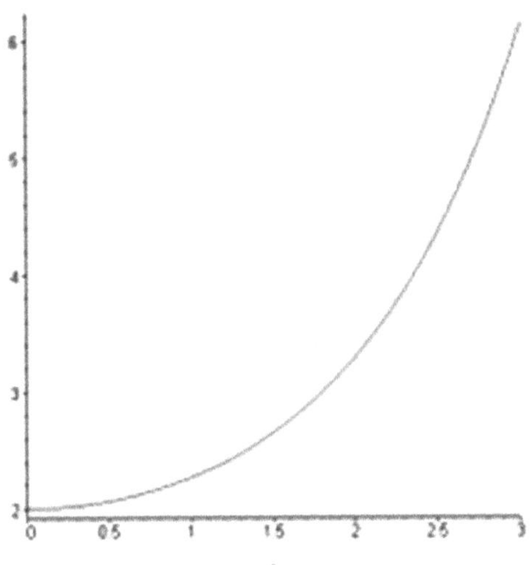

FIGURE 6.6
Numerical plot of Example 6.3.

Numerical Plot:

```
> data:=({seq([i,fy2(i)],i=0..3)}):

> points:=pointplot(data, symbol=diamond, color=black):

> display( points);
```

Exercises

6.1 **Euler's Method.** Given the following differential equation: with the following initial condition:
$y(0) = 2$, estimate $y(2)$ using Euler's method with a step size $h = 0.5$.

$$\frac{dy}{dt} = 0.3ty(t)$$

6.2 **Step Size Impact in Euler's Method.** Solve the same problem as Exercise 6.1, but this time using a step size $h = 0.25$. Compare the results and discuss how the step size affects the accuracy of the solution.

6.3 **Improved Euler's Method.** Consider the equation:

$$\frac{dy}{dt} = 0.2ty(t)$$

with initial condition $y(0) = 3$. Use Improved Euler's Method with a step size of $h = 1$ to estimate $y(3)$.

6.4 **Bungee Jump Modeling.** A bungee jumper is modeled by the first order differential equation:

$$m\frac{dv}{dt} = mg - kv$$

where $m = 70$ kg, $g = 9.8\text{m/s}^2$, and $k = 0.9$. Solve this equation numerically using Euler's method for $v(t)$ with step size $h = 0.5$, and initial velocity $v(0) = 0$.

6.5 **Runge-Kutta Method.** Using the differential equation:

$$\frac{dy}{dt} = ty + 2t$$

with $y(0) = 1$, estimate $y(2)$ using the Runge-Kutta fourth order method (RK4) with a step size of $h = 0.5$.

6.6 **Modeling Disease Spread.** Solve the given ODE for a contagious disease model:

$$\frac{dN}{dt} = N(100 - N)$$

where $N(0) = 5$, using Euler's method with a step size of $h = 1$. Estimate the value of $N(10)$.

6.7 **Comparing Numerical Methods.** Compare Euler's method, Improved Euler's method, and the Runge-Kutta method for the differential equation:

$$\frac{dy}{dt} = -2y + t^2$$

with $y(0) = 1$ over the interval $t \in [0,3]$, using a step size of $h = 0.5$.

PROJECTS

6.1 **Optimal Battery Capacity.** In the modern world, renewable energy sources like solar and wind power are becoming increasingly vital. However, one major challenge is the intermittent nature of these energy sources. Solar energy is only available during the day, and wind energy can fluctuate based on weather conditions. This project focuses on **optimizing energy storage in a battery system** to ensure a reliable power supply, even when renewable sources are not actively generating energy.

The energy storage system can be modeled using *ordinary differential equations (ODEs)*, and numerical methods like *Euler's Method* and *Runge-Kutta Methods (RK4)* will be used to solve these equations. The aim is to determine the best charging and discharging strategies that maximize battery life and ensure a steady energy supply.

Problem Statement

Design an energy storage system for a small community that relies on solar power. The energy produced from the solar panels varies throughout the day, and the goal is to manage the charging and discharging of the battery system efficiently.

The differential equation governing the charge $Q(t)$ of the battery can be written as:

$$\frac{dQ}{dt} = P(t) - C(t)$$

where:

$Q(t)$ = battery charge at time t (in kilowatt-hours, kWh).

$P(t)$ = power generated by the solar panels at time t (in kW). This varies based on time of day and weather conditions.

$C(t)$ = power consumption by the community (in kW). This can also vary based on time of day and usage patterns.

Objectives

- Use **numerical methods** to solve for $Q(t)$, the battery charge, over a 24-hour period.
- Simulate different scenarios (e.g., sunny day vs. cloudy day, high vs. low consumption) and determine the optimal battery size and charging strategy.
- Implement **Improved Euler's Method** and **Runge-Kutta (RK4)** to model the system and compare their accuracy.

Supporting Data

The following real-world data will be used in the project:

1. **Solar Power Generation (P(t)).** The solar panel system generates power according to the following pattern (in kW), based on sunlight intensity:
   - **00:00–06:00**: P(t) = 0kW (no sunlight at night)
   - **06:00–08:00**: P(t) increases gradually from 0 to around 3 kW (sunrise)
   - **08:00–12:00**: P(t) increases sharply from 3 kW to around 8 kW (morning to noon)
   - **12:00–14:00**: P(t) peaks at around 10 kW (midday, maximum solar output)
   - **14:00–18:00**: P(t) decreases from 10 kW to around 3 kW (afternoon)
   - **18:00–20:00**: P(t) decreases gradually to 0 kW (sunset)
   - **20:00–24:00**: P(t) = 0kW (nighttime, no solar generation)
2. **Power Consumption (C(t)).** The community consumes power in the following **pattern** (in kW):
   - **00:00–06:00**: C(t) ≈1 kWC(t)
   - **06:00–08:00**: C(t) increases to around 3 kW (morning activity)
   - **08:00–12:00**: C(t) stays relatively high around 4 kW (morning peak)
   - **12:00–14:00**: C(t) decreases to around 2 kW (midday lull)
   - **14:00–18:00**: C(t) increases again to around 4 kW (afternoon activity)
   - **18:00–20:00**: C(t) peaks around 5 kW (evening peak)
   - **20:00–24:00**: C(t) decreases back to around 1.5 kW (nighttime consumption)
3. **Battery Specifications**
   - Maximum battery capacity: 50kWh.
   - Initial charge $Q(0) = 25$kWh.

Steps to Solve.

1. **Formulate the ODE.** Based on the data for $P(t)$ and $C(t)$, write the ODE for the rate of change of the battery charge:

$$\frac{dQ}{dt} = P(t) - C(t)$$

2. **Solve Using Numerical Methods**
   - Apply **Euler's Method** to solve for $Q(t)$ over the 24-hour period with a step size of $h = 1$ hour.
   - Repeat the process using **Improved Euler's Method** and **RK4** to improve accuracy.
   - Compare results and discuss which method provides better estimates for $Q(t)$.

3. **Analyze Different Scenarios.** Simulate different conditions, such as cloudy days where the power generation is halved and days with higher energy consumption (e.g., during special events).
4. **Optimize Battery Capacity.** Based on the results, suggest an optimal battery capacity that would prevent power shortages.

Expected Output

- A table and graph showing the battery charge over 24 hours under different scenarios.
- Comparison of the accuracy between Euler's Method, Improved Euler's Method, and RK4.
- Recommendations for battery capacity and charging strategies.
- The output will include a graph of $Q(t)$ (battery charge) over time under different weather and consumption scenarios, generated using Python or another plotting tool.

References and Additional Readings

Burden, R., and Faires, J. (2015). *Numerical Analysis*. 10th edition. Boston: Brooks/Cole, Cengage Learning.

Butcher, J. (2008). *Numerical Methods for Ordinary Differential Equations*. 2nd edition. Hoboken: John Wiley & Sons.

Chapra, S., and Canale, R. (2015). *Numerical Methods for Engineers*. 7th edition. New York: McGraw-Hill.

Hairer, E., Nørsett, P., and Wanner, P. (1993). *Solving Ordinary Differential Equations I: Nonstiff Problems*. 2nd edition. Berlin: Springer-Verlag.

7

++++++Numerical Output for Analysis: Graphical and Percent Error

ODEs are fundamental for mathematical modeling and provide an excellent approach for understanding how a system develops over time, in terms of its state and its rate of change. Although analytic solutions exist for many ODEs, real-world problems often introduce complexities such as non-linearity, variable coefficients, or external-forcing terms that make finding solutions to these problems unrealistic or impossible to achieve; numerical methods will be a fitting option for such cases. Numerical techniques like Euler's method, Runge-Kutta methods, and finite difference methods allow us to approximate solutions over discrete time steps, producing numerical outputs that help analyze the system's behavior. These numerical outputs provide insight into trends, stability, and systems dynamics, enabling the analysis of systems in engineering, physics, biology, and economics when exact solutions are unattainable.

First Get the Numerical Output as a Procedure

We will start this chapter by providing a Maple procedure that will allow us to develop a numerical output a problem. The Maple code is

Example 7.1: Numerical analysis for the problem

```
y(t)=2.0e^0.125t^2
> dsol2 := dsolve({eqn, y(0) = 2},
       numeric,method=classical[heunform],stepsize=1,
output=listprocedure);

                dsol2 := [t = (proc(t) … end proc), y(t) = (proc(t) …
end proc)]
```

Prepare for numerical output

```
> fy2 := eval(y(t),dsol2);

                fy2 := proc(t) … end proc
```

View output to compare to the above output (numerical)

```
> seq(fy2(i),i=0..3);

                    2., 2.25, 3.234375, 5.8623046875
```

Prepare the exact solution for the output format:

```
> actual_y2:=evalf(subs(t=i,2*exp(1/8*t^2)));
```

$$\text{actual}_{y2} := 2.0e^{0.125^2}$$

DOI: 10.1201/9781003582274-7

View actual output

```
> seq(actual_y2(i),i=0..3);
```

```
2., 2.26629, 3.2974425, 6.160433698
```

Put into an array that lists **[t, numerical y(t), actual y(t), Percent Error]**

```
> array([seq([i,fy2(i),evalf(subs(t=i, 2*exp(1/8*t^2))),evalf(100*abs
(fy2(i)-actual_y2(i))/actual_y2(i))],i=0..3)]);
```

0	2	2	0
1	2.25	2.266296906	0.7190984592
2	3.234375	3.297442542	1.9126198745
3	5.862304	6.160433698	4.8394159835

Graphical Comparisons

First, we plot (Figure 7.1) the actual solution (if we have one) then the numerical solution (Figure 7.2). Second, we will overlay them on one plot (Figure 7.3). Otherwise, compare numerical solutions to either qualitative plots or slope field plots.

```
> plot(2*exp(1/8*t^2),t=0..3);
```

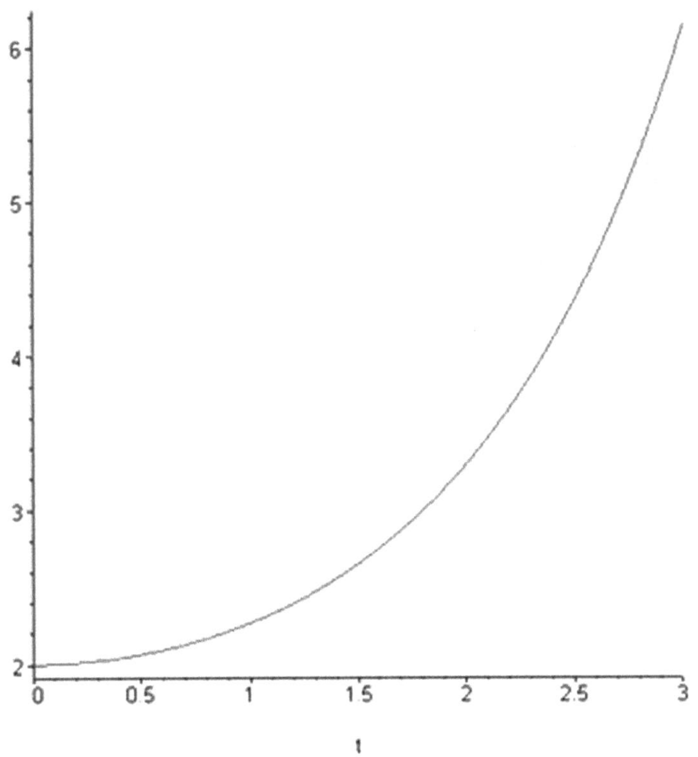

FIGURE 7.1
Graphical plot for Example 7.1.

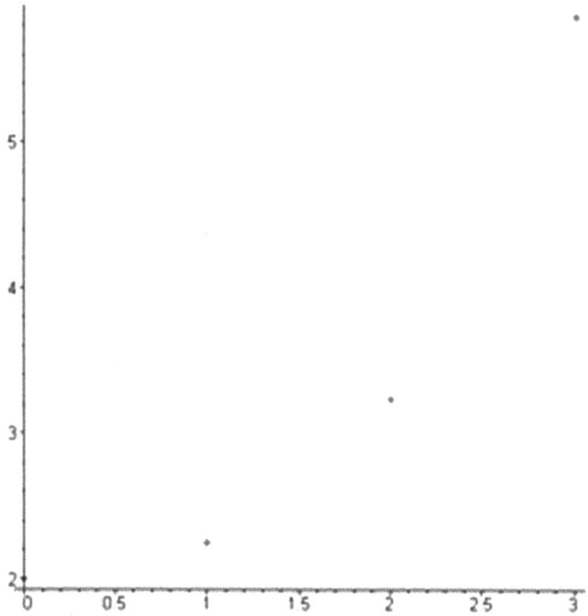

FIGURE 7.2
Graph for Example 7.1.

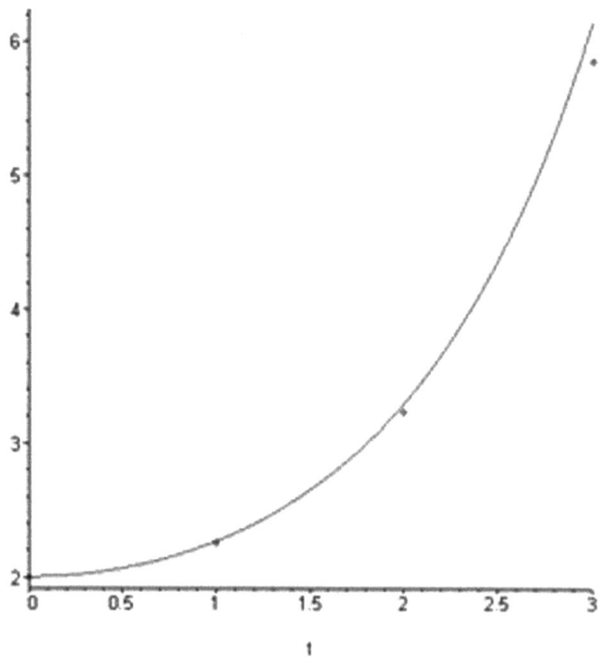

FIGURE 7.3
Graph for Example 7.1.

Numerical Plot:

```
> data:=({seq([i,fy2(i)],i=0..3)}):
> points:=pointplot(data, symbol=diamond, color=black):
> display( points);
```

Overlay

```
> with(plots):
> data:=({seq([i,fy2(i)],i=0..3)}):
> curve:=plot(2*exp(1/8*t^2),t=0..3, color=red):
> points:=pointplot(data, symbol=diamond, color=black):
> display(curve, points);
```

Runge-Kutta Methods

When we use the average of the estimates of the derivatives at the endpoints we can improve the approximation to the solution. A class of approximation techniques that estimate derivatives at various points within an interval and then computes a ***weighted average*** is the Runge-Kutta Method, named after two German mathematicians.

The Runge-Kutta Methods are classified by order, where the order depends upon the number of slope estimates used at each step. A very popular method is the fourth order Runge-Kutta Method.

Fourth Order Runge-Kutta Method

For solving dy/dt = g(t,y), $y(t_o) = y_o$ over an interval.

Step 1. First, divide the interval $x_0 \le x \le b$ into p sub-intervals using equally spaced points. This yields the step size, $h = (b - x_0)/p$.

Step 2. For n = 1,2,3..p obtain the following sequence of approximations

$$y_{n+1} = y_n + \frac{K_1 + 2K_2 + 2K_3 + K_4}{6},$$

where

$$K_1 = g(t_n, y_n)h$$

$$K_2 = g(t_n + h/2, y_n + K_1/2)h$$

$$K_3 = g\left(t_n + h/2, y_n + K_2/2\right)h$$

$$K_4 = g\left(t_n + h, y_n + K_3\right)h$$

We begin with an example by hand:

Example 7.2: RK4 for y′ = 0.25 t y(t), y(0) = 2

We want to estimate the solution to y(3) using a numerical approach.
A few steps of Runge-Kutta Method by hand:

Step size of h = 1.
When t = 0, y = 2 This is the initial condition: (0,2)
When t = 0 + h = 1,

K1 = h*g(t(n),y(n) = 1*0.25*0*2 = 0
K2 = h*g(t(n + h/2),y(n) + K1/2) = 1*0.25*(0 + 1/2)*(2 + 0) = 0.25
K3 = h*g(t(n + h/2),y(n) + K2/2) = 1*0.25*(0 + 1/2)*(2 + 0.125) = 0.265625
K4 = h*g(t(n + h),y(n) + K3) = 1*0.25*(0 + 1)*(2 + 0.265625) = 0.56640625
Y(1) = 2 + (1/6)*(0+2*0.25 + 2*0.265625 + 0.56640625) = 2.266276

When t =1 + h = 2

K1 = 0.566569
K2 = 0.956085
K3 = 1.029119
K4 = 1.647698
Y(2) = 2.266276 + (1/6)*(0.566569 + 2*0.956085 + 2*1.029119 + 1.647698)
= 3.297055

When t = 2 + h = 3

K1 = 1.648528
K2 = 2.575825
K3 = 2.865605
K4 = 4.621995
Y(2) = 3.297055 + (1/6)* (1.648528 + 2*2.575825 + 2*2.86560 + 4.621995)
= 6.155952

The exact solution is $y(t) = 2 \cdot e^{\frac{t^2}{8}}$

Table 7.1 provides an indication of how close our estimate is or how we did.

Notice that as we move farther away from the Initial Condition y(0) = 2, the worse our estimate becomes. Also, note that these estimates are much better estimates than either Euler's Method or Improved Euler's. Think about why.

The RK4 plot is shown in Figure 7.4.

In Maple, we again change the call from within dsolve to RK4 as seen in this sequence of commands.

TABLE 7.1
RK 4 Method for Example 7.2

t	Y (approximate, Ya)	Y (exact, Ye)	Error = $\lvert Ye - Ya \rvert$	Percent Error = $\frac{100 \cdot \lvert Ye - Ya \rvert}{Ye}$
0	2	2	0	0
1	2.266276	2.266297	0.00002	0.002%
2	3.297055	3.297443	0.000388	0.0388%
3	6.155952	6.160434	0.004482	0.4482%

Set UP Maple

```
> restart;
> with(DEtools):with(plots):with(linalg):
```

Enter your ODE

```
> eqn:=diff(y(t),t)=.25*t*y(t);
```

$$\text{eqn} := \frac{d}{dt}y(t) = 0.25ty(t)$$

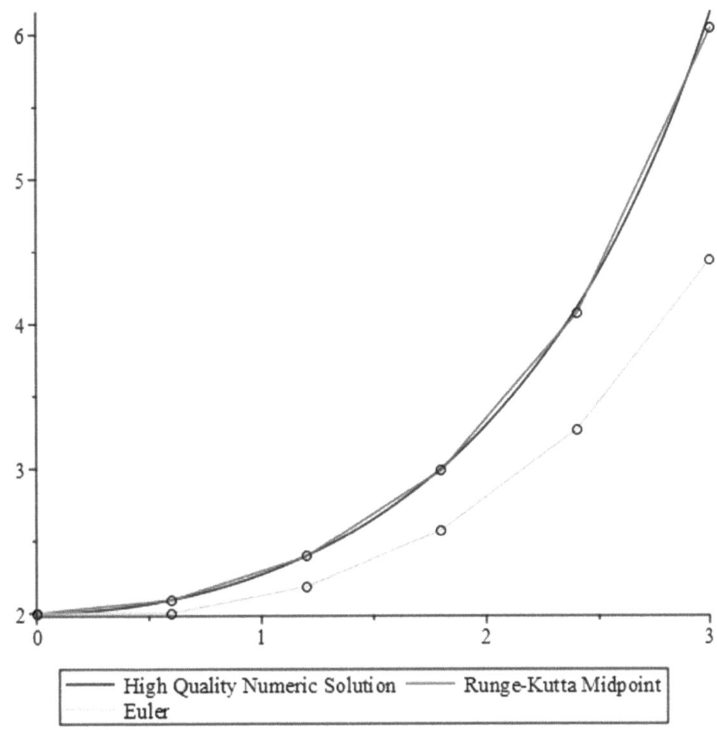

FIGURE 7.4
RK-4 plot for Example 7.2.

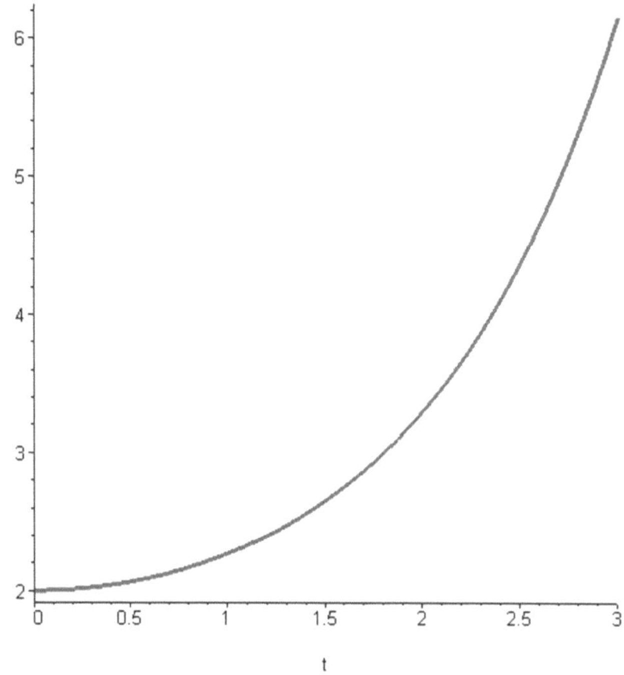

FIGURE 7.5
Maple graph to Example 7.2.

Solve the ODE Analytically & Plot (Figure 7.5)

```
> dsolve({eqn,y(0)=2},y(t));
```

$$y(t) = 2e^{\left(\frac{t^2}{8}\right)}$$

```
> plot(2*exp(1/8*t^2),t=0..3, thickness=3);
```

Note you have to copy and paste the output equation from **dsolve** into the plot command.

Numerical Method–Internal and Classic-OUTPUT only

You will need to specify both the step size you want and the output array you want to see.

```
> Digits := 20:
ans2 := dsolve({eqn, y(0)=2}, numeric,
    method=classical[rk4],
```

```
output=array([0,0.5,1,1.5,2,2.5,3]),
stepsize=0.5);
```

$$ans2 := \begin{bmatrix} [t,\ y(t)] \\ \begin{bmatrix} 0 & 2 \\ 0.5 & 2.06348673 \\ 1 & 2.26629594 \\ 1.5 & 2.64956434 \\ 2 & 3.29741947 \\ 2.5 & 4.36830973 \\ 3 & 6.16009786 \end{bmatrix} \end{bmatrix}$$

Numerical Output for Analysis: graphical and percent error
First, get the numerical output as a procedure as follows:

```
> dsol2 := dsolve({eqn, y(0) = 2},
          numeric,method=classical[rk4],stepsize=1,
output=listprocedure);
          dsol2 := [t = (proc(t) ... end proc), y(t) = (proc(t) ... end
proc)]
```

Prepare for numerical output

```
> fy2 := eval(y(t),dsol2);
                    fy2 := proc(t) ... end proc
```

View output to compare to the above output (numerical)

```
> seq(fy2(i),i=0..3);
               2., 2.66276, 3.290755, 6.1559523
```

Prepare the exact solution for the output format:

```
> actual_y2:=evalf(subs(t=i,2*exp(1/8*t^2)));
```

$$actual_{y2} := 2.0e^{0.125^2}$$

View actual output

```
> seq(actual_y2(i),i=0..3);
               2., 2.26629, 3.2974425, 6.160433698
```

Put into an array that lists [t, numerical y(t), actual y(t), Percent Error]

```
> array([seq([i,fy2(i),evalf(subs(t=i, 2*exp(1/8*t^2))),evalf(100*
abs(fy2(i)-actual_y2(i))/actual_y2(i))],i=0..3)]);
```

0	2	2	0
1	2.266276042	2.266296906	0.0009206413
2	3.297055403	3.297442542	0.0117405535
3	6.155952320	6.160433698	0.0727445029

Graphical Comparisons

First, we plot (Figure 7.6) the actual solution (if we have one) and then the numerical solution. Second, we will overlay them on one plot. Otherwise, compare numerical solution to either qualitative plots or slope field plots.

```
>plot(2*exp(1/8*t^2),t=0..3);
```

Numerical Plot (Figure 7.7):

```
> data:=({seq([i,fy2(i)],i=0..3)}):

> points:=pointplot(data, symbol=diamond, color=black):

> display( points);
```

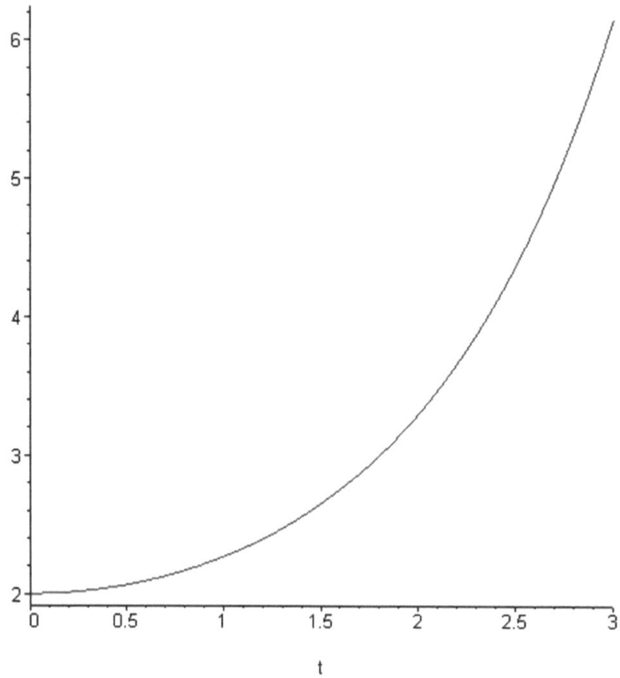

FIGURE 7.6
Graph for Example 7.2.

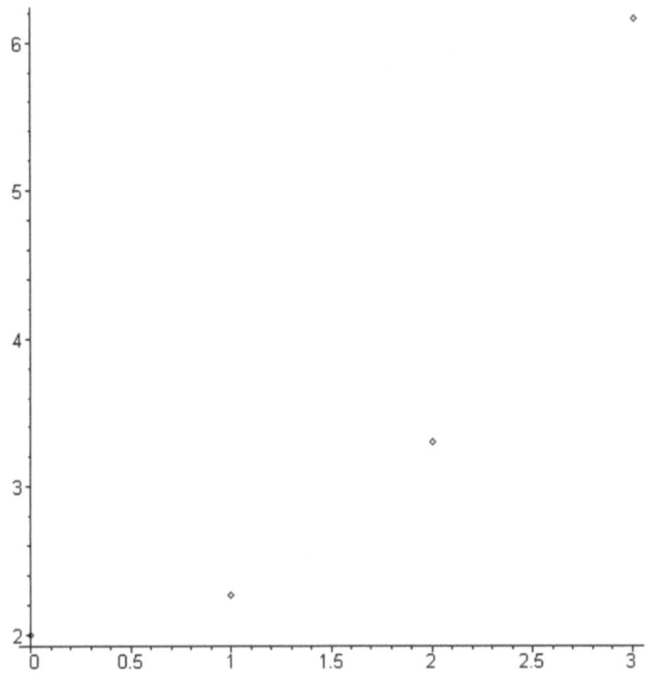

FIGURE 7.7
Maple plot for Example 7.2.

Overlay (Figure 7.8):

```
> with(plots):

> data:=({seq([i,fy2(i)],i=0..3)}):

> curve:=plot(2*exp(1/8*t^2),t=0..3, color=red):

> points:=pointplot(data, symbol=diamond, color=black):

> display(curve, points);
```

Python and Numerical Methods for ODEs (IVP)

Runge-Kutta and Euler's method are two of the most popular numerical techniques for solving ordinary differential equations (ODEs). They will provide an approximate solution to first order ODEs, where an exact solution might be difficult or impossible to find analytically. This section provides Python code for finding a numerical solution using both the Euler's method and Runge-Kutta for a problem that we can solve exactly. This will provide some evidence of how effective these techniques are at finding a numerical solution.

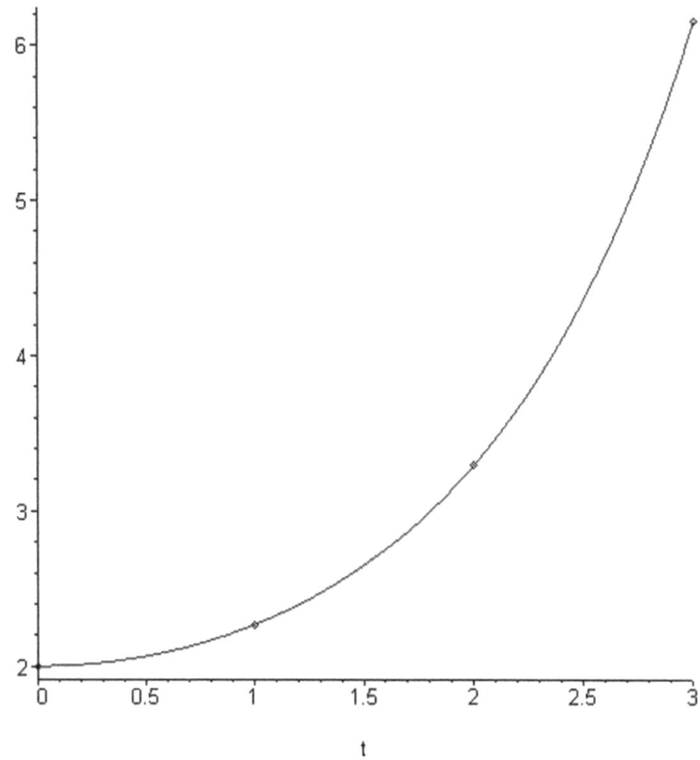

FIGURE 7.8
Maple graph for Example 7.2.

Euler's Method and Plot with Python Code (Figure 7.9)

```
import numpy as np
import matplotlib.pyplot as plt

plt.style.use('seaborn-poster')
%matplotlib inline

# Define parameters
f = lambda t, s: np.exp(-t) # ODE
h = 0.1 # Step size
t = np.arange(0, 1 + h, h) # Numerical grid
s0 = -1 # Initial Condition

# Explicit Euler Method
s = np.zeros(len(t))
s[0] = s0

for i in range(0, len(t) - 1):
    s[i + 1] = s[i] + h*f(t[i], s[i])
```

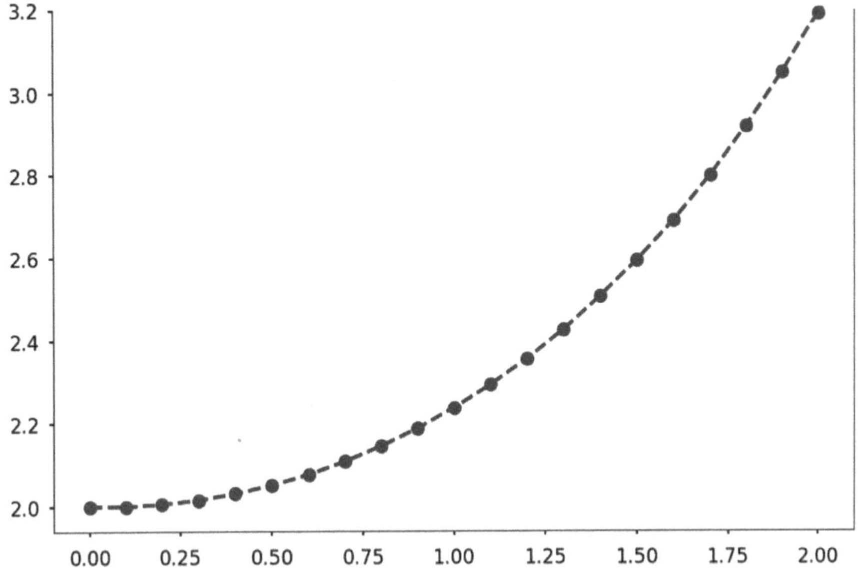

FIGURE 7.9
Graph of Euler's method.

```
plt.Figure(figsize = (12, 8))
plt.plot(t, s, 'bo--', label='Approximate')
plt.plot(t, -np.exp(-t), 'g', label='Exact')
plt.title('Approximate and Exact Solution \
for Simple ODE')
plt.xlabel('t')
plt.ylabel('f(t)')
plt.grid()
plt.legend(loc='lower right')
plt.show()
h = 0.01 # Step size
t = np.arange(0, 1 + h, h) # Numerical grid
s0 = -1 # Initial Condition

# Explicit Euler Method
s = np.zeros(len(t))
s[0] = s0

for i in range(0, len(t) - 1):
    s[i + 1] = s[i] + h*f(t[i], s[i])

plt.Figure(figsize = (12, 8))
plt.plot(t, s, 'b--', label='Approximate')
plt.plot(t, -np.exp(-t), 'g', label='Exact')
plt.title('Approximate and Exact Solution \
for Simple ODE')
plt.xlabel('t')
plt.ylabel('f(t)')
```

```
plt.grid()
plt.legend(loc='lower right')
plt.show()
```

```
Output
Euler:

        Enter initial conditions:

        x0 = 0
        y0 = 2

        Enter calculation point:

        xn = 3

        Enter number of steps:
        Number of steps = 25

                    --------SOLUTION--------
                    -------------------------
                      x0       y0       yn
                    -------------------------
                    0.0000  2.0000  2.0000
                    -------------------------
                    0.1200  2.0000  2.0072
                    -------------------------
                    0.2400  2.0072  2.0217
                    -------------------------
                    0.3600  2.0217  2.0435
                    -------------------------
                    0.4800  2.0435  2.0729
                    -------------------------
                    0.6000  2.0729  2.1102
                    -------------------------
                    0.7200  2.1102  2.1558
                    -------------------------
                    0.8400  2.1558  2.2101
                    -------------------------
                    0.9600  2.2101  2.2738
                    -------------------------
                    1.0800  2.2738  2.3475
                    -------------------------
                    1.2000  2.3475  2.4320
                    -------------------------
                    1.3200  2.4320  2.5283
                    -------------------------
                    1.4400  2.5283  2.6375
                    -------------------------
                    1.5600  2.6375  2.7609
                    -------------------------
```

```
1.6800  2.7609  2.9001
-------------------------
1.8000  2.9001  3.0567
-------------------------
1.9200  3.0567  3.2327
-------------------------
2.0400  3.2327  3.4306
-------------------------
2.1600  3.4306  3.6529
-------------------------
2.2800  3.6529  3.9027
-------------------------
2.4000  3.9027  4.1837
-------------------------
2.5200  4.1837  4.5000
-------------------------
2.6400  4.5000  4.8564
-------------------------
2.7600  4.8564  5.2586
-------------------------
2.8800  5.2586  5.7129
-------------------------

At x = 3.0000, y = 5.7129
```

RK 4 Python Code

```
# RK-4 method python program

# function to be solved
def f(x,y):
    return x+y

# or
# f = lambda x: x+y

# RK-4 method
def rk4(x0,y0,xn,n):

    # Calculating step size
    h = (xn-x0)/n

    print('\n--------SOLUTION--------')
    print('-------------------------')
    print('x0\ty0\tyn')
    print('-------------------------')
    for i in range(n):
        k1 = h * (f(x0, y0))
        k2 = h * (f((x0+h/2), (y0+k1/2)))
        k3 = h * (f((x0+h/2), (y0+k2/2)))
        k4 = h * (f((x0+h), (y0+k3)))
        k = (k1+2*k2+2*k3+k4)/6
```

```
        yn = y0 + k
        print('%.4f\t%.4f\t%.4f'% (x0,y0,yn) )
        print('-------------------------')
        y0 = yn
        x0 = x0+h

    print('\nAt x=%.4f, y=%.4f' %(xn,yn))

# Inputs
print('Enter initial conditions:')
x0 = float(input('x0 = '))
y0 = float(input('y0 = '))

print('Enter calculation point: ')
xn = float(input('xn = '))

print('Enter number of steps:')
step = int(input('Number of steps = '))

# RK4 method call
rk4(x0,y0,xn,step)
```

```
Enter initial conditions:
x0 = 0
y0 = 2

Enter calculation point:
xn = 3

Enter number of steps:
Number of steps = 25

                   --------SOLUTION--------
                  -------------------------
                     x0       y0       yn
                  -------------------------
                    0.0000  2.0000  2.0036
                  -------------------------
                    0.1200  2.0036  2.0145
                  -------------------------
                    0.2400  2.0145  2.0327
                  -------------------------
                    0.3600  2.0327  2.0584
                  -------------------------
                    0.4800  2.0584  2.0921
                  -------------------------
                    0.6000  2.0921  2.1339
                  -------------------------
                    0.7200  2.1339  2.1844
                  -------------------------
```

```
0.8400  2.1844  2.2442
-------------------------
0.9600  2.2442  2.3139
-------------------------
1.0800  2.3139  2.3944
-------------------------
1.2000  2.3944  2.4867
-------------------------
1.3200  2.4867  2.5918
-------------------------
1.4400  2.5918  2.7111
-------------------------
1.5600  2.7111  2.8461
-------------------------
1.6800  2.8461  2.9986
-------------------------
1.8000  2.9986  3.1707
-------------------------
1.9200  3.1707  3.3647
-------------------------
2.0400  3.3647  3.5835
-------------------------
2.1600  3.5835  3.8303
-------------------------
2.2800  3.8303  4.1089
-------------------------
2.4000  4.1089  4.4236
-------------------------
2.5200  4.4236  4.7796
-------------------------
2.6400  4.7796  5.1828
-------------------------
2.7600  5.1828  5.6404
-------------------------
2.8800  5.6404  6.1604
-------------------------

At x = 3.0000, y = 6.1604
```

Table 7.2 provides an indication of how good both methods are in terms of providing a decent numerical estimate of the exact solution.

TABLE 7.2

Summary of Python Results

Method	Approximate y(3)	Exact y(3)	|Error|
Euler	5.7129	6.16043	0.44753
RK4	6.1604	6.16043	0.00003

Population Modeling with Numerical Methods

Example 7.3:

Suppose a society has a population of $x\ (t)$ individuals at time t, in years, and that all left-handed people who mate with other left-handed people have offspring who are also left-handed, while a fixed proportion r of all other offspring are also left-handed. If the birth and death rates for all individuals are assumed to be the constants B and D, respectively, and if right-handed and left-handed people mate at random, the problem can be expressed by the following differential equation.

$$\frac{dx(t)}{dt} = (B - D)x(t)$$
$$\frac{dx(t)}{dt} = (B - D)x(t) + rB(x(t) - x_n(t),$$

where $x_n(t)$ denotes the number of left-handed people in the population at time t.

To simplify our modeling and equation building process, let the variable, $p(t) = x_n(t)/x(t)$ be introduced to represent the proportion of left-handed people in the population and now may be combined into single differential equations:

$$\frac{\mathrm{d}}{\mathrm{dt}}(\mathrm{t}) = \mathrm{r}(\mathrm{B}(1 - \mathrm{p}(\mathrm{t}))$$

Historical data was collected and analyzed to obtain the following coefficient values: $p(0) = 0.01$, $B = 0.02$, $D = 0.015$, and $r = 0.1$. We use these values to approximate the solution $p\ (t)$ from $t = 0$ to $t = 50$ with a step size (arbitrarily chosen) of $h = 1$ year. We will illustrate several numerical methods in obtaining our solution: Euler's Method, Improved Euler's Method, Runge-Kutta, and the Runge-Kutta-Fehlberg method.

Methods of Numerical Solutions Using the Numerical Analysis Package in Maple

Based on our answer for $y(50)$, Maple appears to be using a Runga-Kutta or Modified Runga-Kutta Method for its numeric algorithm. We can obtain an exact solution to this differential equation for $p\ (t)$ exactly, and compare our results from the numerical methods when t = 50 with the exact value at that time. We can make a recommendation of which method is most accurate. We provide an error bound for each method. The exact solution to

$$\frac{\mathrm{d}}{\mathrm{dt}}(\mathrm{t}) = 0.002(1 - \mathrm{p}(\mathrm{t}), \ \mathrm{p}(0) = 0.01$$

can be found by separation of variables or variation of parameters (Figures 7.10 – 7.12) as

```
p(t) = 1. - 0.9900000000*exp(-0.002000000000*t). So p(50) = 0.10241.

Euler(diff(p(t), t) = 0.002*(-p(t) + 1), p(0) = 0.01, t = 50);
                      0.1051
```

Digits := 20 :
ansp := *dsolve*({*eqnp1*, *icp1*}, *numeric*, *method* = *classical*[*foreuler*], *output* = *Array*([10, 20, 30, 40, 50]), *stepsize* = 1);

ansp := (

t	$p(t)$
10.	0.02962274708157006959 5
20.	0.04885655254392067789 4
30.	0.06770912558202499643 3
40.	0.08618802258723684214 6
50.	0.10430065017600459226

ans2 := *dsolve*({*eqnp1*, *icp1*}, *numeric*, *method* = *classical*[*heunform*], *output* = *Array*([10, 20, 30, 40, 50]), *stepsize* = 1)

ans2 := (

t	$p(t)$
10.	0.029603300468266216578
20.	0.048818429836280790261
30.	0.067653074401731933619
40.	0.086114768263516878847
50.	0.10421089633547786816

ans2 := *dsolve*({*eqnp1*, *icp1*}, *numeric*, *method* = *classical*[*rk4*], *output* = *Array*([10, 20, 30, 40, 50]), *stepsize* = 1)

ans2 := (

t	$p(t)$
10.	0.029603313426309658760
20.	0.048818455239194941224
30.	0.067653111751586306342
40.	0.086114817077220810550
50.	0.10421095614438805879

FIGURE 7.10
Maple solutions for Euler, Heunform, and RK4 techniques for Example 7.3.

```
Euler(diff(p(t), t) = 0.002*(-p(t) + 1), p(0) = 0.01, t = 50, output =
Error);
                        0.0009075

Euler(diff(p(t), t) = 0.002*(-p(t) + 1), p(0) = 0.01, t = 50, output =
plot);

RungeKutta(diff(p(t), t) = 0.002*(-p(t) + 1), p(0) = 0.01, t = 50,
submethod = rk4);
                              0.1042
```

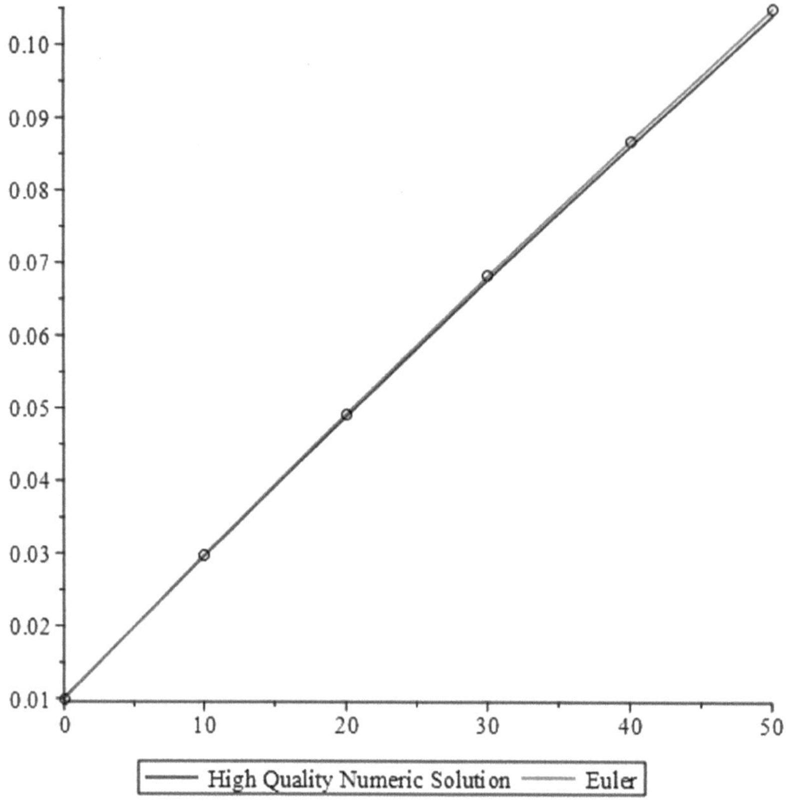

FIGURE 7.11
Maple output for Example 7.3.

```
RungeKutta(diff(p(t), t) = 0.002*(-p(t) + 1), p(0) = 0.01, t = 50,
submethod = rk4, output = plot);

RungeKutta(diff(p(t), t) = 0.002*(-p(t) + 1), p(0) = 0.01, t = 50,
submethod = rk4, output = Error);
                         -10
                   1.214 10
```

It appears in Table 7.3 that the RK4 method did a better job in approximating the solution.

Euler's error bound is found by

$$|p_i - w_i| \leq \frac{hM}{2L}\left(e^{(t_{i+1}-\alpha)L} - 1\right) \leq 0.000396$$

which we see with our estimates is true.

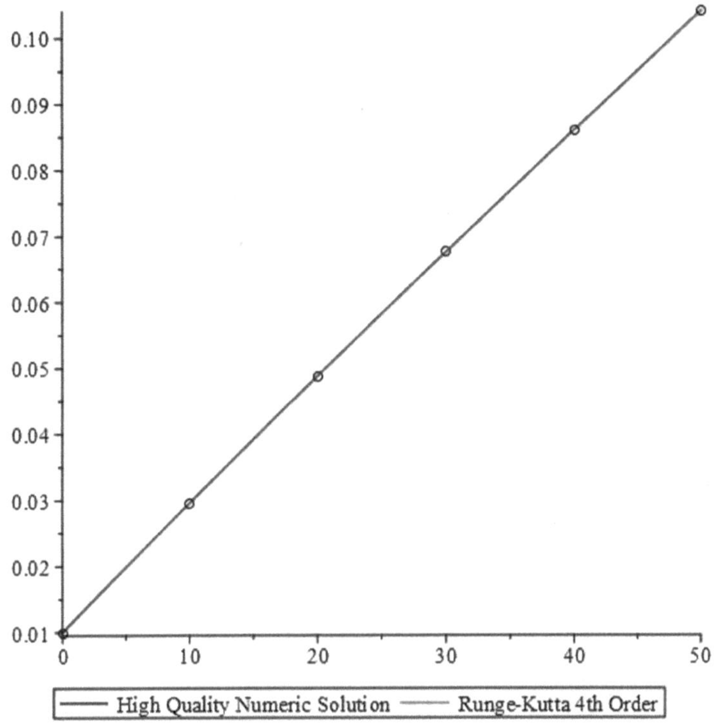

FIGURE 7.12
RK4 graph for Example 7.3.

Improved Euler's method error bound is $O(h^3)$. By using Taylor polynomials (see Chapter 1) we find the local truncation error for $p(50)$ to be

$$0.104210894 - (0.1024175266 + 0.5(0.00179564947 + 0.0017958212)) = -6.1 \times 10^{-9}.$$

For RK4 the error bound is, $y^{(5)}(c)h^5/5!$ or $O(h^5)$. In our problem,

$$y^{(5)}(c)h^5/5! = 2.38877 \times 10^{-16}.$$

We note our errors are less than our bounds.

TABLE 7.3
Comparison of Numerical Techniques for Example 7.3

Method	Approximate for p(50)	Exact Solution	Absolute Error
Euler	0.1043006502	0.10421	8.6988×10^{-4}
Improved Euler	0.1042108964	0.10421	8.6019×10^{-4}
RK 4	0.1042109562	0.10421	9.1354×10^{-6}

Simple Spread of a Contagious Disease Model with Numerical Methods

Example 7.4: We return to our spread of a contagious disease. Consider modeling the spread of a contagious disease.

Given the following ODE model for the spread of a communicable disease as:

$$\frac{dN}{dt} = 0.25N(10 - N),\ N(0) = 2$$

where N is in 100's

We need to analyze the results. We will do this using Python.

Enter initial conditions:

$x0 = 0$

$y0 = 2$

Enter calculation point:

$xn = 15$

Enter number of steps:
Number of steps = 25

```
--------SOLUTION--------
-------------------------
   x0       y0       yn
-------------------------
 0.0000   2.0000   5.2639
-------------------------
 0.6000   5.2639   8.3116
-------------------------
 1.2000   8.3116   9.5181
-------------------------
 1.8000   9.5181   9.8668
-------------------------
 2.4000   9.8668   9.9635
-------------------------
 3.0000   9.9635   9.9900
-------------------------
 3.6000   9.9900   9.9973
-------------------------
 4.2000   9.9973   9.9993
-------------------------
 4.8000   9.9993   9.9998
-------------------------
 5.4000   9.9998   9.9999
-------------------------
```

```
 6.0000   9.9999  10.0000
-------------------------
 6.6000  10.0000  10.0000
-------------------------
 7.2000  10.0000  10.0000
-------------------------
 7.8000  10.0000  10.0000
-------------------------
 8.4000  10.0000  10.0000
-------------------------
 9.0000  10.0000  10.0000
-------------------------
 9.6000  10.0000  10.0000
-------------------------
10.2000  10.0000  10.0000
-------------------------
10.8000  10.0000  10.0000
-------------------------
11.4000  10.0000  10.0000
-------------------------
12.0000  10.0000  10.0000
-------------------------
12.6000  10.0000  10.0000
-------------------------
13.2000  10.0000  10.0000
-------------------------
13.8000  10.0000  10.0000
-------------------------
14.4000  10.0000  10.0000
-------------------------
```

At x = 15.0000, y = 10.0000

We find that 100% have the disease in 6.6 time periods. This means that we need to quickly get a vaccine or some medications for these people.

Bungee Jumping Model with Numerical Methods

Example 7.5

In bridge jumping, a participant attaches one end of a bungee cord to himself (or herself), attaches the other end to a bridge railing, and then drops off the bridge. In this project, the jumper will be dropping off of the Royal Gorge Bridge, a suspension bridge that is 1053 feet above the floor of the Royal Gorge in Colorado. The jumper will use a 200-foot-long bungee cord. It would be nice if the jumper has a safe jump, meaning that the jumper does not crash into the floor of the gorge or run into the bridge on the rebound. In this project, you will do some analysis of the fall.

Assume the jumper weighs 160 lbs. The jumper will free fall until the bungee cord begins to exert a force that acts to restore the cord to its natural (equilibrium) position. In order to determine the spring constant of the bungee cord, you found that a MASS weighing 4 lbs stretches the cord 8 feet. Hopefully, this spring force will help slow down the descent sufficiently so that the jumper does not hit the bottom of the gorge. Throughout this problem, we assume that DOWN is the POSITIVE direction.

Before the bungee cord begins to retard the fall of the jumper, the only forces that act on the jumper are his weight and the force due to wind (air) resistance.

If the force due to the wind resistance is 0.9 times the velocity of the jumper, then use Newton's second law (ΣF = MA) to write a differential equation that models the fall of the jumper.

$$\sum F = MA$$

$$m x'' = -mg - kx'$$

Since k = 0.5 since from Hooke's law ($F = Ks$) 4 = k * 8 so k = 4/8 = 0.5

$$g = 32.17 \text{ ft/s}^2$$

$$W = mg \text{ so } W/g = m. \text{ This implies } m = 160/32.17 = 4.9735$$

We divide the entire equation by m to get

$$x'' = -g/m - k/mx' - 0.9/m\,x'$$

$-k/m\ x'$ only after bungee has stretched out. Otherwise, delete during free fall.

$$4.9735\,x'' = 160\ \ 0.9\,x'$$

$$v' + 0.18095\,v = 32.17\,v(0) = 0$$

Dealing only with velocity we have

$$dv/dt = x'',\ x' = v(t) \text{ and we get}$$

$$4.975\ dv/dt = -32.17/(w/g) - 0.9\ v(t) = v' = -160 - 0.9/dv/dt + 0.18095\ v(t) = -160$$

$$160 - 0.9v = 0.5\ dv/dt =$$

$$320 - 1.8\ \ v = d$$

$$v/dt$$

We solve using numerical methods (Figures 7.13 and 7.14).

First, we see we reach a terminal velocity of 177.7777 in about 50 s. If we estimate the velocity of our jumper after 4 s, we see v(4) = 97.405 ft/s.

If the jumper has fallen 200 feet, we find the speed is 177.7777.

Be sure to include the initial conditions for the jumper for your differential equation. The initial condition will be x(0) = 0, x′(0) = 0

$$160 - 0.9v = 0.5dv/dt$$

```
> Digits := 20 :
> ans2 := dsolve({eq3, ic3}, numeric, method = classical[foreuler], output = Array([1, 2, 3, 4,
    5, 6, 7, 8, 9, 10]), stepsize = 1);
```

$$ans2 := \begin{bmatrix} [\ t \quad v(t)\ ] \\ \begin{bmatrix} 1. & 320.00000000000000000 \\ 2. & 64.00000000000000000 \\ 3. & 268.80000000000000000 \\ 4. & 104.96000000000000000 \\ 5. & 236.03200000000000000 \\ 6. & 131.17440000000000000 \\ 7. & 215.06048000000000000 \\ 8. & 147.95161600000000000 \\ 9. & 201.63870720000000000 \\ 10. & 158.68903424000000000 \end{bmatrix} \end{bmatrix}$$

FIGURE 7.13
Euler method solution for Example 7.5.

Now we will use the following Maple to look at the exact solution (Figures 7.15 and 7.16).

$$> eq3 := diff(v(t), t) = \frac{160}{0.5} - \frac{0.9}{0.5}v(t)$$

$$eq3 := diff(v(t), t) = 320.0 - 1.8v(t)$$

$$> \mathrm{ic3} := v(0) = 0$$

$$\mathrm{ic3} := \mathrm{v}(\mathrm{o}) = 0$$

$$> dsolve(eq3)$$

$$v(t) = \frac{1600}{9} + e^{-\frac{9t}{5}} c_1$$

$$> dsolve(\{eq3, ic3\})$$

$$v(t) = \frac{1600}{9} + \frac{1600e^{-\frac{9t}{5}}}{9}$$

$$> f := \frac{1600}{9} - \frac{1600}{9} exp\left(-\frac{9t}{5}\right)$$

$$f := \frac{1600}{9} + \frac{1600e^{-\frac{9t}{5}}}{9}$$

= *dsolve*({*eq3*, *ic3*}, *numeric*, *method* = *classical*[*heunform*], *output* = *Array*([1, 2, 3, 4,
6, 7, 8, 9, 50]), *stepsize* = 1)

ans2 :=

t	v(t)
1.	32.000000000000000000
2.	58.240000000000000000
3.	79.756800000000000000
4.	97.400576000000000000
5.	111.86847232000000000
6.	123.73214730240000000
7.	133.46036078796800000
8.	141.43749584613376000
9.	147.97874659382968320
50.	177.76905670135150858

(3

:= *dsolve*({*eq3*, *ic3*}, *numeric*, *method* = *classical*[*rk4*], *output* = *Array*([1, 2, 3, 4, 5, 6,
8, 9, 50]), *stepsize* = 1)

ans2 :=

t	v(t)
1.	127.04000000000000000
2.	163.29721600000000000
3.	173.64502544640000000
4.	176.59829026240256000
5.	177.44115204088969062
6.	177.68170479246991770
7.	177.75035854777091451
8.	177.76995232953381900
9.	177.77554439484895194
50.	177.77777777777777778

(3

FIGURE 7.14
Heunform and RK4 numerical technique solutions for Example 7.5.

The plot is seen in Figure 7.17.

We will now look at finding an approximation for Example 7.5 (Figure 7.18).

```
xdis<-cumtrapz(t,v)
> xdis
             [,1]
 [1,]     0.00000
 [2,]    26.37468
```

```
> Digits := 20 :
> ans2 := dsolve({eq3, ic3}, numeric, method = classical[foreuler], output = Array([1, 2, 3, 4,
    5, 6, 7, 8, 9, 10]), stepsize = 1);
```

ans2 :=

t	$v(t)$
1.	320.00000000000000000
2.	64.0000000000000000
3.	268.80000000000000000
4.	104.96000000000000000
5.	236.03200000000000000
6.	131.17440000000000000
7.	215.06048000000000000
8.	147.95161600000000000
9.	201.63870720000000000
10.	158.68903424000000000

FIGURE 7.15
Euler method numerical technique solutions for updated Example 7.5.

```
 [3,]    89.84719
 [4,]   168.40255
 [5,]   253.09014
 [6,]   340.27091
 [7,]   428.46533
 [8,]   517.07187
 [9,]   605.84596
[10,]   694.68818
[11,]   783.55809
[12,]   872.43926
[13,]   961.32502
[14,]  1050.21263
[15,]  1139.10100
[16,]  1227.98968
[17,]  1316.87848
[18,]  1405.76733
[19,]  1494.65621
[20,]  1583.54509
[21,]  1672.43398
>

xdis<-cumtrapz(t,v)
> xdis
             [,1]
 [1,]     0.00000
 [2,]    26.37468
 [3,]    89.84719
```

= *dsolve*({*eq3*, *ic3*}, *numeric*, *method* = *classical*[*heunform*], *output* = *Array*([1, 2, 3, 4, 6, 7, 8, 9, 50]), *stepsize* = 1)

ans2 :=

t	$v(t)$
1.	32.000000000000000000
2.	58.240000000000000000
3.	79.756800000000000000
4.	97.400576000000000000
5.	111.86847232000000000
6.	123.73214730240000000
7.	133.46036078796800000
8.	141.43749584613376000
9.	147.97874659382968320
50.	177.76905670135150858

(3

:= *dsolve*({*eq3*, *ic3*}, *numeric*, *method* = *classical*[*rk4*], *output* = *Array*([1, 2, 3, 4, 5, 6, 8, 9, 50]), *stepsize* = 1)

ans2 :=

t	$v(t)$
1.	127.04000000000000000
2.	163.29721600000000000
3.	173.64502544640000000
4.	176.59829026240256000
5.	177.44115204088969062
6.	177.68170479246991770
7.	177.75035854777091451
8.	177.76995232953381900
9.	177.77554439484895194
50.	177.77777777777777778

(3

FIGURE 7.16
Heunform and RK4 numerical technique solutions for updated Example 7.5.

```
[4,]   168.40255
[5,]   253.09014
[6,]   340.27091
[7,]   428.46533
[8,]   517.07187
```

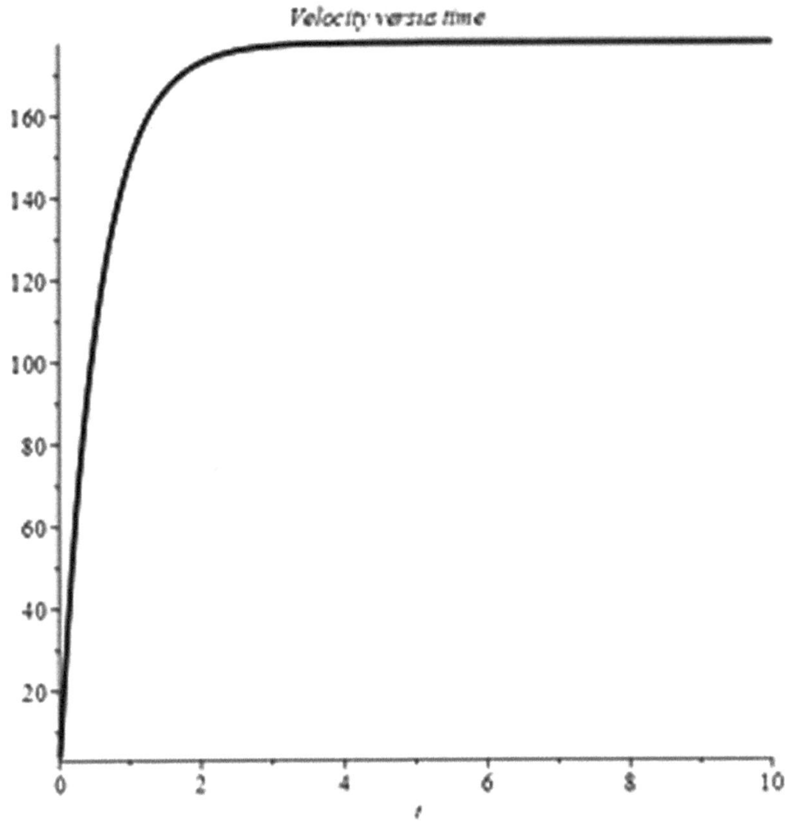

FIGURE 7.17
Plot of velocity versus time with a constant terminal velocity = 177.777.

```
 [9,]   605.84596
[10,]   694.68818
[11,]   783.55809
[12,]   872.43926
[13,]   961.32502
[14,] 1050.21263
[15,] 1139.10100
[16,] 1227.98968
[17,] 1316.87848
[18,] 1405.76733
[19,] 1494.65621
[20,] 1583.54509
[21,] 1672.43398
>
```

We find somewhere between the 4th and 5th timestep we reach 200 feet.

ApproximateInt(*f*, *t* = 0 ..5, *method* = *trapezoid*, *output* = *plot*)

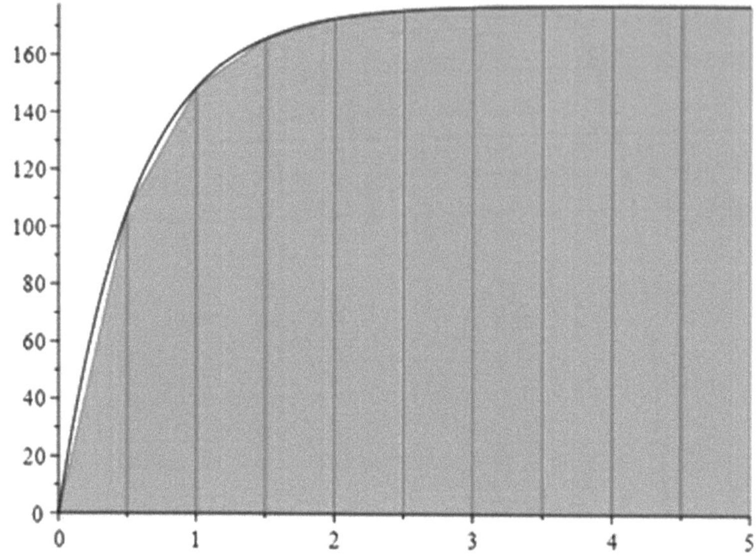

An approximation of $\int_0^5 f(t)\,dt$ using trapezoid rule, where

$f(t) = 177.77777777777777778 - 177.77777777777777778\,e^{-1.8000000000000000000\,t}$

and the partition is uniform. The approximate value of the integral is 783.55808931061237839. Number of subintervals used: 10.

ApproximateInt(*f*, *t* = 0 ..1.6568, *method* = *trapezoid*, *output* = *value*)

200.08827026796965368

FIGURE 7.18
Numerical approximation of an integral.

Harvesting a Species with Numerical Methods

Example 7.6

Consider harvesting the blue crab in Virginia from the Chesapeake Bay. There have been many newspaper reports about the declining populations of blue crabs and the difficulty in harvesting these crabs. Let's model the situation and analyze some "what ifs." (Figure 7.19)

The basic Balance law for Harvesting is:

$$P'(t) = r\left(1 - \frac{P(t)}{k}\right)P(t) - H(t)$$

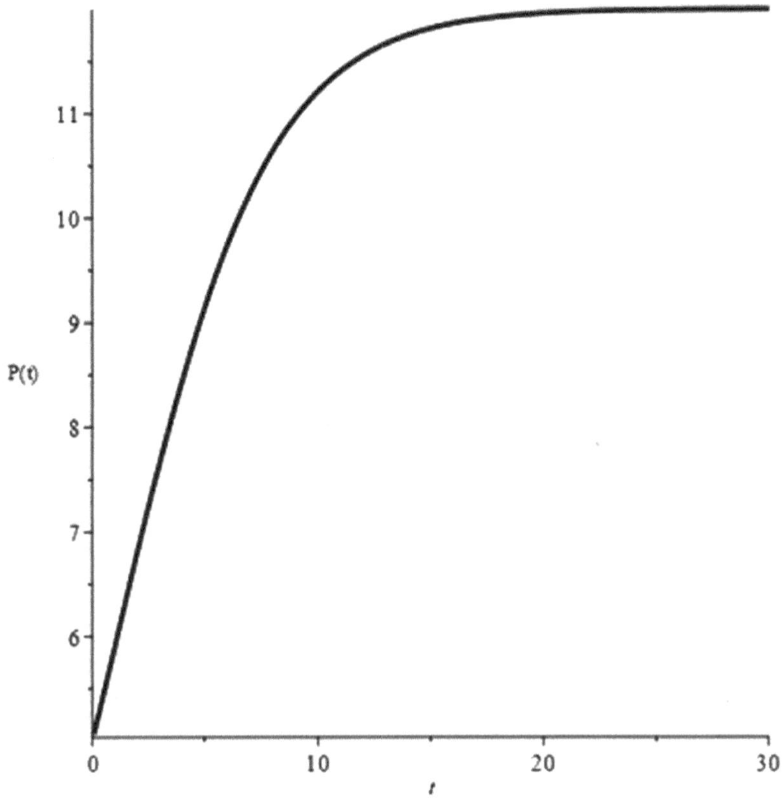

FIGURE 7.19
Graph of harvesting species.

where r is the intrinsic rate coefficient (growth rate if > 0) and k represents the carrying capacity (or saturation level).

In the absence of harvesting (H(t) = 0), the ODE is autonomous. Let r = 0.3 k = 12, and P(0) = 5.

$$> woh := diff\left(P(t),t\right) = 0.3 \cdot \left(1 - \frac{P(t)}{12}\right) \cdot P(t);$$

$$woh := diff\left(P(t),t\right) = 0.3 \cdot \left(1 - \frac{P(t)}{12}\right) \cdot P(t);$$

$$> dsolve\left(\{who, ich\}, \mathrm{P(t)}\right);$$

$$P(t) = \frac{60}{5 + e^{-\frac{3t}{10}}}$$

Now, let's consider harvesting with $h(t) = 0.4$.

$$> harvest := diff\left(P(t),t\right) = 0.3\cdot\left(1-\frac{P(t)}{12}\right)\cdot P(t) - 0.4;$$

$$harvest := \frac{d}{dt}\left(P(t)\right) = 0.3\cdot\left(1-\frac{P(t)}{12}\right)\cdot P(t) - 0.4;$$

$$> harvest := \left(\left\{harvest,\ ich\right\}\right),\ P(t));$$

$$P(t) = 2\sqrt{5}tanh\left(\frac{\left(-4\sqrt{5}arctanh\left(\frac{\sqrt{5}}{10}+t\right)\right)}{20}\right)+6$$

$$> plot\left(\left(\left(2\sqrt{5}tanh\left(\frac{\left(-4\sqrt{5}arctanh\left(\frac{\sqrt{5}}{10}+t\right)\right)}{20}\right)\right)+6,\ t = 0.200, thickness = 3, color = black\right);$$

We still reach an equilibrium of $P(t) = 10.47213595$ (Figure 7.20).

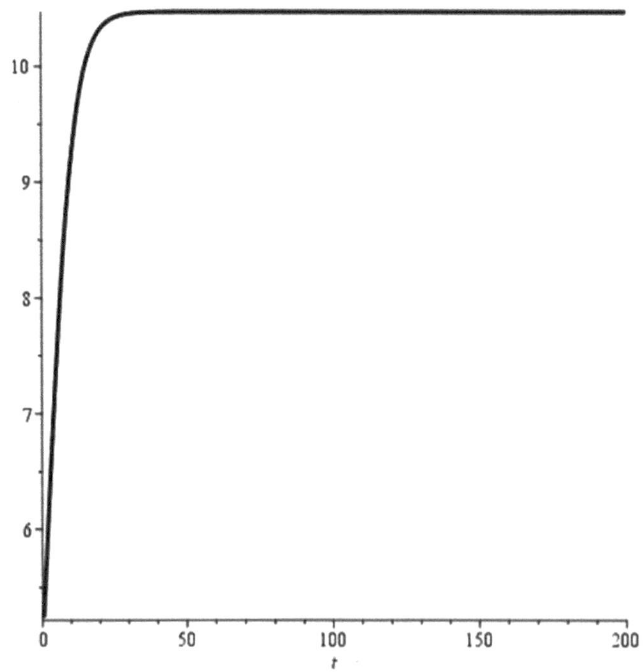

FIGURE 7.20
Harvesting with h(t) = 0.4.

t	$p(t)$
0.	5.
1.	4.50000000000000
2.	4.10625000000000
3.	3.78737402343750
4.	3.52345332418821
5.	3.30117157409959
6.	3.11125751941100
7.	2.94704087359399
8.	2.80359064766800
9.	2.67717740086741
10.	2.56492486499622
15.	2.15106492834448
20.	1.88636605706183
25.	1.70333349210895
30.	1.56998773160613
35.	1.46915392065279
40.	1.39075888698673
45.	1.32849378416310
50.	1.27820076204226
55.	1.23702553154162
60.	1.20294215480541
65.	1.17447207211523
70.	1.15051052537237
75.	1.13021528892347
80.	1.11293304897355
100.	1.06506226353485
200.	1.00487392955600
300.	1.00038578638956
400.	1.00003066551926

FIGURE 7.21
Euler's method approximation.

and we will solve numerically using Euler's method and obtain a plot of the numerical solution. We want to predict 400 days (Figure 7.21).

By Euler's method, our approximate solution is P(400) = 1.00003066551926.

Compare to the analytical solution which is

$$p(t) = -5./\left(4.*\exp\left(-0.02500000000*t\right)-5.\right)$$

$$p(400) = 1.000036322.$$

t	$p(t)$
0.	5.
1.	4.18861631936901
2.	3.51465557922920
3.	2.93545246693939
4.	2.42293896655056
5.	1.95758961923879
6.	1.52506631852378
7.	1.11422679678759
8.	0.715854209553363
9.	0.321772079138831
10.	−0.0758492648140658

FIGURE 7.22
Harvesting over time.

Our absolute error is 5.6654808×10^{-6}

Now consider light harvesting where H(t) = 0.04 (Assume r and k are the same values as before.)

Solve numerically using RK 4 method with P(0) = 5.

We see from our tabulated values that we ran out of crabs before time period 10 (Figure 7.22).

Exercises

7.1 Get Euler estimates for y(3) using the differential equation $y' = y + t$ for step sizes of 0.1, 0.05, and 0.01. Compute the percent error for each.

7.2 Use Euler's method by hand y′ = y + 1, y(0) = 1, with step size of 0.25 to estimate y(0.5). Then do Euler's method using Maple or any other program and compare your results to make sure you have the correct solution. Compute the % error since you can find this exact solution. Change the step sizes to 0.1 and then 0.05 and use Maple or any other program to estimate y(0.5).

7.3 Consider v′ = 32 − 1.6 v, v(0) = 0. Using Euler's method in Maple or any other program and a step size of 0.05 estimate v(2). What is the terminal velocity? Keep stepping out until you approximate the terminal velocity to 3 decimal places of accuracy.

7.4 Try **by hand** y′ = y + 1, y(0) = 1, with step size of 0.25 to estimate y(0.5). Then do Improved Euler's method using Maple or any other program and compare your results to make sure you have the correct solution. Compute the % error since you can find this exact solution.

7.5 Consider v′ = 32 − 1.6 v, v(0) =, Using Heunform in Maple or any other program and a step size of 0.05 estimate v(2). What is the terminal velocity? Keep stepping out until you approximate the terminal velocity to 3 decimal places of accuracy.

7.6 Consider P′ = P*(15 − 3*P), P(0) = 2, Using Heunform in Maple or any other program and a step size of h = 1, h = 0.5, h = 0.1. Discuss the results as compared to qualitative solution.

7.7 Try **by hand** $y' = y + 1$, $y(0) = 1$ with step size of 0.25 to estimate y(0.5). Then do the RK4 method using Maple or any other program and compare your results to make sure you have the correct solution. Compute the % error since you can find this exact solution.

7.8 Consider $v' = 32 - 1.6\,v$, $v(0) = 0$. Using RK4 in Maple or any other program and a step size of 0.05 estimate v(2). What is the terminal velocity? Keep stepping out until you approximate the terminal velocity to 3 decimal places of accuracy.

7.9 Consider $P' = P*(15 - 3*P)$, $P(0) = 2$, Using RK4 in Maple or any other program and a step size of h = 1, h = 0.5, h = 0.1. Discuss the results as compared to qualitative solution.

7.10 In the bungee jumping example assume the drag force is now $0.9\,v + 0.0009\,v^2$ due to wind resistance. Resolve to find the new terminal velocity using numerical methods.

PROJECTS

7.1 **The Spread of a Contagious Disease.** Given the following ODE model for the spread of a communicable disease as:

$$\frac{dN}{dt} = 0.25(10 - N),\ N(0) = 2$$

where N is in 100's

Let's completely analyze the behavior of this ODE.

a. Since this is an autonomous ODE perform a complete graphical analysis:
   I. Plot dN/dt versus *N*. Find and label all rest points (equilibrium points).
   II. Find the value where the rate of change of the disease is the fastest. Why is it this value that you provided?
   III. Plot *N* versus *t* from the following initial conditions:

$$N(0) = 2,\ N(0) = 7,\ \text{and}\ N(0) = 14.$$

   IV. Describe the *stability* of each rest point (equilibrium point).

b. Obtain a slope field plot of this ODE from Maple. Briefly discuss how you would analyze the slope field plot. Compare it to the qualitative plot in part (a).

c. Solve this ODE using separable variables. (Hint: You will need partial fraction decomposition as well.) Ensure you find the value of arbitrary constant C using the initial condition *N(0)* = 2.

d. Simplify your solution in (b) and plot *N* versus *t*.

e. Compare your actual plot with the graphical solution in (a.3) and the slope field plot in (b).

f. From your graph of the analytical solution in (d), estimate the solution values of *N(0.5)* and *N(5)*.

g. Find the actual values for these two solutions *N(.5)* & *N(5)* using your analytical solution from part (c) and substitution.

h. Compute the time, *t*, when N is changing the fastest using the initial condition *N*(0) = 2. Recall, you have found the value of N already.

i. Use Euler's method with step sizes of $h = 0.1$ and then $h = 1$ to approximate the solutions to the ODE for $N(.5)$ & $N(5)$. Find the relative error or absolute differences.
j. Plot Euler's approximations for h = 0.1 and h = 1.0 and compare these graphs to your graphical analysis in part (a) and your other plots. Briefly discuss these plots…you may compare and contrast these plots.
k. What happened with Euler's method? (Any opinions?)

7.2 **Chemical Reactions**

Purpose: To model the changing amounts of salt dissolved in brine as an initial value problem, and to use the computer to obtain graphical and numerical solutions.

Background: A brine solution is a solution of salt in water. If the brine is in a tank equipped with fill and drain pipes, then the total amount of dissolved salt in the tank varies as the concentration in the inflow stream changes and the inflow and outflow rates are adjusted. The amount of salt in the tank can be modeled by appealing to the

$$\textit{Balance law}: \textit{New rate of change} = \textit{rate in} - \textit{rate out}$$

The term "rate in" refers to the rate at which salt is added to the brine by means of the inflow stream and the term "rate out" is the rate at which salt leaves the tank through the outflow pipe. It is assumed throughout that the inflow stream is instantaneously mixed with the brine in the tank so that at any time the concentration of salt in the tank is uniform. Each of these terms (rate in and rate out) is a product of the appropriate brine flow rate with corresponding salt concentrations.

Consider a tank with a capacity of 4000 liters holds 2000 liters of brine that contains 50 kg of dissolved salt. Brine with a salt concentration of 0.2 kg/liter is piped into the tank at a rate of 40 liters/min. Well-mixed brine is drawn off at a rate of a liter/min.

$$\frac{dx}{dt} = (0.2)(40) + \frac{x}{2000 + (40 - a)t}a$$

Requirement 1. Show that the above is the general ODE; find the appropriate initial condition for this Brine problem.

Requirement 2. Solve the ODE leaving a as a parameter.

Requirement 3. Solve the ODE and obtain plots when the "rate out" $a = 30$, 40, and 50.

Requirement 4. (BULK SALT) Suppose that 200 kg of bulk salt is placed in a tanking holding 2000 liters of brine already containing x_0 kg of dissolved salt. Suppose that the saturation level of brine is 300 kg of salt. The model representing this is:

$$\frac{dx}{dt} = 0.01(200 - x)(300 - x),\ x(0) = x_o$$

Solve this ODE, and obtain a plot for various choices of x_0, ($x_0 = 25, 50, 100, 200$).

Requirement 5. Bimolecular Chemical Reactions.

Background: In a bimolecular chemical reaction, two species interact and create one or more products. Consider the reaction,

$$A + B -> C + D$$

The Law of Conservation: $\frac{dC}{dt} = \frac{dD}{dt} = -\frac{dA}{dt} = -\frac{dB}{dt}$

The Law of Mass Action: The rate of an elementary reaction is proportional to the product of the concentrations of the reactants.

The ODE,

$$\frac{dx}{dt} = k_1(a+c-x)(b+c-x), x(0)=c$$

Consider the following reversible chemical reaction,

$$\frac{dx}{dt} = (0.7-x)(0.4-x)-0.1x^2, x(0)=c$$

Solve qualitatively (autonomous). Solve this ODE (at least numerically) and show the curve and discuss the equilibrium values. Let c = 0.2. This reaction is only valid for $0 \le c \le 0.4$.

7.3 **Harvesting a Species.** Consider harvesting the blue crab in South Carolina. There have been many newspaper reports about the declining populations of blue crabs and the difficulty in harvesting these crabs. Let's model the situation and analyze some "what ifs."

The basic Balance law for Harvesting is:

$$P'(t) = r\left(1-\frac{P(t)}{k}\right)P(t)-H(t)$$

where r is the intrinsic rate coefficient (growth rate if > 0) and k represents the carrying capacity (or saturation level).

Requirement 1. In the absence of harvesting (H(t) = 0), the ODE is autonomous. Let r = 0.3, k =12, and P(0) = 5 and

a. Perform a qualitative assessment of this situation since it is autonomous. List and classify each equilibrium value.
b. Solve this ODE.
c. Plot the solution to this ODE and compare to your qualitative solution. Briefly discuss similarities and/or differences.
d. Solve numerically using Euler's Method and obtain a plot of the numerical solution. Compare to the analytical solution.

Requirement 2. Now consider light harvesting where H(t)=0.4 (Assume r and k are the same values as before.)

a. Obtain a slope field plot and discuss any equilibrium values observed.
b. Find all equilibrium values of this ODE.
c. Solve this ODE. Assume initial conditions of P(0) = 19, P(0) = 8, and P(0) = 0.8.
d. Solve numerically using Euler's Method with P(0) = 5.

Requirement 3. Now consider a heavier harvesting where H(t) = 1.5.

a. Obtain a slope field plot and discuss any equilibrium values observed.
b. Find all equilibrium values of this ODE.
c. Solve this ODE. Assume initial conditions of P(0) = 19, P(0) = 8, and P(0) = .8.
d. Solve numerically using Euler's Method with P(0) = 5.

Requirement 4. Determine which of the models is more likely to represent the current situation for blue crabs in the Chesapeake Bay.

7.4 **Rural Water Supplies.** Consider a cylindrical water tank discharging into a pipe located in the middle of its base. The potential energy given by the pressure of the column of liquid is converted into kinetic energy of the stream of water, as it exists in the tank.

If the level of the water in the tank is h ft, then it follows that the exit velocity in ft/s is $v = \sqrt{2gh}$ where $g = 32.2$ ft/s^2 is the acceleration of gravity. Since a convergent flow is set up from all sides to the orifice, inertia from the particles in the jet forces them to overshoot the edge and to converge to a smaller cross-section termed the *vena contracta*. The ratio of the diameter of the vena contracta to the diameter of the orifice is called the contraction coefficient, α; and it ranges from 0.5 to 1.0 depending on the smoothness of the edge of the orifice. The diameter of the vena cross-section and the exit velocity of the stream determine the volume rate of the discharge of the stream. Thus, if the tank has cross-section A and is being filled from the top at a rate K while draining for the bottom through an orifice of area a, the height in the tank is governed by the differential equation:

$$\frac{dh}{dt} = \frac{\left(K - a\alpha\sqrt{2gh}\right)}{A}$$

Now, consider we are in the mountains outside of Stowe, VT. The water for the cabin is supplied from an aquifer some 125 feet below the surface at a maximum rate of 7 gallons per minute. Since the flow rate is insufficient to meet daily needs, you must design a more complicated system. In your design, this water is pumped to a cylindrical tank with a diameter of 10 feet and height 35 feet which is located on a nearby rise about 75 feet from the cabin, with a base 25 feet above the level of the cabin floor. The water is stored in this gravity tank until needed (water is turned on by opening the spicket).

Assume that the coefficient of contraction at the exit is $\alpha = 0.63$ and that the water main consists of 6-inch diameter PVC pipe, so that the friction losses are less than 2 psi.

Requirement:

a. Determine the minimum flow rate in the system and the minimum water pressure during the first two hours.
b. The local fire code demands that water supplies maintain a pressure of at least 20 psi at a flow rate of 30 gallons per minute for a minimum of two hours. Will your design pass inspection?
c. At continuous maximum flow, how long would it take for the level of the top of the water in the tank to fall to 10 ft above the base?

7.5 **Falling Bodies** In bridge jumping, a participant attaches one end of a bungee cord to himself (or herself), attaches the other end to a bridge railing, and then drops off the bridge. In this project, the jumper will be dropping off of the Columbia Gorge, a suspension bridge that is 400 feet above the floor of the Gorge in Oregon. The jumper will use a 100-foot long bungee cord. It would be nice if the jumper has a safe jump, meaning that the jumper does not crash into the floor of the gorge or run into the bridge on the rebound. In this project, you will do some analysis of the fall.

Assume the jumper weighs 200 lbs. The jumper will free fall until the bungee cord begins to exert a force that acts to restore the cord to its natural (equilibrium) position. In order to determine the spring constant of the bungee cord, you found that a MASS weighing 4 lbs stretches the cord 8 feet. Hopefully, this spring force will help slow down the descent sufficiently so that the jumper does not hit the bottom of the gorge.

Throughout this project, we will assume that DOWN is the POSITIVE direction. Assume the force of gravity is 32.17 ft/s^2.

Requirement 1. Before the bungee cord begins to retard the fall of the jumper, the only forces that act on the jumper are his weight and the force due to wind (air) resistance.

a. If the force due to the wind resistance is 0.9 times the velocity of the jumper, then use Newton's second law ($\Sigma F = MA$) to write a differential equation that models the fall of the jumper. Be sure to include the initial conditions for the jumper for your differential equation. (Hint: This problem can be formulated as a 2nd order DE in position, or as a 1st order DE in velocity – use the first order DE model in this project.)
b. Solve this differential equation and find a function that describes the jumper's velocity (as a function of time). Knowing integral calculus, find a function that describes the jumper's position (as a function of time).
c. What is the velocity of the jumper after the jumper has fallen 200 feet?
d. What is the terminal velocity of the jumper, if any?

Requirement 2. After a bit more research, you have found that the force due to wind resistance is not linear, as assumed above. Apparently, the force due to wind resistance is more closely modeled by

$$0.9\,v + 0.0009\,v^2.$$

a. Write a new differential equation governing the velocity of the jumper (prior to the bungee cord coming into effect).
b. You should notice that this DE is no longer easy to solve. Nonetheless, you are determined to find the velocity of the jumper after 4 s by using a numerical technique. Use Euler's method with a step size of 0.5, and estimate the velocity of the jumper after 4 s.
c. Find the terminal velocity, if any exists.
d. How do your results compare with those found in requirement one under the linear assumption?
e. What is the velocity of the jumper after 75 feet? (Hint: You can find this from your numerical Table.)

References and Additional Readings

Burden, R., and Faires, J. (2010). *Numerical Analysis*. 9th edition. Cengage Learning.

Chapra, S. (2017). *Applied Numerical Methods with MATLAB for Engineers and Scientists*. 4th edition. McGraw-Hill Education.

Chapra, S. C., and Canale, R. (2015). *Numerical Methods for Engineers*. 7th edition. McGraw-Hill Education.

8

Higher Order Differential Equations

Introductory Model

Higher order differential equations are a class of differential equations with the highest derivative of order two or greater. These equations typically appear in more complex real-world situations where systems are described by rates of change dependent on more than just the first derivative. These include modeling physical systems that involve oscillations, such as springs and pendulums, and electrical circuits, using second order differential equations. Higher order differential equations model are more complicated dynamic systems in which acceleration, curvature, or even higher-dimensional changes play an important role. Solutions, either analytically or computationally, begin to present a lot of difficulty numerical methods such as Runge-Kutta are usually applied to obtain approximate solutions.

Bungee Jumping

Let's consider the example problem from Chapter 5 concerning bungee jumping. We examined the free fall phase as a first ODE. We discovered that the time to reach terminal velocity was y and the terminal velocity speed was 90.5214. We also found the bungee jumper traveled 200 feet before reaching terminal velocity and 320 feet before the bungee cord becomes active.

Now, let's treat this bungee jumping problem as a second ODE.

Much of the act of bungee jumping can be modeled with the proper understanding of some physics and second ODEs. The importance of doing this type of analysis is to ensure the safety of the jumpers on a bungee jump. In order for the bungee jumpers to be safe, they need to not hit the ground below or the bridge that they just jumped off of for fun. Also, the bungee jumpers need to not experience too much acceleration. The furthest stretch and the highest acceleration occur on the first bounce, so if the bungee jumpers are safe during that time, then the bungee jumpers are safe for the entire jump.

Modeling the motion with a bungee jump: The model of the motion in one dimension with respect to time can be broken into two different sections: the free fall from the ledge and the time when the bungee cord is pulling on the jumper. We solved the free fall portion in Chapter 4. To model both sections we need Newton's second law of

 DOI: 10.1201/9781003582274-8

motion, $\sum F = ma$. It says that the total of all the forces acting on an object is equal to the mass of the object multiplied by its acceleration.

$$\sum F = ma.$$

To describe how the bungee cord will pull on a bungee jumper we will use Hooke's law. Hooke's law says that the force exerted by a spring is proportional to the distance that it has been stretched from its equilibrium point. The relationship is said to be $F = -kx$, where k is a constant that is specific to the spring. The negative is in the equation so that the force is in the opposite direction of the stretch (pushing/pulling toward the resting point of the spring). For our model, we assume that a bungee cord will behave like a spring described by Hooke's law. The spring constant k can be found experimentally by hanging a known weight on a spring and measuring how far it stretches. This was exhibited in Chapter 1 under proportionality. In order to keep the math initially doable, we will also assume that the wind resistance is linearly proportional to the velocity. There are more accurate and complicated formulas to account for wind resistance that will be considered later, but for now:

$$F_{wind} = -\beta v.$$

We will define the positive x direction to be going down.

When the bungee jumper first jumps off of the bridge, they will have the forces of gravity and wind resistance acting on them. The force of gravity is positive because we've defined the positive x direction to be going down. The wind resistance is $-\beta v$ because the force of wind resistance is always in the opposite direction of the velocity. Again, this portion was previously viewed in Chapter 4. We must now concern ourselves with the bungee cord action.

Using Newton's second law again: $mg - \beta v = ma$ and using the fact that $a = dv/dt$, $v = dx/dt$ then with a little algebra, we can put this into the form of a nonhomogeneous equation second order differential equation in position, x, shown in Equation 8.1.

$$x'' + \beta m x' = g \tag{8.1}$$

First, we must solve $x'' + \beta m x' = 0$. This is the homogeneous equation. To find the general solution of the second ODE equation, we must first find the solutions to the equivalent homogeneous equation. In other words, we will set the driving force g to zero. (Note: there are other easier ways, but we will solve our nonhomogeneous equation this way so that the process is the same as the process needed in the latter part of this section.) $x'' + \beta\ mx' = 0$. We use the characteristic equation for the homogeneous equation is: $p(\lambda) = \lambda^2 + \beta m \lambda$. We set $p(\lambda)$ equal to 0 to find that the roots of the equation are: $\lambda_1 = 0$, $\lambda_2 = -\beta m$. The solution to the homogeneous equation is shown in Equation 8.1.

$$x_h = C_1 e^{\lambda 1t} + C_2 e^{\lambda 2t} \tag{8.2}$$

Substituting in our values for λ_0 and $= -\beta$ m, we have $x_h = C_1 + C_2 e^{-\beta t/m}$.

Now that the homogeneous equation has been solved, the only thing left to find is the particular solution. If we look at Equation 8.1 and look at the driving force that is on the right-hand side, then we see that the driving force is a polynomial of zeroth degree (a constant).

The particular solution to the equation should also be a polynomial. If we say a particular solution is: $x_p = at + b$, then x′ = a and x″ = 0. Then we can substitute these into Equation 8.1 and solve for

$$a(0) + \beta m(a) = g\beta ma = ga = mg\beta$$

Our general solution is:

$$x = C_1 + C_2 e - \beta t/m + mg\beta t + bC_1 \text{ and } b \text{ are both constants.}$$

because C_1 is an arbitrary constant, we say that it accounts for *b*.

$$x = C_1 + C_2 e - \beta t/m + mg\beta t(3)$$

Below is the equilibrium point for the cord so far, an equation for the motion of the bungee jumper has been found. Now, we need to find an equation of motion for when the bungee jumper is below the equilibrium point. To do this, we look back to Newton's second law: ΣF = ma. When the jumper is below the equilibrium point, they still have gravity and wind resistance acting on them. But, now they also have the force of the cord pulling on them. We will say that the force applied by the bungee cord is –*kx* (from Hooke's law). Using this with Newton's second law:

$$mg - \beta dx/dt - kx = mx''.$$

Once again, this needs to be rearranged algebraically to look like an inhomogeneous equation, as shown in Equation 8.3.

$$x'' + \beta m x' dt + kmx = g \qquad (8.3)$$

Next, we need to solve the homogeneous equation. The first step to solving the inhomogeneous equation is to solve the equivalent homogeneous equation.

$$x'' + \beta m x' + kmx = 0$$

The characteristic equation is

$$P(\lambda) = \lambda^2 + \beta m\lambda + km.$$

We set P(λ) to 0 and use the quadratic formula to find that the roots of the characteristic equation are:

$$\lambda = -\beta/m \pm q(\beta/m)^2 - 4(k/m)2(4)$$

At this point, the solution to the homogeneous equation can go in three directions, depending on the discriminant. There are three cases, when *(b/m)*^{2} – *4(k/m)* is greater than zero, equal to zero, or less than zero.

Case 1. $(\beta/m)^2 - 4(k/m) > 0$. This is referred to as over damped, and the solution will not oscillate. If bungee cord and jumper system fall into this category then there is a problem, because the bungee jumper will likely experience too much acceleration. In this case, the characteristic equation has two distinct real roots. The solution will take the form:

$$x_h = C_1 e^{\lambda 1t} + C_2 e^{\lambda 2t}$$

We can substitute in our expression for λ to get:

$$x_h = C1e\left(-\beta/m + \surd\,(\beta/m)2 - 4k/m\right)t/2 + C2e\left(-\beta/m + \surd(\beta/m)2 - 4k/m\right)t/2$$

Case 2. $(\beta/m)2 - 4(k/m) = 0$. This case is said to be critically damped, and its solution appears very similar to the over damped case. The characteristic equation has one real root, and the solution will

$$x'' - .5(x + 320) \text{ with } x(0) = 320,\ x' = \text{velocity at 200 feet}$$

$> ode2 := 5diff\left(x(t),\, t,\, t\right) = 160 - 0.9diff\left(x(t), t\right) - 0.5\left(x(t) + 320\right);$

$$ode2 := 5\frac{d^2}{dr^2}x(t) = -0.9\frac{d}{dr}x(t) - 0.5x(t)$$

$> ics1 := \mathrm{x}(0) = 320, \mathrm{D}(\mathrm{x})(0) = 90.5214;$

$$ics1 := \mathrm{x}(0) = 320, \mathrm{D}(\mathrm{x})(0) = 90.5214;$$

The solution using Maple is:

solbj := x(t) = 0.00002176278564 * exp(−0.09000000000 * t) *
(1.808614883 * 10^7 * sin(0.3031501278 * t) + 1.4704000 *
10^7 * cos(0.3031501278 * t))

$$x(t) = 0.0000217627e^{(-0.09t)}(1.808614883 * 10^7(\sin(0.3031501278\ t) + 1.4704 * 10^7\left(\cos(0.3031501278\ t)\right)$$

The plot that shows the oscillations over time is shown in Figure 8.1.

The bungee jumper will be safe and not either hit the ground or rebound back up and hit the bridge. But how did we obtain the analytical solution?

We will discuss higher order odes and their solution techniques in subsequent sections.

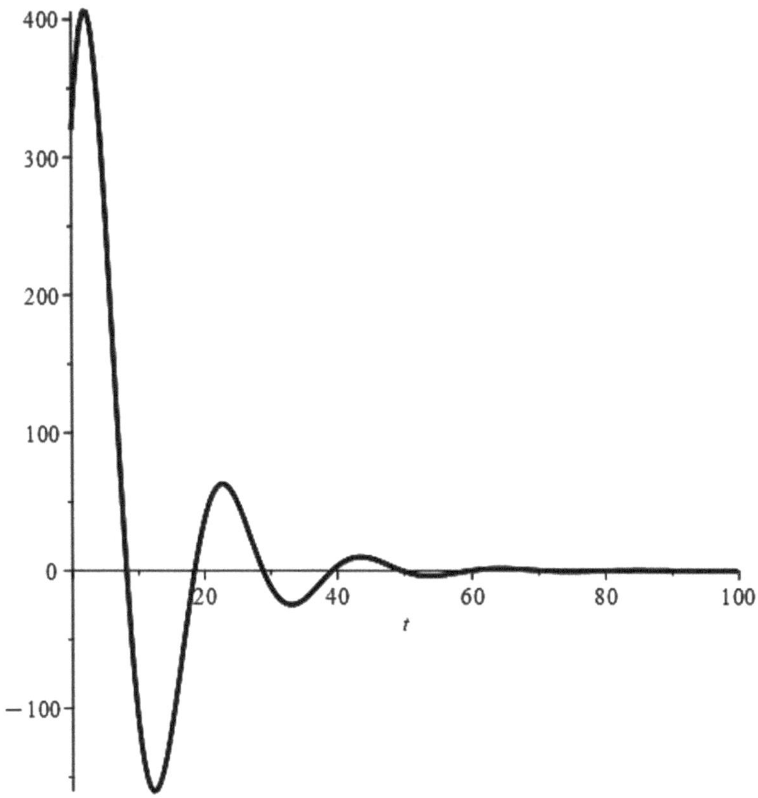

FIGURE 8.1
The bungee jump of the bridge.

Mixtures

There are many types of mixture problems. Such problems are standard in a first course on differential equations as examples of first ODEs. Typically, these examples consist of a tank of brine, water containing a specific amount of salt with pure water entering and the mixture leaving, or the flow of a pollutant into, or out of, a lake.

In general, one has a rate of flow of some concentration of mixture entering a region and a mixture leaving the region. The goal is to determine how much stuff is in the region at a given time. This is governed by the following equation:

$$\text{Rate of change of substance} = \text{Rate In} - \text{Rate Out}.$$

This can be generalized to the case of two interconnected tanks. We will provide an example.

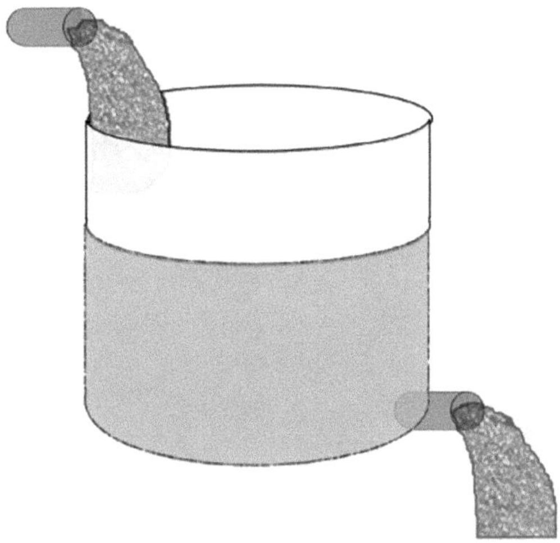

FIGURE 8.2
A typical mixing problem.

Example 8.1: Single Tank Problem

A 50-gallon tank of pure water has a brine mixture with a concentration of 2 pounds per gallon entering at the rate of 5 gallons per minute, as shown in Figure 8.2. At the same time, the well-mixed contents drain out at the rate of 5 gallons per minute. Find the amount of salt in the tank at time t. In all such problems, one assumes that the solution is well mixed at each instant of time.

Let $x(t)$ be the amount of salt at time t.

Then the rate at which the salt in the tank increases is due to the amount of salt entering the tank less than leaving the tank. To compute or model these rates, one notes that dx/dt has units of pounds per minute. The amount of salt entering per minute is given by the product of the entering concentration times the rate at which the brine enters. This gives the correct units:

$$(2\,\text{pounds gal})(5\,\text{gal min}) = 10\,\text{pounds min.} = 10\,\text{pounds min.}$$

Similarly, one can determine the rate out as (x pounds 50 gal) (5 gal min) = 10 x pounds min.

Thus, we have

$$dx/dt = 10 - 10x$$

This equation is easily solved using the methods for first ODEs.

Example 8.2: Double Tank Problems

One has two tanks connected together, labeled tank x and tank y, as shown in Figure 8.3.

Let tank x initially have 100 gallons of brine made with 100 pounds of salt. Tank y initially has 100 gallons of pure water. Pure water is pumped into tank X at a rate of 2.0 gallons per minute. Some of the mixture of brine and pure water flows into tank Y at 3 gallons per minute. To keep the tank levels the same, 1 gallon of the Y mixture

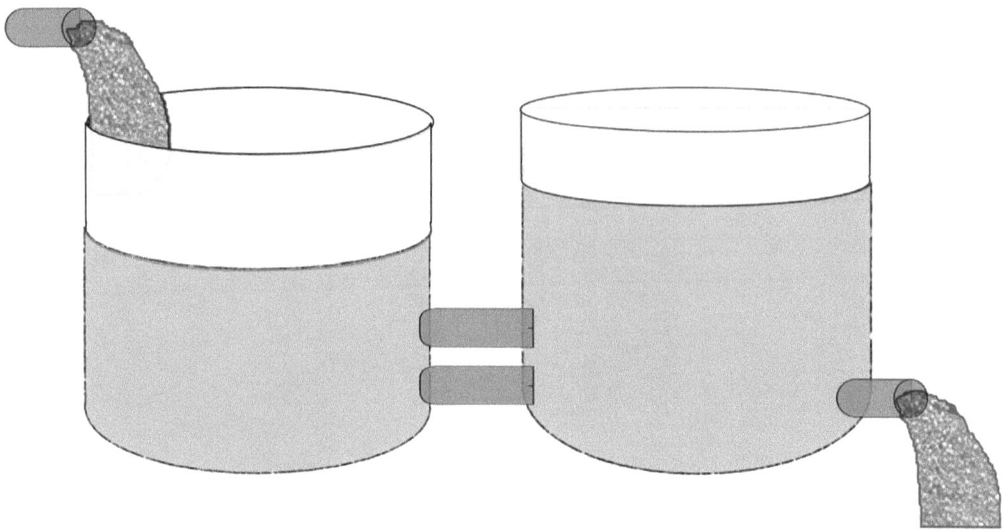

FIGURE 8.3
The two tank problem.

flows back into tank X at a rate of 1 gallon per minute and 2.0 gallons per minute drain out. Find the amount of salt at any given time in the tanks. What happens over a long period of time?

In this problem, we set up two equations. Let $x(t)$ be the amount of salt in tank x and $y(t)$ the amount of salt in tank y. Again, we carefully look at the rates into and out of each tank in order to set up the system of differential equations. We obtain the system

$$dx/dt = y/100 - 3x/100 \quad dy/dt = 3x/100 - 3y/100$$

We put this in matrix form

$$x' = \begin{bmatrix} -0.03 & 0.01 \\ 0.03 & -0.03 \end{bmatrix} \begin{bmatrix} x(t) \\ y(t) \end{bmatrix} = \begin{bmatrix} 100 \\ 0 \end{bmatrix}$$

First, we find the eigenvalues:

λ are –1.26 and –4.73 as our approximate eigenvalues. Our eigenvectors are $\begin{bmatrix} 1 \\ \sqrt{3} \end{bmatrix}$ and $\begin{bmatrix} 1 \\ -\sqrt{3} \end{bmatrix}$.

We solve for c_1 and c_2 and find $c_1 = c_2 = 50$.
The solution is

$$x(t) = 50\left(e^{-3+\sqrt{3}t} + e^{-3-\sqrt{3}t}\right)$$

$$y(t) = 50\sqrt{3}\left(e^{-3+\sqrt{3}t} - e^{-3-\sqrt{3}t}\right)$$

We plot as shown in Figure 8.4.

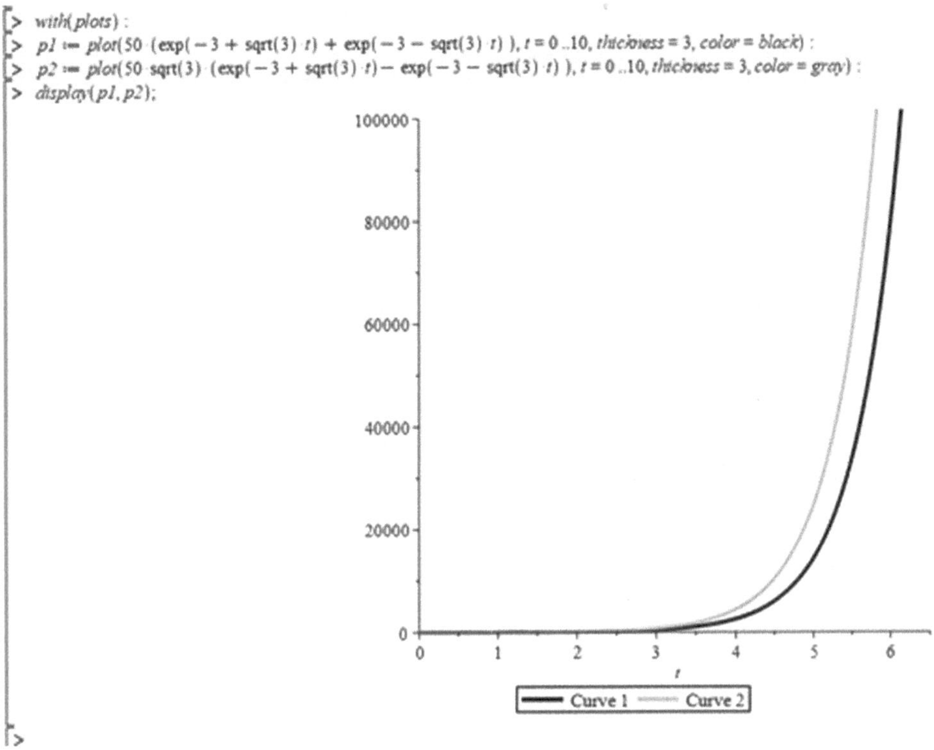

FIGURE 8.4
Plot of x(t) black and y(t) gray.

Solution Techniques to Second ODEs

Second order differential equations can be classified into various types based on their characteristics and properties. Some common types that we will discuss are:

- Homogeneous and Non-homogeneous Equations
- Linear and Non-Linear Differential Equations
- Second ODE with Constant Coefficients
- Second ODE with Variable Coefficients

Homogeneous and Non-Homogeneous Equations

We begin with discussions of homogeneous and non-homogeneous differential equations.

Homogeneous equations have the form $ay'' + by' + cy = 0$

$$\text{An example is} : 5y'' + 4y' + y = 0$$

Non-homogeneous equations have an additional function on the right-hand side, such as f(x), making them: $ay'' + by' + cy = f(x)$

An example of a second order differential equation is $5y'' + 4y' + y = \cos(x)$

Linear and Non-Linear Differential Equations

Linear differential equations are equations where the dependent variable and its derivatives appear linearly. They can be represented as x′ + P(x)y = Q(x), where P(x) and Q(x) are functions of x. An example is:

$$y' + 2y = sin\, f_0(x)$$

Non-linear differential equations are those where the dependent variable and its derivatives appear in a non-linear manner. These equations do not follow the superposition principle and can lead to diverse solutions, sometimes exhibiting chaotic behavior. A common example is:

$$t' = y^2$$

Second Order Differential Equation with Constant Coefficients

Constant coefficient equations have coefficients that are constants throughout the equation, like $y'' + 2y' + 3y = 0$. Variable coefficient equations have coefficients that vary with the independent variable.

For example, consider the following:

$$2y'' + 11y' - 5y = x$$
$$5y'' - 6y' + 10y = \cos(x)$$

Second Order Differential Equation with Variable Coefficients

Second ODE with variable coefficients are those in which the coefficient of differential equation is a variable.

For example, consider the following:

$$xy'' + y' - e^{xy} = x$$
$$\ln(x)y'' - xy' + y = \cos x$$

Solutions of Second Order Differential Equations

The solutions of second order differential equations have two components: a General Solution and a Particular Solution.

General solution of a second order differential equation includes both the complementary function (homogeneous solution) and the particular integral (non-homogeneous solution).

For example, $y'' + 4y' + 4y = 8°$.

We will use the characteristic equation, explained in more detail later in this chapter to find the two repeating roots: –2, –2. Thus, the general solution is

$$y = c_1e^{-2x} + c_2xe^{-2x} + 2$$

before considering any initial conditions.

Initial Value Problems and Boundary Value Problems

Initial value problems involve finding a solution that satisfies the equation and given initial conditions, like y(0) = 1, dy/dx(0) = 2. Boundary value problems involve finding a solution that satisfies the equation and given conditions at different points, such as y(0) = 1, y(1) = 3.

Solving Second-Order Differential Equations

Different types for solving second order differential equations are:

- Analytical method
- Numerical method

Solving Homogeneous Second Order Differential Equation

To solve a homogeneous second order differential equation of the form $ay'' + by' + cy = 0$, where *a*, *b*, and *c* are constants, we can follow these steps:

Find the characteristic equation by substituting $y = e^{rx}$ where r is a root into the equation and simplifying.

- Solve the characteristic equation to find the roots r_1 and r_2.
- If the roots are real and distinct, the general solution is $y = c_1e^{r1x} + c_2e^{r2x}$.
- If the roots are complex conjugates, $f \pm g\ i$ then (r = the general solution is $y = e^{f(x)}\{c_1 \cos f_0(gx) + c_2 \sin f_0(gx)]$, where α and β are determined from the roots.
- If the roots are repeated, the general solution is $y = (c_1 + c_2x)e^{rx}$, where r is the repeated root.

Example 8.3

a. $y'' + 3y + 2 = 0$, roots are 2 and 1. As distinct roots, the solution is $y = c_1e^{2x} + c_2e^{x}$.
b. $y'' - 4y = 0$, roots are 2 and 2 are repeated roots. As repeated roots, the solution is $y = (c_1 + c_2x)e^{2x}$
c. $y'' + 4y = 0$, roots are complex $\pm 2\ i$. The solution is $e^0x\{c_1 \cos(2x) + c_2 \sin(2x)]$.

Solving Non-Homogeneous Second Order Differential Equations

To solve a non-homogeneous second order differential equation of the form a y "+ by' + cy = f(x), where a, b, and c are constants, and f(x) is a function of x, we can use the method of undetermined coefficients or the method of variation of parameters.

- Solve the associated homogeneous equation to find the general solution complementary solution, y_c.

- Find a particular solution y_p using the method of undetermined coefficients or variation of parameters.
- General solution of the non-homogeneous equation is $y = y_c + y_p$.

Second Order Differential Equation Examples

Second order differential equations are widely used in various fields to model and analyze complex systems. Here are a few examples of how second order differential equations are applied in different domains:

Mechanical Vibrations

In mechanical systems, second order differential equations are used to model the motion of masses connected to springs and dampers. For example, the equation of motion for a mass-spring-damper system is:

$$mx'' + cx' + kx = F(t)$$

This equation can be used to analyze the vibrations and oscillations of mechanical systems, such as suspension systems in vehicles.

An analytical way of solving differential equations with such methods as undetermined coefficients and the method of variation of parameters is applied. The particular solution for an undetermined coefficients method is approximated in a particular form and the coefficients are derived by substituting the solution back into the differential equation.

Let's consider an example to illustrate the undetermined coefficients method for solving second order linear differential equations with constant coefficients:

Example 8.4: Solve the differential equation $y'' - 3y' + 2y = 2e^x$

Solution:

Step 1: Solve the associated homogeneous equation

$$y'' - 3y' + 2y = 0$$

Characteristic equation is $r^2 - 3r + 2 = 0$, which has roots $r_1 = 1$ and $r_2 = 2$
Therefore, the general solution of the homogeneous equation is:

$$yc = c_1e^x + c_2e^{2x}$$

Step 2: Assume a particular solution of the form $yp = Ae^x$
Substitute yp into the original equation:

$$(Ae^x)'' - 3(Ae^x)' + 2(Ae^x) = 2e^x$$

We simplify and solve for A:

$$Ae^x - 3Ae^x + 2Ae^x = 2e^x$$
$$-Ae^x = 2e^x$$
$$A = -2$$

Step 3: Write the general solution

$$y = yc + yp$$

$$y = c1e^x + c2e2^x - 2e^x$$

Now, if we have initial conditions for y(0) = 2 and y′(0) = 3.
We substitute into the solution for y and its derivative to find c_1 and c_2.

$$2 = c1 + c2 - 2 \text{ or } 4 = c1 + c2$$

$$Y' = 3 = c_1 + 2c_2 - 2 \text{ or } 5 = c1 + 2c2$$

Solving for c1 and c2 we find c1 = 3 and c2 = 1.
The solution with these initial conditions is

$$y = 3e^x + e2^x - 2e^x$$

Second Order Equations with Constant Coefficients

Now that we have a better feel for second order equations and linear differential equations, we are going to concentrate on solving second order equations of the form

$$ay'' + by' + cy = 0, \tag{8.4}$$

where a, b, and c are constants.

Since all the coefficients are constants, the solutions are probably going to be functions with derivatives that are constant multiples of themselves. We need all the terms to cancel out, and if taking a derivative introduces a term that is not a constant multiple of the original function, it is difficult to see how that term cancels out. Exponential functions have derivatives that are constant multiples of the original function, so let's see what happens when we try a solution of the form $y(x) = e^{\lambda x}$, where λ (the lowercase Greek letter lambda) is some constant.

If $y(x) = e^{\lambda x}$, then $y'(x) = \lambda e^{\lambda x}$ and $y''(x) = \lambda^2 e^{\lambda x}$. Substituting these expressions into Equation 8.4, we get

$$ay'' + by' + cy = a\left(\lambda^2 e^{\lambda x}\right) + b\left(\lambda e^{\lambda x}\right) + ce^{\lambda x} = e^{\lambda x}\left(a\lambda^2 + b\lambda + c\right).$$

Since $e^{\lambda x}$ is never zero, this expression can be equal to zero for all xx only if

$$a\lambda^2 + b\lambda + c = 0.$$

We call this the *characteristic equation* of the differential equation.

Definition: The characteristic equation of a second ODE,

$$ay'' + by' + cy = a\lambda^2 + b\lambda + c = 0.$$

The characteristic equation is very important in finding solutions to differential equations of this form. We can solve the characteristic equation either by factoring or by using the quadratic formula

$$\lambda = \frac{-b \pm -= \sqrt{b^2 - 4ac}}{2a}$$

This gives three cases. The characteristic equation has either

1. distinct real roots when $\sqrt{b^2 - 4ac} > 0$;
2. a single, repeated real root when $\sqrt{b^2 - 4ac} = 0$; or
3. complex conjugate roots when $\sqrt{b^2 - 4ac} < 0$.

We consider each of these cases separately.

Case 1: Distinct Real Roots

If the characteristic equation has distinct real roots λ_1 and λ_2, then $e^{\lambda_1 x}$ and $e^{\lambda_2 x}$ are linearly independent solutions to our example, and the general solution is given by

$$y(x) = c_1 e^{(\lambda_1 x)} + c_2 e^{(\lambda_2 x)},$$

where c_1 and c_2 are constants.

For example, the differential equation $y'' + 9y' + 14y = 0$ has the associated characteristic equation $\lambda^2 + 9\lambda + 14 = 0$. This factors into $(\lambda + 2)(\lambda + 7) = 0$, which has roots $\lambda_1 = -2$ and $\lambda_2 = -7$. Therefore, the general solution to this differential equation is

$$y(x) = c_1 e^{-2x} + c_2 e^{-7x}$$

Case 2: Single Repeated Real Root

Things are a little more complicated if the characteristic equation has a repeated real root, λ. In this case, we know $e^{\lambda x}$ is a solution to equation but it is only one solution and we need two linearly independent solutions to determine the general solution. We might be tempted to try a function of the form $ke^{\lambda x}$, where k is some constant, but it would not be linearly independent of $e^{\lambda x}$. Therefore, let's try $xe^{\lambda x}$ as the second solution. This works.

$$y' = e^{\lambda x} + \lambda x e^{\lambda x} \quad y'' = 2\lambda e_{\lambda x} + \lambda 2 x e^{\lambda x}.$$

Substituting both expressions into equation, we see that it works. This shows that $xe^{\lambda x}$ is a solution to Equation 8.4. Since $e^{\lambda x}$ and $xe^{\lambda x}$ are linearly independent, when the characteristic equation has a repeated root λ, the general solution to equation is given by the following:

$$y(x) = c_1 e^{\lambda x} + c_2 x e^{\lambda x},$$

where c_1 and c_2 are constants.

For example, the differential equation $y'' + 12y' + 36y = 0$ has the associated characteristic equation $\lambda^2 + 12\lambda + 36 = 0$.

This factors into $(\lambda + 6)^2 = 0$, which has a repeated root $\lambda_1 = -6$ and $\lambda_2 = -6$. Therefore, the general solution to this differential equation is

$$y(x) = c_1 e^{-6x} + c_2 x e^{-6x}.$$

Case 3: Complex Conjugate Roots

The third case we must consider is when $b^2 - 4ac < 0$. In this case, when we apply the quadratic formula, we are taking the square root of a negative number. We must use the imaginary number $I = -\sqrt{i} = -1$ to find the roots, which take the form

$$\lambda_1 = \alpha + \beta i \lambda_1 = \alpha + \beta i \text{ and } \lambda_2 = \alpha - \beta i . \lambda_2 = \alpha - \beta i.$$

The **complex number** $\alpha + \beta i\alpha + \beta i$ is called the *conjugate* of $\alpha - \beta i\alpha - \beta i$. Thus, we see that when the discriminate $b^2 - 4ac$ is negative, the roots of our characteristic equation are always **complex conjugates**.

This creates a little bit of a problem for us. If we follow the same process we used for distinct real roots – using the roots of the characteristic equation as the coefficients in the exponents of exponential functions – we get the functions

$$e(\alpha + \beta i) xe(\alpha + \beta i)x \text{ and } e(\alpha - \beta i) xe(\alpha - \beta i)x \text{ as our solutions.}$$

However, there are problems with this approach. First, these functions take on complex (imaginary) values, and a complete discussion of such functions is beyond the scope of this text. Second, even if we were comfortable with complex-value functions, in this course we do not address the idea of a derivative for such functions. So, if possible, we'd like to find two linearly independent *real-value* solutions to the differential equation. For purposes of this development, we are going to manipulate and differentiate the functions

$$e(\alpha + \beta i) xe(\alpha + \beta i)x \text{ and } e(\alpha - \beta i) xe(\alpha - \beta i)x \text{ as if they were real-value functions.}$$

For these particular functions, this approach is valid mathematically, but be aware that there are other instances when complex-value functions do not follow the same rules as real-value functions. Those of you interested in a more in-depth discussion of complex-value functions should consult a complex analysis text.

Based on the roots $\alpha \pm \beta i$ of the characteristic equation, the functions $e^{(\alpha+\beta i)x}$ and $e^{(\alpha-\beta i)x}$ are linearly independent solutions to the differential equation and the general solution is given by

$$y(x) = c_1 e^{(\alpha+\beta i)x} + c_2 e^{(\alpha-\beta i)x}.$$

Using some smart choices for c_1 and c_2, and a little bit of algebraic manipulation, we can find two linearly independent, real-value solutions to Equation 8.4 and express our general solution in those terms.

We encountered exponential functions with complex exponents earlier. One of the key tools we used to express these exponential functions in terms of sines and cosines was Euler's formula, which tells us that

$$e^{i\theta} = \cos\theta + i\sin\theta$$

Euler's formula for all real numbers θ. Going back to the general solution,if we were to choose $c1 = c2 = 12$, the second term is zero and we get

$$y(x) = e^{\alpha x}\cos(\beta x)$$

as a real-value solution to equation. Similarly, if we choose $c1 = -i^2$ and $c2 = i^2$, the first term of the equation is zero and we get

$$y(x) = e^{\alpha x}\sin(\beta x)$$

as a second, linearly independent, real-value solution to Equation 8.4. Based on this, we see that if the characteristic equation has complex conjugate roots $\alpha \pm \beta i$, $\alpha \pm \beta i$, then the general solution is given by

$$y(x) = c_1 e^{\alpha x}\left[\cos(\beta x) + c_2\sin(\beta x)\right],$$

where c_1 and c_2 are constants.

For example, the differential equation $y'' - 2y' + 5y = 0$ has the associated characteristic equation $\lambda^2 - 2\lambda + 5 = 0$. By the quadratic formula, the roots of the characteristic equation are $1 \pm 2i$. Therefore, the general solution to this differential equation is

$$y(x) = e^x(c_1\cos(2x) + c_2\sin(2x)).$$

Summary of Results

We can solve second order, linear, homogeneous differential equations with constant coefficients by finding the roots of the associated characteristic equation. The form of the general solution varies, depending on whether the characteristic equation has

TABLE 8.1

Summary of Characteristics Equation

Cases	
Characteristic Equation Roots	
Distinct real roots, λ_1 and λ_2	$y(x) = c_1 e^{\lambda 1x} + c_2 e^{\lambda 2x}$
A repeated real root, λ	$y(x) = c_1 e^{\lambda x} + c_2 x e^{\lambda x}$
Complex conjugate roots $\alpha \pm \beta i$	$y(x) = e^{\alpha x}(c_1 \cos \beta x + c_2 \sin \beta x)$

distinct, real roots; a single, repeated real root; or complex conjugate roots. The three cases are summarized in Table 8.1.

Strategy for equations with constant coefficients.

Step 1. Write the differential equation in the form $a\,y'' + by' + cy = 0$

Step 2. Find the corresponding characteristic equation, $ay^2 + by + c = 0$.

Step 3. Factor or use the quadratic formula or technology to find the roots.

Step 4. Select the appropriate general solution form based on distinct roots, repeated roots, or complex roots.

Example 8.5: Find the general solutions to the following:

a. $y'' + 3y' - 4y = 0$
b. $y'' + 2y' + y = 0$
c. $y'' + 6y' + 12y = 0$
d. $y'' - 5y' = 0$
e. $y'' - 16y = 0$
f. $y'' + 16y = 0$

SOLUTIONS

a. $\lambda^2 + 3\lambda - 4 = 0$ with roots $\lambda_1 = -4$ and $\lambda_2 = 1$, so the general solution is $y(x) = c_1 e^{-4x} + c_2 e^{x}$
b. $\lambda^2 + 2\lambda + 1 = 0$ with roots –1, –1 (repeated), so the general solution is $y(t) = c_1 e^{-t} + c_2 t^{e-t}$
c. $\lambda^2 + 6\lambda + 12 = 0$ with complex roots $-3 \pm 1.7320i$. The general solution is $y(t) = e - 3t(c_1 \cos(1.732t) + c_2 \sin(1.7320t))$
d. $\lambda^2 - 5\lambda = 0$ with roots 0 and 5. The general solution is $y(t) = c_1 + c_2 e^{5t}$.
e. $\lambda^2 - 16 = 0$ so this factors into $(\lambda + 4)(\lambda - 4) = 0$, a repeated root. Our general solution is $y(t) = c_1 e^{4t} + c_2 e^{-4t}$
f. $\lambda^2 + 16 = 0$ so this factors into complex roots, $\pm 4i$. The general solution is $y(t) = c_1 \cos(4t) + c_2 \sin(4t)$

Initial-Value Problems and Boundary-Value Problems

So far, we have been finding general solutions to differential equations. However, differential equations are often used to describe physical systems, and the person studying that physical system usually knows something about the state of that system at one or more points in time. For example, if a constant-coefficient differential equation is representing how far a motorcycle shock absorber is compressed, we might know that the rider is sitting still on his motorcycle at the start of a race, time $t = t_0$. This means the system is at equilibrium, so $y(t_0) = 0$, and the compression of the shock absorber is not changing, so $y'(t_0) = 0$. With these two initial conditions and the general solution to the differential equation, we can find the *specific* solution to the differential equation that satisfies both initial conditions. This process is known as *solving an initial-value problem.* Note that second order equations have two arbitrary constants in the general solution, and therefore we require two initial conditions to find the solution to the initial-value problem.

Sometimes we know the condition of the system at two different times. For example, we might know $y(t_0) = y_0$ and $y(t_1) = y_1$. These conditions are called boundary conditions, and finding the solution to the differential equation that satisfies the boundary conditions is called solving a boundary-value problem.

Mathematicians, scientists, and engineers are interested in understanding the conditions under which an initial-value problem or a boundary-value problem has a unique solution. Although a complete treatment of this topic is beyond the scope of this text, it is useful to know that, within the context of constant-coefficient, second order equations, initial-value problems are guaranteed to have a unique solution as long as two initial conditions are provided. Boundary-value problems, however, are not as well behaved. Even when two boundary conditions are known, we may encounter boundary-value problems with unique solutions, many solutions, or no solution at all.

Example 8.6: Solve the initial-value problem, $y'' + 3y' - 4y = 0, y(0) = 1, y'(0) = -9$

Our general solution to the $y(t) = e^{-4t} + c2e^t$ found in Example 8.6 (a). We apply the initial conditions, when t = 0 we have $c_1 + c_2 = 1$. We apply y′(0) = 9 to get the equation,

$$-4c_1 + c_2 = -9.$$

We solve two equations and two unknowns to *get* $c_1 = 2$ and $c_2 = -1$. *Our* solution is $(t) = 2e^{-4t} - e^t$.

Models Using Second Order ODE

Bungee Jumping

We refer to the first model in this chapter for bungee jumping. Using Newton's second law again: $mg - \beta v = ma$ and using the fact that $a = dv/dt$, $v = dx/dt$ then with a little algebra, we can put this into the form of a nonhomogeneous equation Second order differential equation in position, x, shown in Equation 8.1

First, we must solve $x'' + \beta m x' = 0$.

This is the homogeneous equation. To find the general solution of the second order ODE equation, we must first find the solutions to the equivalent homogeneous equation. In other words, we will set the driving force g to zero. (Note: there are other easier ways, but we will solve our nonhomogeneous equation this way so that the process is the same as the process needed in the latter part of this section.) $x'' + \beta m x' = 0$. We use the characteristic equation for the homogeneous equation is: $p(\lambda) = \lambda^2 + \beta m\lambda$. We set $p(\lambda)$ equal to 0 to find that the roots of the equation are: $\lambda_1 = 0$, $\lambda_2 = -\beta$ m The solution to homogeneous equation is shown in Equation 8.2.

$$x_h = C_1 e^{\lambda 1t} + C_2 e^{\lambda 2t} \tag{8.2}$$

Substituting in our values for λ 0 and $= -\beta$ m, we have $x_h = C_1 + C_2 e^{-\beta t/m}$.

Now that the homogeneous equation has been solved, the only thing left to find is the particular solution. If we look at Equation 8.1 and we look at the driving force that is on the right-hand side, then we see that the driving force is a polynomial of zero'th degree (a constant). The particular solution to the equation should also be a polynomial. If we say a particular solution is:

$$x_p = at + b, \text{then } x' = a \text{ and } x'' = 0.$$

Then we can substitute these into Equation 8.1 and solve for

$$a(0) + \beta m(a) = g\,\beta m a = g\, a = mg\,\beta$$

Our general solution is:

$$x = C_1 + C_2 e - \beta t/m + mg\,\beta t + bC_1 \text{ and } b \text{ are both constants.}$$

because C_1 is an arbitrary constant, we say that it accounts for

$$x = C_1 + C_2 e - \beta t/m + mg\beta t(3)$$

Below is the equilibrium point for the cord. So far, an equation for the motion of the bungee jumper has been found. Now, we need to find an equation of motion for when the bungee jumper is below the equilibrium point. To do this we look back to Newton's second law: $\Sigma F = ma$. When the jumper is below the equilibrium point, they still have gravity and wind resistance acting on them. But, now they also have the force of the cord pulling on them. We will say that the force applied by the bungee cord is $-kx$ (from Hooke's law). Using this with Newton's second law:

$$mg - \beta\, dx\, dt - kx = m x''.$$

Once again, this needs to be rearranged algebraically to look like an inhomogeneous equation, as shown in Equation 8.3.

$$x'' + \beta m x' dt + kmx = g.$$

Next, we need to solve the homogeneous equation. The first step to solving the inhomogeneous equation is to solve the equivalent homogeneous equation.

$$x'' + \beta m x' + kmx = 0$$

The characteristic equation is

$$P(\lambda) = \lambda^2 + \beta m\lambda + km.$$

We set P(λ) to 0 and use the quadratic formula to find that the roots of the characteristic equation are:

$$\lambda = -\beta/m \pm q(\beta/m)^2 - 4(k/m)2(4)$$

At this point, the solution to the homogeneous equation can go in three directions, depending on the discriminant. There are three cases, when *(b/m)*^{2} – *4(k/m)* is greater than zero, equal to zero, or less than zero.

Case 1. $(\beta/m)^2 - 4(k/m) > 0$. This is referred to as over damped, and the solution will not oscillate. If bungee cord and jumper system fall into this category then there is a problem, because the bungee jumper will likely experience too much acceleration. In this case, the characteristic equation has two distinct real roots. The solution will take the form:

$$x_h = C_1 e^{\lambda 1t} + C_2 e^{\lambda 2t}$$

We can substitute in our expression for λ to get:

$$x_h = C1e\left(-\beta/m + \sqrt{}(\beta/m)2 - 4k/m\right)t/2 + C2e\left(-\beta/m + \sqrt{}(\beta/m)2 - 4k/m\right)t/2$$

Case 2. $(\beta/m)2 - 4(k/m) = 0$. This case is said to be critically damped, and its solution appears very similar to the over damped case. The characteristic equation has one real root, and the solution will

$$x'' - .5(x + 320) \text{ with } x(0) = 320, x' = \text{velocity at 200 feet}$$

$$> ode2 := 5diff\left(x(t),\, t,\, t\right) = 160 - 0.9diff\left(x(t), t\right) - 0.5\left(x(t) + 320\right);$$

$$ode2 := 5\frac{d^2}{dr^2}x(t) = -0.9\frac{d}{dr}x(t) - 0.5x(t)$$

$$> ics1 := x(0) = 320, D(x)(0) = 90.5214;$$

$$ics1 := x(0) = 320, D(x)(0) = 90.5214;$$

The solution (In Maple formatted output) is:

solbj := x(t) = 0.00002176278564 * exp(−0.09000000000 * t)
* (1.808614883 * 10^7 * sin(0.3031501278 * t) +
1.4704000 * 10^7 * cos(0.3031501278 * t))

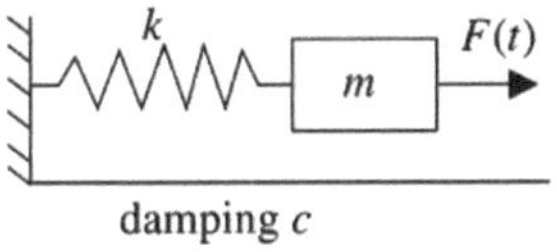

FIGURE 8.5
Spring-mass with damping.

$$x(t) = 0.0000217627e^{(-0.09t)}(1.808614883 * 10^7(\sin(0.3031501278\ t) + 1.4704 * 10^7 \\ (\cos(0.3031501278\ t))$$

The plot that shows the oscillations over time is seen in Figure 8.5.

The bungee jumper will be safe and not either hit the ground or rebound back up and hit the bridge.

But how did we obtain the analytical solution? The roots of second order ODE is

$-0.09 \pm 3031501278t$. This yields the complex solution:

$$y(t) = e^{-0.09t}\left[c_1\, cos(0.3031501278t) + c_2\, sin(0.3031501278t)\right].$$

Now, we solve two equations and two unknowns for the values of c_1 and c_2. We find our solution is

$$y(t) = 0.00002176278564 * \exp(-0.09000000000 * t) \\ *(1.808614883 * 10\text{^}7 * \sin(0.3031501278 * t) + \\ 1.4704000 * 10\text{^}7 * \cos(0.3031501278 * t)).$$

Spring damping

Consider a mass on a spring, like in Figure 8.5, The second ODE for a damped spring is

$$mx'' + cx' + kx = f(t).$$

How about undamped force motion and resonance? For our second order ODE,

$$mx'' + cx' + kx = f(t)\ for\ undamped\ c = 0.$$

So lets' consider when $w \neq w_0$

$$mx'' + kx = F_o\, cos(wt)$$

The complementary solution to the homogeneous solution will be

$$x(t) = c_1 \cos(wt) + c_2 \sin(wt) \text{ where}$$

$$\varpi_o = \sqrt{\frac{k}{m}}$$

The particular solution with $F_o\ cos(wt)$ is

$$x_p(t)=\frac{F_o}{\left(\varpi_o^2-\varpi_0^2\right)}\cos(\varpi t)$$

$$0.5x''+8x=10\cos(wt)\,\text{with}\,x(0)=0\,\text{and}\,x'(0)=0$$

The homogeneous part, we solve the characteristic equation

$$0.5\lambda^2+8=0$$

$$\lambda^2=-16=\pm 4i\lambda$$

$$x_c(t)=c_1cos(4t)+c_2\,sin(4t)$$

The particular solution, we begin with a guess of $A\cos(\pi t)$

$$x'=-A\pi\sin(\pi t)\text{and}$$

$$x''=-A\pi^2\cos(\pi t)$$

so, $-0.5\left(A\pi^2\cos(\pi)\right)+8A\cos(\pi t)=10\cos(\pi t)$

$$0.5\pi^2A+8A=10$$

$$A=20/\left(-\pi^2+16\right)$$

The complete solution is

$$x(t)=c_1\,cos(4t)=c_2\,sin(4t)+20/\left(-\pi^2+16\right)\cos(\pi t).$$

Now we use our initial conditions $x(0)=0$ and $x'(0)=0$ to find c_1 and c_2.

$$x(0)=c_1+20/\left(-\pi^2+16\right)=0$$

$$c_1=-20/\left(-\pi^2+16\right)$$

Taking the derivative of the general solution:

$$x'(t)=-4c_1\sin(4t)+4c_2\cos(4t)-20/\left(-\pi^2+16\right)\sin(\pi t).$$

Substitute $t=0$.

$$x'(0)=0=4c_2\,\text{so}\,c_2=0$$

Our complete solution,

$$x(t)=-20/\left(-\pi^2+16\right)cos(4t)+20/\left(-\pi^2+16\right)\cos(\pi t).$$

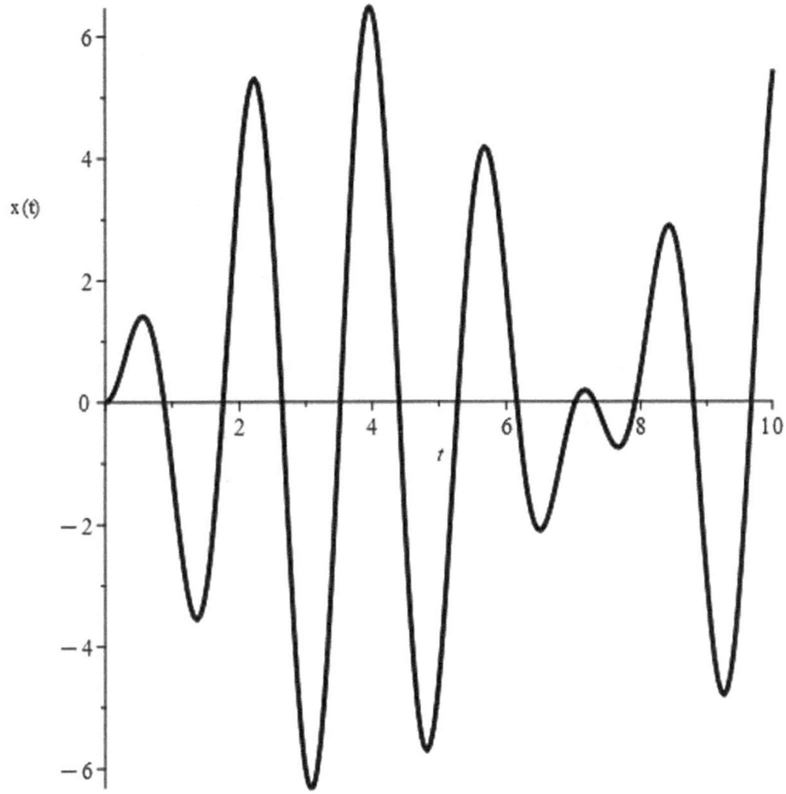

FIGURE 8.6
Plot of $x(t) = -20/(-\pi^2 + 16)cos(4t) + 20/(-\pi^2 + 16)\cos(\pi t)$ from t = 0.10.

Our plot is shown in Figure 8.6

But what is $w = w_0$, In this case, our solution is simply, 2/π*t*sin(πt). We plot this case in Figure 8.7.

This could actually be used to model the effects of an earthquake.

Nonhomogeneous Equations and Undetermined Coefficients

In the previous sections, we solved the homogeneous higher order differential equation. We now turn our attention to the nonhomogeneous equation:

$$ay'' + by' + cy = f(x)$$

First, we state that there are two components to the solution y_c and y_p and the solution is

$$y = y_c + y_p,$$

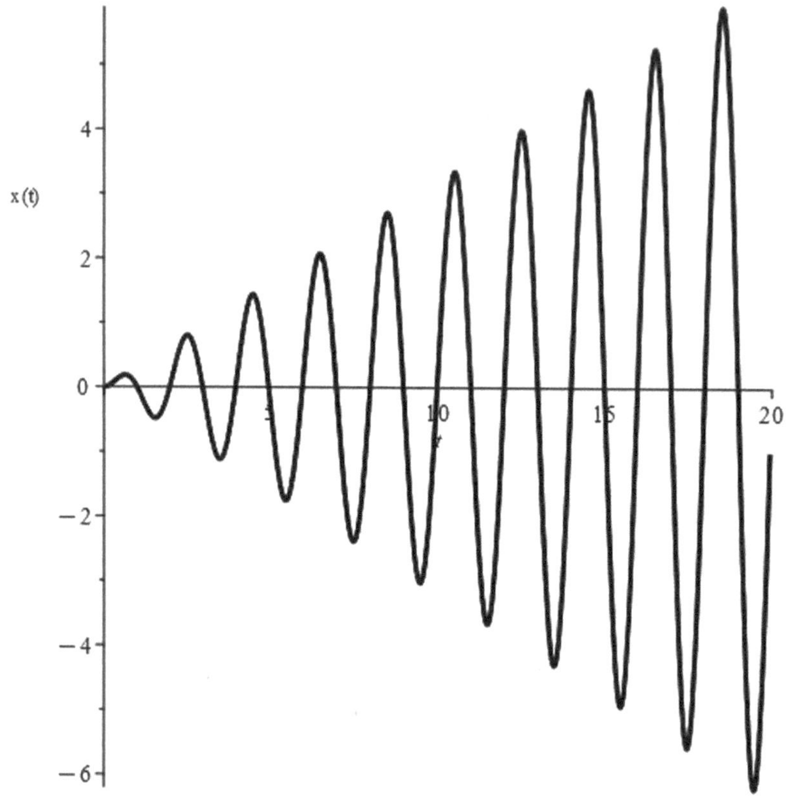

FIGURE 8.7
Plot of 2/π*t*sin(πt).

where, y_c is the solution to the homogeneous part and y_p is the solution to the nonhomogeneous part. We present two methods for solving the nonhomogeneous part: Method of variation of parameters and annihilators

Variation of Parameters

We begin with the method of undetermined coefficients. We will assume that the function, f(x), is a linear combination of a polynomial in x, and exponential in x or a cos(kx) or sin (kx) function. The procedure states that we obtain our generalized guess of one of these functions, take the appropriate derivatives, substitute back into our second order ODE and solve for the coefficients.

We provide an example of each. f(x) is a constant or polynomial function

$$y'' - 3y' + 2y = 2x + 3$$

Our guess might be by $y_p(x) = Ax + B$.

We find $y_p'(x) = A$ and $y_p''(x) = 0$. We substitute into our second ODE and get

$$0 - 3\mathrm{A} + 2(\mathrm{Ax} + \mathrm{b}) = 2\mathrm{x} + 3$$

$$-\mathrm{A} + 2\mathrm{B} + 2 + 2\mathrm{Ax} = 2\mathrm{x} + 3$$

We set 2Ax = 2x, to find that A = 1, then –A + 2B + 2 = 3 becomes –1 + 2B + 2 = 3 or 2B = 2, so B = 1

Therefore, $y_p(x) = x + 1$.

What if f(x) is exponential

$$\mathrm{y}'' - 3\mathrm{y}' + 2\mathrm{y} = 2\mathrm{e}^{3\mathrm{x}}$$

It is reasonable to try Ae^{3x} as our guess. We find

$$y_p'(x) = 3Ae^{3x} \text{ and } y_p''(x) = 9Ae^{3x}$$

By substitution,

$$9\mathrm{Ae}^{3\mathrm{x}} - 3(3\mathrm{Ae}^{3\mathrm{x}}) + 2(\mathrm{Ae}^{3\mathrm{x}}) = 2\mathrm{e}^{3\mathrm{x}}$$

A = 1 so

$$y_p(x) = e^{3x}$$

Example 8.7: $y'' - 4y = 2e^{3x}$

Again, we let our guess be Ae^{3x}. We find $y_p'(x) = 3Ae^{3x}$ and $y_p''(x) = 9Ae^{3x}$

By substitution,

$$9Ae^{3x} - 4(Ae^{3x}) = 2e^{3x}$$

$$5Ae^{3x} = 2e^{3x}, \text{so } A = 2/5.$$

$$y_p(x) = 2/5e^{3x}$$

What if the *f(x)* is either A *cos(kx)* + r or B *sin(Ax),* with *A* and *B* as constants.

Example 8.8: $3y'' + y' - 2y = 2\cos(x)$

Our guess might be $A\cos(x) + B\sin(x)$

$$y_p'(x) = -A\sin(x) + B\cos(x)$$

$$y_p''(x) = -A\cos(x) - B\sin(x)$$

By substitution,

$$3*(-A\cos(x)-B\sin(x))+(-A\sin(x)+B\cos(x))-2*(A\cos(x)+B\sin(x))$$
$$=2\cos(x)=3A+B-2A=2 \text{ for } \cos(x)$$
$$=-3B - A-2B=0 \text{ for } \sin(x)$$

So,

$$-5A+B=2$$
$$-A-5B=0$$

We solve the two equations and two unknowns to find $A = -5/13$ and $B = 1/13$. Therefore, $y_p(x) = -5/13\ cos(x) + 1/13\ sin(x)$

Example 8.9: $y'' - 4y = 2\ e^{2x}$

We begin by guessing $y_p(x) = Ae^{2x}$.

$$y'p(x)=2A^{e2x}$$

$$yp''(x)=4\,Ae^{2x}$$

By substitution,

$$4Ae^{2x}-4Ae^{2x}=2e^{2x}.$$

This is not possible so we change our guess to

$$y_p(x)=Axe^{2x}.$$

$$y_p'(x)=Ae^{2x}+2Axe^{2x} \text{ and}$$

$$y_p''(x)=2Ae^{2x}+2Ae^{2x}+4Ae^{2x}$$

By substitution and simplification, we obtain A = ½ and all the xe^{2x} terms cancel out.

$$y_p(x)=1/2xe^{2x}$$

Method of Variation of Parameters

If a nonhomogeneous differential equation is of the form $y'' + P(x)y' + Q(x)y = f(x)$ with complementary solution, $y_c = c_1\,y_1(x) + c_2\,y_2\,(x)$ then the particular solution is given by

$$y_p(x)=-y_1(x)\int\frac{y_2(x)f(x)}{W(x)}+y_2(x)\int\frac{y_1(x)f(x)}{W(x)}$$

where $W = W(y_1,y_2)$ is the Wronskian of y_1 and y_2.

Example 8.10: $y'' - 5y' + 6y = e\wedge(2t)$

The characteristic equation, $\lambda^2 - 5\lambda + 6 = 0$ or yields two distinct roots, 2, 3 so our complementary solution is

$$y_c = c_1 e^{2x} + c_2 e^{3x} \text{ where } y_1 = e^{2x} \text{ and } y_2 = e^{3x}$$

The Wronskian is

$$W = \begin{bmatrix} e^{2x} & e^{3x} \\ 2e^{2x} & 3e^{3x} \end{bmatrix}$$

The determinant is e^{5x}.
We use Maple and obtain the following (Figure 8.8).

>S1:=matrix(2,2,[[0, exp(3*t)], [exp(2*t),3*exp(3*t)]]);

$$s1 := \begin{bmatrix} 0 & e^{3t} \\ e^{2t} & e^{3t} \end{bmatrix}$$

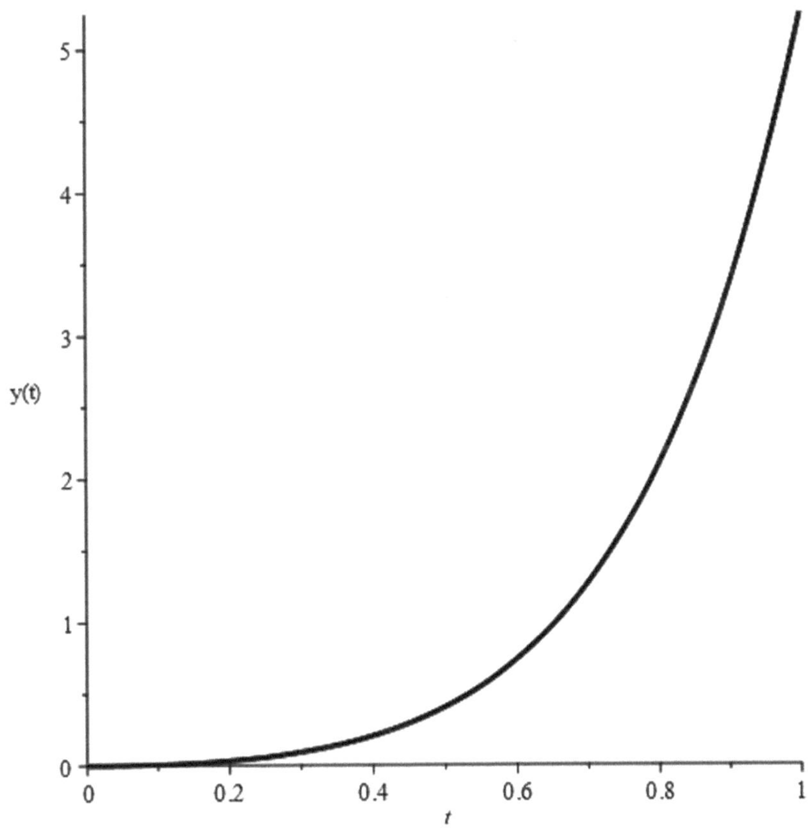

FIGURE 8.8
Plot of output.

```
> w1 := det(S1);
```

$$w1 := -e^{2t}e^{3t}$$

```
> w1 := simplify(%);
```

$$w1 := -e^{5t}$$

```
> S2:=matrix(2,2,[[exp (2*t),0],[2*exp(2*t),exp(2*t)]]);
```

$$s1 := \begin{bmatrix} e^{2t} & 0 \\ 2e^{2t} & e^{2t} \end{bmatrix}$$

```
> w2:=det(S2);
```

$$w2 := \left(e^{2t}\right)^2$$

```
> w2:=simplify(%);
```

$$w2 := e^{4t}$$

```
> u1:=int(w1/step_2,t);
```

$$u1 := -t$$

```
> u2:=int(w2/step_2,t);
```

$$u2 := -\frac{1}{e^t}$$

```
> yp:=u1*exp(2*t)+u2*exp(3*t);
```

$$yp := -te^{2t} - \frac{e^{3t}}{e^t}$$

```
> yp:=simplify(%);
```

$$yp := -e^{2t}(t+1)$$

```
> Yg:=c1*exp(2*t)+c2*exp(3*t)+yp;
```

$$Yg := -c1e^{2t} + c2e^{3t} - e^{2t}(t+1)$$

```
> eq1:=simplify(subs(t=0, Yg));
```

$$eq1 := c1 + c2 - 1$$

```
> eq2:=simplify(subs(t=0,diff(Yg,t)));
```

$$eq2 := 2c1 + 3c2 - 3$$

```
> solve({0= c1+c2-1,2*c1+3*c2-3=0},{c1,c2});
```

$$\{c2 = 1,\ c1 = 0\}$$

```
> sol:=subs({c2 = 1, c1 = 0},Yg);
```

$$sol := e^{3t} - e^{2t}(t+1)$$

> simplify(%);

$$e^{3t} - te^{2t} - e^{2t}$$

> plot(exp(3*t)-t*exp(2*t)-exp(2*t),t=0.1);

Method of Annihilators

Method #2 Annihilators using Maple to find the solution (Figure 8.9):

> f:=A1*t*exp(2*t);

$$f := A1te^{2t}$$

$$solbj := x(t) = 0.00002176278564\,e^{-0.09000000000t}\,(1.808614883 \times 10^{7}\sin(0.3031501278\,t) + 1.4704000 \times 10^{7}\cos(0.3031501278\,t))$$

$plot(0.00002176278564\,e^{-0.09000000000t}\,(1.808614883 \times 10^{7}\sin(0.3031501278\,t) + 1.4704000 \times 10^{7}\cos(0.3031501278\,t)), t = 0..100, thickness = 3, color = black);$

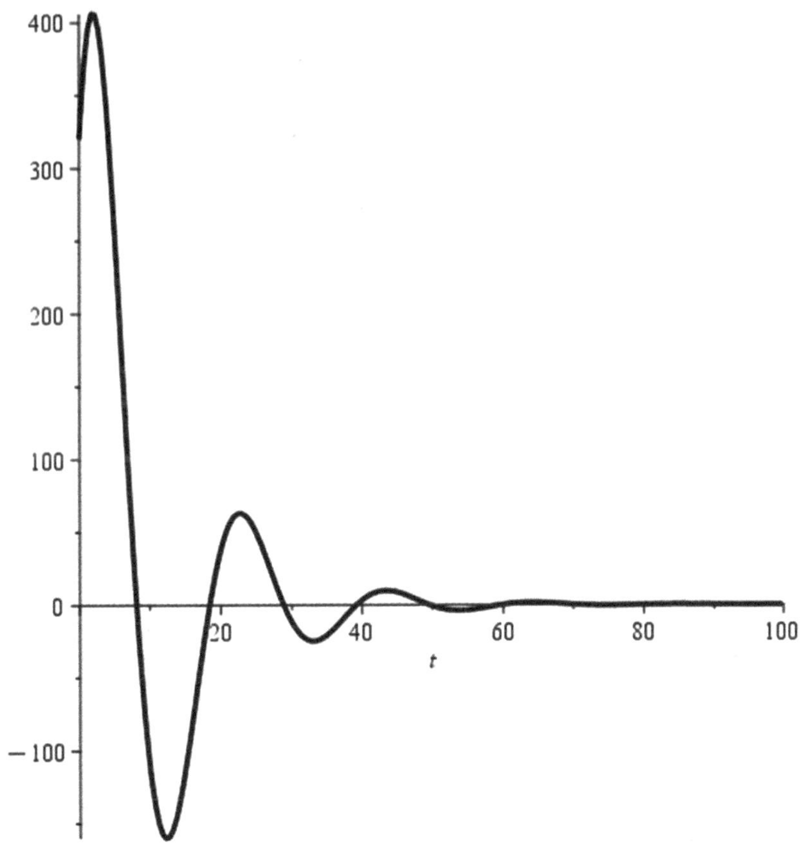

FIGURE 8.9
Plot of bungee jumping solution over time.

```
> fp:=diff(f,t);
```

$$fp := A1te^{2t} + 2A1te^{2t}$$

```
> fpp:=diff(fp,t);
```

$$fpp := 4A1e^{2t} + 4A1te^{2t}$$

```
> fg:=fpp-5*fp+6*f;
```

$$fg := -A1te^{2t}$$

```
> A:=–1;
```

$$A := 1$$

```
> C:=–1/2;
```

$$C := -\frac{1}{2}$$

```
> B:=–1;
```

$$B := 1$$

```
> Yf:=subs({A1 = –1,B = 1,C = –1/2},f);
```

$$Yf := -te^{2t}$$

```
> Yg:=c1*exp(2*t)+c2*exp(3*t)+Yf;
```

$$Yg := c1e^{2t} + c2e^{3t} - te^{2t}$$

```
> eq1:=simplify(subs(t=0, Yg));
```

$$eq1 := c1 + c2$$

```
> eq2:=simplify(subs(t=0,diff(Yg,t)));
```

$$eq2 := 2c1 + 3c2 - 1$$

```
> solve({0= c1 + c2,2*c1+3*c2 – 1= 0},{c1,c2});
```

$$\{c2 = 1, C1 = -1\}$$

```
> sol:=subs({c2 = 1, c1 = –1},Yg);
```

$$sol := e^{3t} + te^{2t} - e^{2t}$$

```
> simplify(%);
```

$$e^{3t} + te^{2t} - e^{2t}$$

Back to our Bungee Jumping model

Our second ODE was

$$ode2 := 5\frac{d^2}{dr^2}x(t) = -0.9\frac{d}{dt}x(t) - 0.5x(t)$$

$$5x'' + 0.9x' + 0.5x = 32.00$$

The characteristic equation is:

$$\lambda^2 + 5\lambda + 0.5 = 32 \text{ with roots } -0.09 \pm .3031501278 \text{ i.}$$

Our complementary solution is

$$y_c = e^{-0.09t} * (c1\cos(.3031501278t) + c2\sin(.3031501278t))yc + yp$$
$$= e^{-0.09t} * (c1\cos(.3031501278t) + c2\sin(.3031501278t))$$

We solve for c1 and c2 with our initial conditions and obtain our solution:

Exercises

8.1 Find the general solution to the following Differential Equations,
   a. $y'' - 2y' + 10y = 0$
   b. $y'' - 5y' + 4y = 0$
   c. $y'' + 5y' + 6y = 0$
   d. $y'' + 4y' + 4y = 0$
   e. $y'' + 14y' + 49y = 0$
   f. $y'' - 6y' - 7y = 0$

8.2 Find the general and particular solutions to the following differential equations and graph the solution
   a. $y'' - 3y' - 10y = 0, y(0) = 0, y'(0) = 7.$
   b. $y'' + 6y' + 13y = 0, y(0) = 0, y'(0) = 2$
   c. $y'' - 2y' + 10y = 0, y(0) = 2, y'(0) = -1$

8.3. Suppose the following initial-value problem models the position (in feet) of a mass in a spring-mass system at any time t. Given our ode is $y''+ 14y' + 49y$, $y(0) = 0$, $y'(0) = 1$. Solve and determine the position of the mass at time, $t = 0$ and $t = 1$. How fast is the mass moving at time $t = 0.1$? What direction is the mass moving?

8.4. Suppose the following initial-value problem models the position (in feet) of a mass in a spring-mass system at any time t. Given our ode is $y'' - 2y' + 410y$, $y(0) = 0$, $y'(0) = 1$. Solve and determine the position of the mass at time, $t = 0$ and $t = 1$. How fast is the mass moving at time $t = 0.1$? What direction is the mass moving?

8.5. Solve the following by variation of parameters and state an interval over which the general solution is defined.
   a. $y'' + y = \cos x$
   b. $y'' + y = \cos^2 x$
   c. $y'' - 4y = e^{2x}/4$
   d. $4y'' - 4y' + y = 8e^{-x} + x$

8.6. Solve the following by variation of parameters of the following initial value problems.
   a. $y'' - y = xe^x$, $y(0) = 1$, $y'(0) = 0$.
   b. $2y'' + y' - y = x + 2$, $y(0) = 1$, $y'(0) = 0$
8.7. Solve the following using the method of annihilators.
   a. $y'' + y = \cos x$
   b. $y'' + y = \cos^2 x$
   c. $y'' - 4y = e^{2x}/4$
   d. $4y'' - 4y' + y = 8e^{-x} + x$
8.8. Solve the following by variation of parameters of the following initial value problems.
   a. $y'' - y = x\,e^x$, $y(0) = 1$, $y'(0) = 0$.
   b. $2y'' + y' - y = x + 2$, $y(0) = 1$, $y'(0) = 0$

PROJECTS

8.1 A mass weighing 2 lbs stretches a spring 6 inches. At $t=0$, the mass is released from a point 8 inches below the equilibrium position with an upward velocity of 4/3 ft/sec. Determine the function $x(t)$ that models the free motion.

8.2 A mass weighing 12 lbs stretches a spring 12 inches. At $t=0$, the mass is released from a point 10 inches below the equilibrium position with an upward velocity of 4/3 ft/sec. Determine the function $x(t)$ that models the free motion.

8.3 An 8-lb weight stretches a spring 2 feet. Assume a damping force numerically equal to two times the instantaneous velocity acts on a system. Determine the equation of motion if the weight is released from the equilibrium position with an upward velocity of 3ft/sec.

8.4 A 16-lb weight is attached to a 5-foot-long spring. At equilibrium the spring measures 8.2 ft. If the weight is pushed up and released from a rest point 2 feet above the equilibrium position, model the displacement, $x(t)$. Further, it is known the surrounding medium offers a resistance equal to the instantaneous velocity.

8.5 A 16-lb weight stretches a spring 8/3 feet. Initially, the weight starts from rest 2 feet below the equilibrium position and the subsequent motion takes place in a medium that offers a damping force numerically equal to 1.2 the instantaneous velocity. Find the equation of motion if the external force is equal to $f(t) = 10\cos 3t$.

References and Additional Readings

Boyce, W., and DiPrima, R. (2012). *Elementary Differential Equations and Boundary Value Problems.* 10th edition. Wiley.

Zill, D. (2017). *A First Course in Differential Equations with Modeling Applications.* 11th edition. Cengage Learning.

9

Higher Order Numerical Methods to Solve IVP and BVP

Often the second order differential equations are too difficult or simply cannot be solved for as a closed form analytical solution. In those cases, we will use numerical methods. We discuss Euler's method, Heun's method, and Fourth Order Runge-Kutta methods. We will repeat several of the modeling examples from Chapter 7 to illustrate numerical methods.

Euler's Method

Most ordinary differential equations arising in real-world applications cannot be solved exactly. These ODEs can be analyzed qualitatively. However, qualitative analysis may not be able to give accurate answers. A *numerical method* can be used to get an accurate approximate solution to a differential equation. There are many programs and packages for solving differential equations. With today's computer, an accurate solution can be obtained rapidly. In this section, we focus on Euler's method, a basic numerical method for solving initial value problems.

Example 9.1: Consider the differential equation for a mechanical device with a forcing function:

$$y'' = -y' + \sin(ty), y(0) = 1, y'(0) = 2$$

The first step is to convert the above second ODE into two first ODE. This is a standard operation. Let v(t) = y′(t). Then v′(t) = y″(t). We then get two differential equations. The first is easy

$$y' = v$$

The second is obtained by rewriting the original ODE. Using the fact that y″ = v′ and y′ = v, so

$$v' = -v + \sin(ty)$$

The initial conditions are y(0) = 1 and y′(0) = v(0) = 2. The algorithm now states that we assign a value to *h* (we start with *h* = *1*). This method incorporates a discrete dynamical system. We will iterate and plot using EXCEL to illustrate the algorithm.

DOI: 10.1201/9781003582274-9

Given the second ODE,

$$y'' + 5y' + 6y = 10\,sin(wt)\text{ with initial conditions } y(0) = 0 \text{ and } y'(0) = 5.$$

We start by transforming the second ODE into two first ODE. Let use $z_1 = y$ and $z_2 = y'$. We assume w = 1.

$$z'_1 = z_2$$
$$z'_2 = y'' = 10\ \sin(wt) = 5z_2 - 6z_1$$

We choose a stepsize, like $\Delta h = 0.1$ and we iterate from t = 0 to t = 3 stepping in increments of 0.1. Our two dynamical systems are

$$z_1(i+1) = z_1(i) + \Delta h\, z_2(i) \text{and}$$
$$z_2(i+1) = z_2(i) + \Delta h\left(10 * \sin(t) - 5z_2(i) - 6z_1(i)\right) \text{with } t \text{ in degrees.}$$

We can step through this using Excel (Figure 9.1).

The solution plot for Example 9.1 is shown in Figure 9.2.
We may use Python to find a better estimate.

Python Code for Example 9.1.

```
# import libraries for numerical functions and plotting
          import numpy as np
          import matplotlib.pyplot as plt

# these lines are only for helping improve the display
          import matplotlib_inline.backend_inline
          matplotlib_inline.backend_inline.set_matplotlib_
formats('pdf', 'png')
          plt.rcParams['Figure.dpi']= 300
          plt.rcParams['savefig.dpi'] = 300

# plot exact solution first
          time = np.linspace(0, 3)

          y_exact = (
              -6*np.exp(-3*time) + 7*np.exp(-2*time) +
              np.sin(time) - np.cos(time)
              )
          plt.plot(time, y_exact, label='Exact')

          omega = 1

          dt = 0.1
          time = np.arange(0, 3.001, dt)

          f = lambda t,z1,z2: 10*np.sin(omega*t) - 5*z2 - 6*z1
```

	h=.1	
time	z1	z2
0	0	5
0.1	0.5	2.5
0.2	0.75	1.049833
0.3	0.854983	0.273586
0.4	0.882342	-0.08068
0.5	0.874274	-0.18033
0.6	0.856242	-0.1353
0.7	0.842712	-0.01675
0.8	0.841036	0.130214
0.9	0.854058	0.277841
1	0.881842	0.409813
1.1	0.922823	0.517272
1.2	0.97455	0.59615
1.3	1.034165	0.645384
1.4	1.098704	0.665751
1.5	1.165279	0.659103
1.6	1.231189	0.627879
1.7	1.293977	0.5748
1.8	1.351457	0.502679
1.9	1.401725	0.414313
2	1.443156	0.312422
2.1	1.474398	0.199615
2.2	1.49436	0.078378
2.3	1.502197	-0.04893
2.4	1.497304	-0.18008
2.5	1.479297	-0.31296
2.6	1.448001	-0.44559
2.7	1.403442	-0.57609
2.8	1.345833	-0.70273
2.9	1.27556	-0.82388
3	1.193172	-0.93803

FIGURE 9.1
Excel solution for Example 9.1.

```
        z1 = np.zeros_like(time)
        z2 = np.zeros_like(time)
        # initial conditions
        z1[0] = 0
        z2[0] = 5

# Forward Euler iterations
        for idx, t in enumerate(time[:-1]):
            z1[idx+1] = z1[idx] + dt*z2[idx]
            z2[idx+1] = z2[idx] + dt*f(t, z1[idx], z2[idx])
```

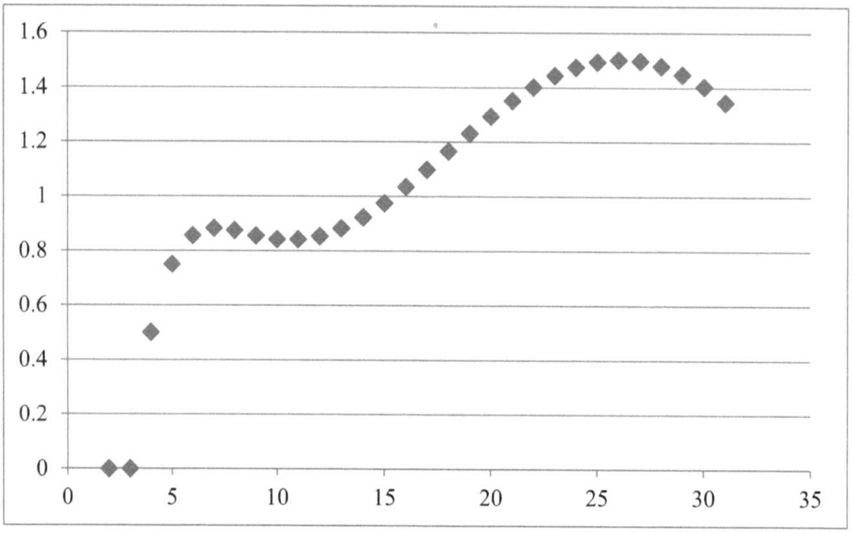

FIGURE 9.2
Euler's approximation to the solution for Example 9.1.

```
plt.plot(time, z1, 'o--', label='Forward Euler solution')
plt.xlabel('time')
plt.ylabel('displacement')
plt.legend()
plt.grid(True)
plt.show()
print(time,z1,z2)
```

Python output (Figure 9.3)
We see the estimate is fair at best as shown in Figure 9.3.

Maple Examples

Maple only does RK4 internally and not Euler's method. From Chapter 7, we have the following Maple Code and solution (Figure 9.4):

> with(DEtools):

$$> woh := diff(P(t),t) = 0.3 \cdot \left(1 - \frac{P(t)}{12}\right) \cdot P(t);$$

$$woh := diff(P(t),t) = 0.3 \cdot \left(1 - \frac{P(t)}{12}\right) \cdot P(t)$$

$$> ich := P(0) = 5;$$

$$ich := P(0) = 5$$

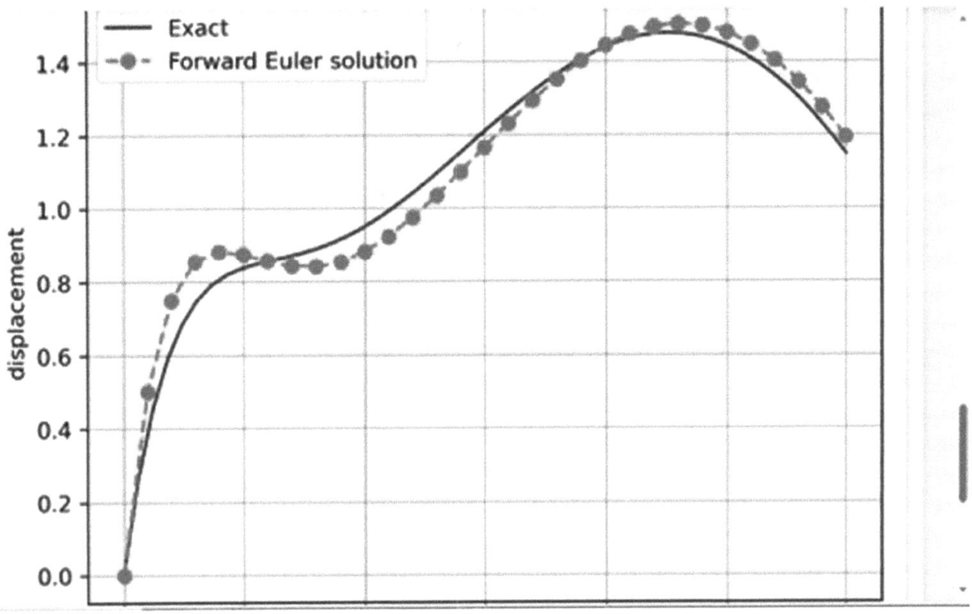

FIGURE 9.3
Python output and comparison to exact solution (Example 9.1).

```
>dsolve({who,ich},P(t));
```

$$P(t)=\frac{60}{5+e^{-\frac{3t}{10}}}$$

$$>eqn:=\left\{iff(P(t),t)-0.3\cdot\left(1-\frac{P(t)}{12}\right)=0,\quad P(0)=5\right\};$$

$$eqn:=\left\{iff(P(t),t)-0.3\cdot\left(1-\frac{P(t)}{12}\right)=0,\quad P(0)=5\right\}$$

```
> egn_euler := dsolve(eqn, numeric,method=classical[foreuler],
stepsize=0.5,output=array ([0,0.5,1,1.5,2,2.5,3.,3.5,4,4.5,5,
5.5,6,6.5,7,7.5,8,8.5,9,9.5,10,20,30,40,200]));
```

Now with harvesting (Figure 9.5):
Testing values of harvesting, we find if h = 1 that we run out of crabs in about 20 days.

Example 9.2: Let's solve a forced damped mass-spring problem given by a second ODE:

$$y''+5y'+6y=10\ \sin f_0\omega t$$

with the initial conditions $y(0) = 0$ and $y'(0) = 5$.

·:=

t	P(t)
0.	5.
0.5	5.43750000000000
1.	5.88354492187500
1.5	6.33337539955974
2.	6.78198616009684
2.5	7.22434238066455
3.	7.65560470235091
3.5	8.07134186572033
4.	8.46771115166176
4.5	8.84159117256131
5.	9.19065816766158
5.5	9.51340442337576
6.	9.80910429034831
6.5	10.0777383466639
7.	10.3198889713658
7.5	10.5366209623042
8.	10.7293593403590
8.5	10.8997738432309
9.	11.0496765467958
9.5	11.1809361314551
10.	11.2954098914774
20.	11.9707603367109
30.	11.9988635536127
40.	11.9999559471530
200.	12.0000000000000

FIGURE 9.4
Solution for RK4 method for Example 9.1.

Forward Euler

Then, let's solve numerically using the forward Euler method. Recall that the recursion formula for forward Euler is:

$$yi + 1 = yi + \Delta x(xi, yi)$$

where $f(x,y) = dydx$.

Let's solve using $\omega = 1$ and with a step size of $\Delta t = 0.1$, over $0 \le t \le 3$.

```
> egn_euler := dsolve(eqn1, numeric,method=classical[foreuler], stepsize=0.5,output=array([0,0.5,1,1.5,2,2.5,3.
  ,3.5,4,4.5,5,5.5,6,6.5,7,7.5,8,8.5,9,9.5,10,20,30,40, 200]));
```

egn_euler :=

t	$P(t)$
0.	5.
0.5	5.23750000000000
1.	5.48023242187500
1.5	5.72685544268413
2.	5.97592284331923
2.5	6.22591559695081
3.	6.47527762373899
3.5	6.72245401349365
4.	6.96592976597349
4.5	7.20426701206357
5.	7.43613877410926
5.5	7.66035759187801
6.	7.87589775021668
6.5	8.08191034560083
7.	8.27773096201183
7.5	8.46288023282049
8.	8.63705799480526
8.5	8.80013205895568
9.	8.95212281461079
9.5	9.09318495070397
10.	9.22358753646320
20.	10.3364331869404
30.	10.4592670437040
40.	10.4709326875258
200.	10.4721359549996

(8)

FIGURE 9.5
Euler solution numerical method for Example 9.1.

We can compare this against the exact solution, obtainable using the method of undetermined coefficients (Figure 9.6):

$$y(t) = -6e - 3t + 7e - 2t + \sin f_0 t - \cos f_0 t$$

```
# plot exact solution first
       time = np.linspace(0, 3)
       y_exact = (
           -6*np.exp(-3*time) + 7*np.exp(-2*time) +
           np.sin(time) - np.cos(time)
           )
       plt.plot(time, y_exact, label='Exact')

       omega = 1

       dt = 0.1
       time = np.arange(0, 3.001, dt)

       f = lambda t,z1,z2: 10*np.sin(omega*t) - 5*z2 - 6*z1
```

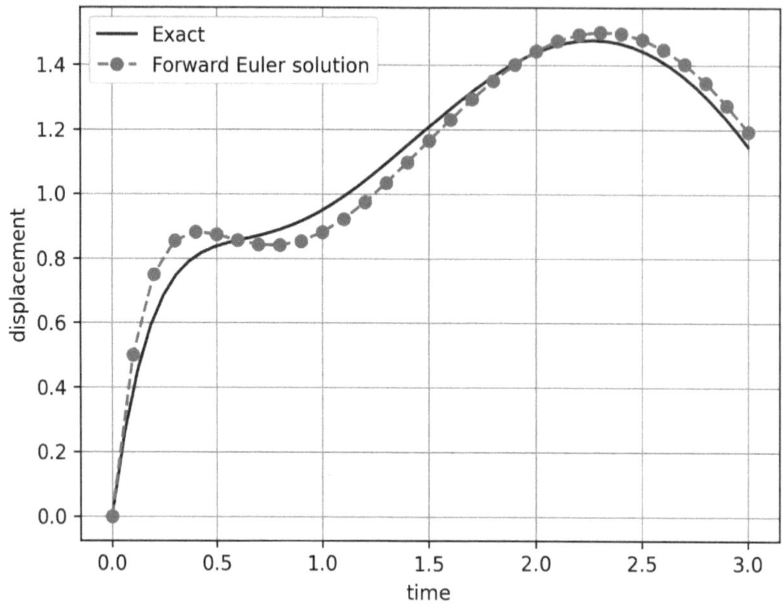

FIGURE 9.6
Python plot of Example 9.2.

```
        z1 = np.zeros_like(time)
        z2 = np.zeros_like(time)

# initial conditions
        z1[0] = 0
        z2[0] = 5

# Forward Euler iterations
        for idx, t in enumerate(time[:-1]):
             z1[idx+1] = z1[idx] + dt*z2[idx]
             z2[idx+1] = z2[idx] + dt*f(t, z1[idx], z2[idx])

plt.plot(time, z1, 'o--', label='Forward Euler solution')
plt.xlabel('time')
plt.ylabel('displacement')
plt.legend()
plt.grid(True)
plt.show()
```

Since the method is only first order accurate, we see both qualitative and quantitative differences compared with the exact solution, but the overall solution agrees. We can use more accurate methods to better capture the exact solution (Figure 9.7).

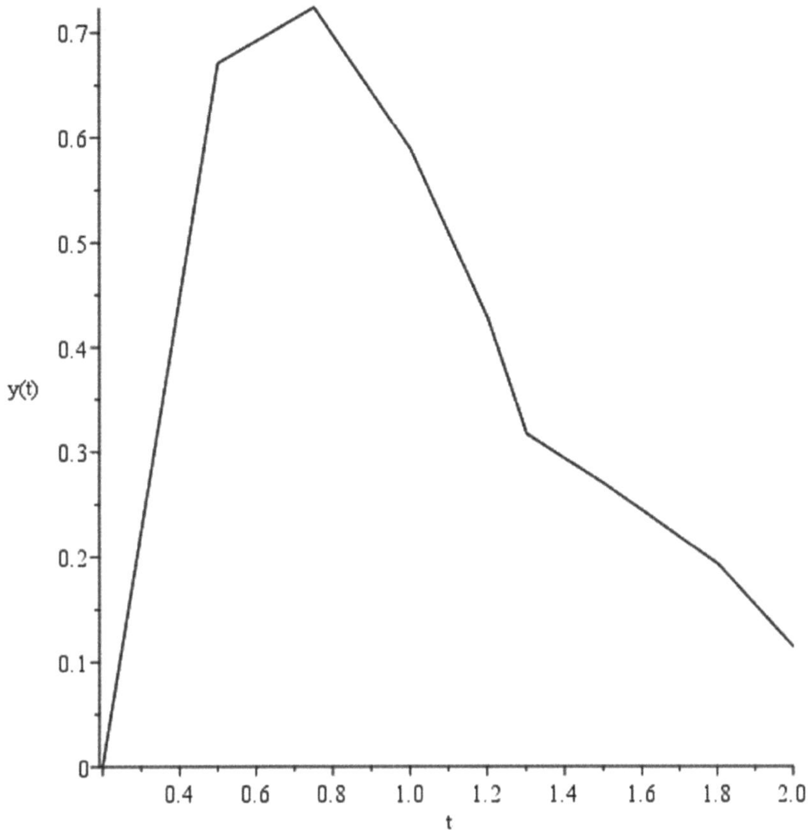

FIGURE 9.7
Maple plot of Example 9.2.

Heun's Method

For schemes that involve more than one stage, like Heun's method, we'll need to implement both stages for each first ODE. For example:

```
# plot exact solution first
time = np.linspace(0, 3)
y_exact = (
    -6*np.exp(-3*time) + 7*np.exp(-2*time) +
    np.sin(time) - np.cos(time)
    )
plt.plot(time, y_exact, label='Exact')

omega = 1
```

```
dt = 0.1
time = np.arange(0, 3.001, dt)

f = lambda t,z1,z2: 10*np.sin(omega*t) - 5*z2 - 6*z1

z1 = np.zeros_like(time)
z2 = np.zeros_like(time)

# initial conditions
z1[0] = 0
z2[0] = 5

# Forward Euler iterations

    for idx, t in enumerate(time[:-1]):
        # predictor
        z1p = z1[idx] + dt*z2[idx]
        z2p = z2[idx] + dt*f(t, z1[idx], z2[idx])

        # corrector
        z1[idx+1] = z1[idx] + 0.5*dt*(z2[idx] + z2p)
        z2[idx+1] = z2[idx] + 0.5*dt*(
            f(t, z1[idx], z2[idx]) + f(time[idx+1], z1p, z2p)
            )

plt.plot(time, z1, 'o--', label='Heuns method solution')
plt.xlabel('time')
plt.ylabel('displacement')
plt.legend()
plt.grid(True)
plt.show()
```

As expected (Figure 9.8), the numerical solution using Heun's method agrees much more closely with the exact solution (vs. forward Euler), though we can still see some error due to the relatively large step size.

Runge-Kutta Method

We can also solve using the fourth order Runge–Kutta method available through the solve_ivp() function, by providing a function that defines the system of first ODEs.

Since there are technically two variables being integrated (z1 representing y, and z2 representing y'), the solution object contained in sol.y will be a two-dimensional array.

The first column, sol.y[0,:], will contain y and the second column, sol.y[1,;], will contain y'. We are typically mostly interested in the solution to y (Figure 9.9).

```
from scipy.integrate import solve_ivp

# plot exact solution first
    time = np.linspace(0, 3)
```

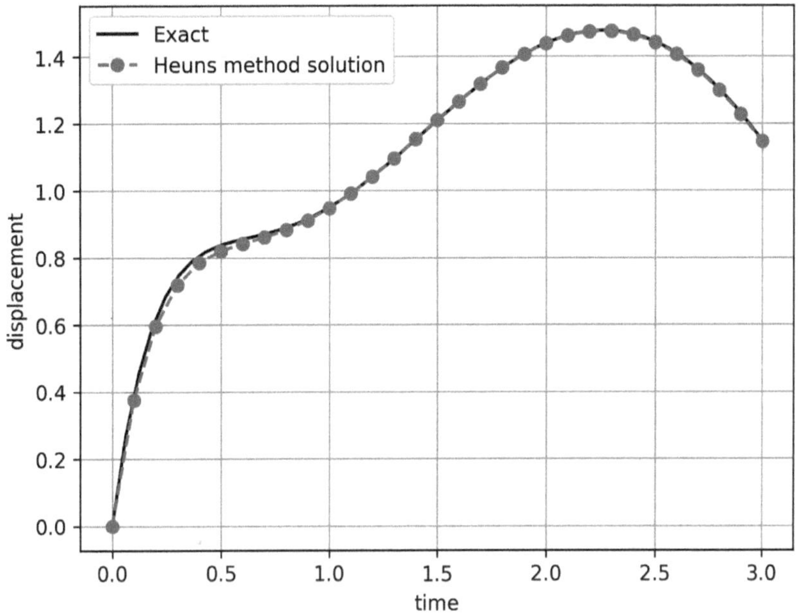

FIGURE 9.8
Python plot of Heun's method for Example 9.2.

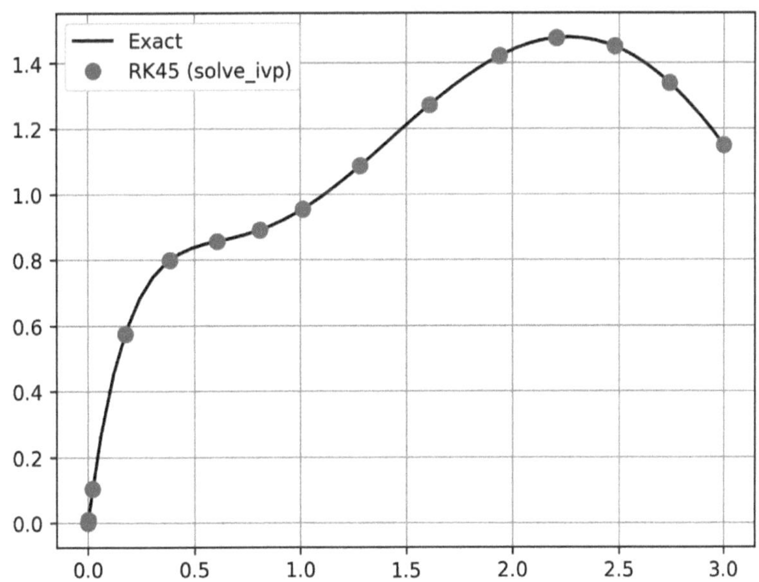

FIGURE 9.9
Plot of Runge-Kutta method.

```
y_exact = (
    -6*np.exp(-3*time) + 7*np.exp(-2*time) +
    np.sin(time) - np.cos(time)
    )
plt.plot(time, y_exact, label='Exact')

def mass_spring(time, z):
    '''Function providing dz/dt derivative system'''
    omega = 1
    return [z[1], 10*np.sin(omega*time) - 6*z[0] - 5*z[1]]

sol = solve_ivp(mass_spring, [0, 3], [0, 5], method='RK45')

plt.plot(
    sol.t, sol.y[0,:], '.',
    markersize=15, label='RK45 (solve_ivp)'
    )
plt.grid(True)
plt.legend()
plt.show()
```

Backward Euler for Second ODEs

We saw how to implement the Backward Euler method for a first ODE, but what about a second ODE? (Or in general a system of first ODEs?). The recursion formula is the same, except now our dependent variable is an array/vector:

$$y_{i+1} = y_i + \Delta(t_i + 1, y_i + 1)$$

In general, we can use Backward Euler to solve second ODEs in a similar fashion as our other numerical methods:

1. Convert the second ODE into a system of two first ODEs.
2. Insert the ODEs into the Backward Euler recursion formula and solve for $yi + 1$.

The main difference is that we will now have a system of two equations and two unknowns: y_1, $+ 1$ and y_2, $i + 1$. Let's illustrate with an example.

Example 9.3

$$y'' + 6y' + 5y = 10 \text{ and } \mathrm{y}(0) = 0, y'(0) = 5$$

where the exact solution is

$$\mathrm{y}(t) = -34e^{-5t} - 54e^{-t+2}$$

To solve numerically,

Step 1. Convert the second ODE into a system of two first ODEs:

$$y_1 = y,\ y1(t=0) = 0 \text{ and } y_2 = y' \text{ and } y2(t=0) = 5$$

Then, for the derivatives of these variables:

$$y1' = y2y2' = 10 - 6y2\ -5y1$$

Step 2. Then substitute these derivatives into the Backward Euler recursion formulas and solve for y_1, $i+1$ and y_2, $i+1$:

$$y_1, i+1 = y1, i + \Delta t,$$
$$y_2, i+1y_2, i+1 = y_2, i + \Delta t(10 - 6y_2, i+1 - 5y1, i+1) \to y1, i+1 - \Delta ty_2,$$
$$i+1 = y_1, i5\Delta ty_1, i+1 + (1+6\Delta t)y_2, i+1 = y_2, i + 10\Delta t$$

or, represented as a product of a coefficient matrix and a vector:

$$[1 - \Delta t5\Delta(1+6\Delta t)][y1, +1y_2, i+1\big] = \big[y_1, iy_2, i + 10\Delta t]Ayi + 1 = b$$

To isolate yi+1 and get a usable recursion formula, we need to solve this system of two equations. We could solve this by hand using the substitution method, or we can use Cramer's rule (from linear algebra):

$$y_1, i+1 = y1, i(1+6\Delta t) + \Delta t(y_2, i + 10\Delta t)1 + 6\Delta t + 5\Delta t2y_2,$$
$$i+1 = y2, i + 10\Delta t - 5\Delta ty1, i1 + 6\Delta t + 5\Delta t2$$

Let's confirm that this gives us a good, well-behaved numerical solution and compare with the Forward Euler method (Figure 9.10):

```
# Exact solution
time = np.linspace(0, 5)
y_exact = lambda t: -0.75*np.exp(-5*t) - 1.25*np.exp(-t) + 2
plt.plot(time, y_exact(time), label='Exact')

dt = 0.1
time = np.arange(0, 5.001, dt)

# Forward Euler
f = lambda t,y1,y2: 10 - 6*y2 - 5*y1
y1 = np.zeros_like(time)
y2 = np.zeros_like(time)
y1[0] = 0
y2[0] = 5
for idx, t in enumerate(time[:-1]):
    y1[idx+1] = y1[idx] + dt*y2[idx]
    y2[idx+1] = y2[idx] + dt*f(t, y1[idx], y2[idx])

plt.plot(time, y1, 's', label='Forward Euler')

# Backward Euler
# use a single array to contain both variables being integrated
```

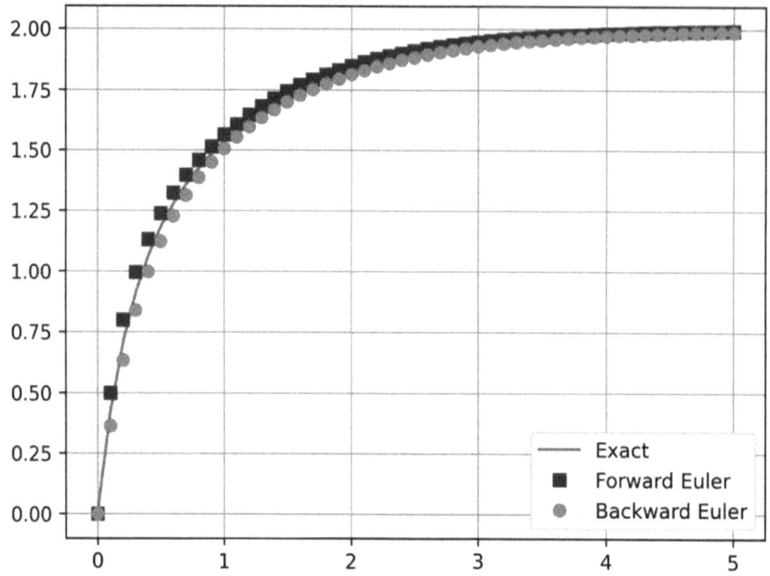

FIGURE 9.10
Plot of solution for Example 9.3.

```
y_vals = np.zeros((len(time), 2))
y_vals[0,0] = 0
y_vals[0,1] = 5
for idx, t in enumerate(time[:-1]):
    denom = 1 + 6*dt + 5*dt**2
    y_vals[idx+1,0] = (
        y_vals[idx,0]*(1 + 6*dt) + dt*(y_vals[idx,1] + 10*dt)
        ) / denom
    y_vals[idx+1,1] = (
        y_vals[idx,1] + 10*dt - y_vals[idx,0]*5*dt
        ) / denom

plt.plot(time, y_vals[:,0], 'o', label='Backward Euler')
plt.legend()
plt.grid(True)
plt.show()

print(
    'Maximum error of Forward Euler: '
    f'{np.max(np.abs(y1 - y_exact(time))):. 3f}'
    )
print(
    'Maximum error of Forward Euler: '
    f'{np.max(np.abs(y_vals[:,0] - y_exact(time))):. 3f}'
    )

Maximum error of Forward Euler:  0.099
Maximum error of Forward Euler:  0.068
```

So, for $\Delta t = 0.1$, we see that the Forward and Backward Euler methods give an error $0(\Delta t)$, as expected since both methods are *first order* accurate.

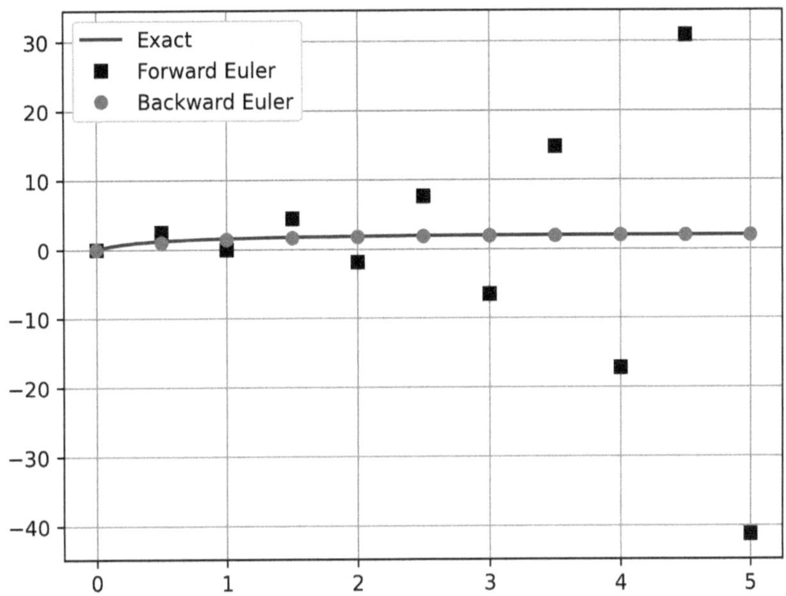

FIGURE 9.11
Plot of larger step size for Example 9.3.

Let's see how they compare for a larger step size (Figure 9.11):

```
# Exact solution
time = np.linspace(0, 5)
y_exact = lambda t: -0.75*np.exp(-5*t) - 1.25*np.exp(-t) + 2
plt.plot(time, y_exact(time), label='Exact')

dt = 0.5
time = np.arange(0, 5.001, dt)

# Forward Euler
f = lambda t,y1,y2: 10 - 6*y2 - 5*y1
y1 = np.zeros_like(time)
y2 = np.zeros_like(time)
y1[0] = 0
y2[0] = 5
for idx, t in enumerate(time[:-1]):
    y1[idx+1] = y1[idx] + dt*y2[idx]
    y2[idx+1] = y2[idx] + dt*f(t, y1[idx], y2[idx])

plt.plot(time, y1, 's', label='Forward Euler')

# Backward Euler
# use a single array to contain both variables being integrated
y_vals = np.zeros((len(time), 2))
y_vals[0,0] = 0
y_vals[0,1] = 5
for idx, t in enumerate(time[:-1]):
    denom = 1 + 6*dt + 5*dt**2
    y_vals[idx+1,0] = (
        y_vals[idx,0]*(1 + 6*dt) + dt*(y_vals[idx,1] + 10*dt)
        ) / denom
```

```
    y_vals[idx+1,1] = (
        y_vals[idx,1] + 10*dt - y_vals[idx,0]*5*dt
        ) / denom

plt.plot(time, y_vals[:,0], 'o', label='Backward Euler')
plt.legend()
plt.grid(True)
plt.show()

print(
    'Maximum error of Forward Euler: '
    f'{np.max(np.abs(y1 - y_exact(time))):. 3f}'
    )
print(
    'Maximum error of Backward Euler: '
    f'{np.max(np.abs(y_vals[:,0] - y_exact(time))):. 3f}'
    )
```

```
Maximum error of Forward Euler:  43.242
Maximum error of Backward Euler:  0.228
```

Backward Euler, since it is unconditionally stable, remains well-behaved at this larger step size, while the Forward Euler method blows up.

One other thing: instead of using Cramer's rule to get expressions for $y1$, i+1 and $y2$, i+1, we could instead use built-in linear algebra routines to solve the linear system of equations at each time step, using the function **np.linalg.solve()**.

To do that, we could replace it with

```
A = np.array([1,  -dt], [5*dt, 1+6*dt])
b = np.array([y_vals[idx,0], y_vals[idx,1]+10*dt])
y_vals[idx+1,:] = np.linalg.solve(A, b)
```

Let's confirm that this gives the same answer (Figure 9.12):

```
# Exact solution
time = np.linspace(0, 5)
y_exact = lambda t: -0.75*np.exp(-5*t) - 1.25*np.exp(-t) + 2
plt.plot(time, y_exact(time), label='Exact')

dt = 0.1

time = np.arange(0, 5.001, dt)

# Backward Euler
y_vals = np.zeros((len(time), 2))
y_vals[0,0] = 0
y_vals[0,1] = 5
for idx, t in enumerate(time[:-1]):
    A = np.array([[1,  -dt], [5*dt, 1+6*dt]])
    b = np.array([y_vals[idx,0], y_vals[idx,1]+10*dt])
    y_vals[idx+1,:] = np.linalg.solve(A, b)

plt.plot(time, y_vals[:,0], 'o', label='Backward Euler')
plt.legend()
plt.grid(True)
plt.show()
```

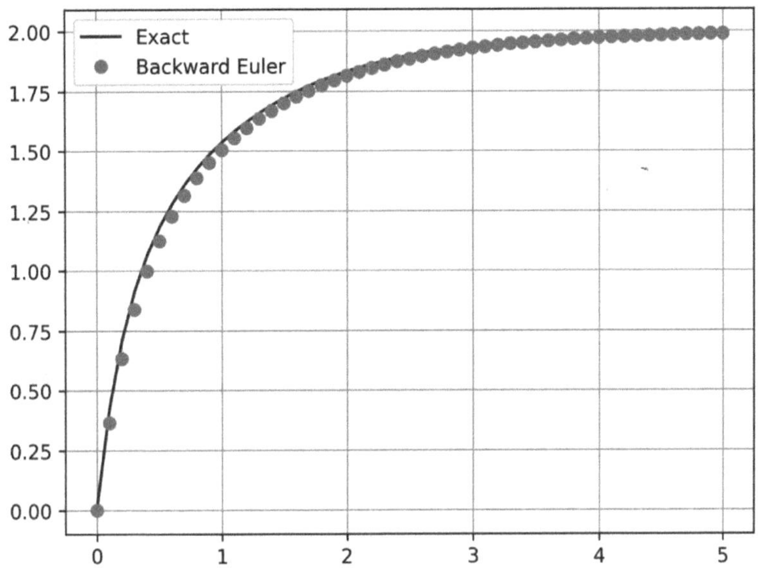

FIGURE 9.12
Backward Euler for Example 9.3.

Numerical Methods

Numerical approaches are used for situations when the analytical solution is not possible. Approximate solutions are calculated by applying methods like Euler's technique and Runge-Kutta methods. The Euler method applies a technique of partitioning the domain, to which the tangent lines then are used for the finite solution approximation purpose.

Runge-Kutta methods are higher order numerical techniques that form a particular type of approximation, known as the higher order Runge-Kutta, by taking multiple intermediate steps.

Let's consider an example to illustrate Euler's method for solving second ODEs:

Example 9.4: Solve the initial value problem y′ = x + y, y(0) = 1 on the interval using Euler's method with step size h = 0.25.

SOLUTION

Step 1: *Discretize the interval using the step size h = 0.25:*

$$x0 = 0,\ x1 = 0.25,\ x2 = 0.50,\ x3 = 0.75,\ x4 = 1.00$$

Step 2: *Use Euler's method to approximate the solution at each point:*

$$yn + 1 = yn + hf\left(xn, yn\right)$$

where $f(x,y) = x + y$
$y0 = 1$

$y1 = y0 + hf(x0, y0) = 1 + 0.25(0 + 1) = 1.25$
$y2 = y1 + hf(x1, y1) = 1 .25 + 0.25(0.25 + 1.25) = 1.625$
$y3 = y2 + hf(x2, y2) = 1.625 + 0.25(0.50 + 1.625) = 2.125$
$y4 = y3 + hf(x3, y3) = 2.125 + 0.25(0.75 + 2.125) = 2.75$

Therefore, using Euler's method with a step size of h = 0.25, the approximate solution at x = 1 is y ≈ 2.75.

Second Order Differential Equation Examples

Second ODEs are widely used in various fields to model and analyze complex systems. Here are a few examples of how second order differential equations are applied in different domains:

Mechanical Vibrations

In mechanical systems, second ODEs are used to model the motion of masses connected to springs and dampers. For example, the equation of motion for a mass-spring-damper system is:

$$md2x/dt2 + c\,dx/dt + kx = F(t)$$

This equation can be used to analyze the vibrations and oscillations of mechanical systems, such as suspension systems in vehicles.

Electrical Circuits

In electrical engineering, second ODEs are employed to analyze the behavior of RLC (resistor-inductor-capacitor) circuits. The equation governing the current i(t) in an RLC series circuit is:

$$Ld2i/dt2 + R\,di/dt + 1/C\,i = V(t)$$

This equation can be used to study the transient and steady-state behavior of electrical circuits, such as the response to step inputs or sinusoidal excitations.

Structural Analysis

In civil and structural engineering, second ODEs are used to analyze the behavior of structures under various loads, such as wind, earthquakes, and impact forces. For example, the equation of motion for a single-degree-of-freedom structure subjected to ground acceleration is:

$$md^2x/dt^2 + c\,dx/dt + kx = -mx\,G$$

This equation can be used to assess the seismic response of buildings and design appropriate structural systems.

Example 9.5: Bridge Jumping

In bridge jumping, a participant attaches one end of a bungee cord to himself (or herself), attaches the other end to a bridge railing, and then drops off the bridge. In this project, the jumper will be dropping off of the Royal Gorge Bridge, a suspension bridge that is 1053 feet above the floor of the Royal Gorge in Colorado. The jumper will use a 200-foot-long bungee cord. It would be nice if the jumper has a safe jump, meaning that the jumper does not crash into the floor of the gorge or run into the bridge on the rebound. In this project, you will do some analysis of the fall.

Assume the jumper weighs 160 lbs. The jumper will free fall until the bungee cord begins to exert a force that acts to restore the cord to its natural (equilibrium) position. In order to determine the spring constant of the bungee cord, you found that a MASS weighing 4 lbs stretches the cord 8 feet. Hopefully, this spring force will help slow down the descent sufficiently so that the jumper does not hit the bottom of the gorge.

Throughout this example, we will assume that DOWN is the POSITIVE direction.

Before the bungee cord begins to retard the fall the jumper, the only forces that act on the jumper are his weight and the force due to wind (air) resistance. If the force due to the wind resistance is 0.9 times the velocity of the jumper, then use Newton's second law ($\Sigma F = MA$) to write a differential equation that models the fall of the jumper. Be sure to include the initial conditions for the jumper for your differential equation. This problem can be formulated as a second ODE in position, or as a first ODE in velocity – we will use the second ODE model in this example.

Requirement 1.

a. Solve this differential equation and find a function that describes the jumper's velocity (as a function of time). Knowing integral calculus, find a function that describes the jumper's position (as a function of time).
b. What is the velocity of the jumper after the jumper has fallen 200 feet?
c. What is the terminal velocity of the jumper, if any?

Requirement 2. After a bit more research, you have found that the force due to wind resistance is not linear, as assumed above. Apparently, the force due to wind resistance is more closely modeled by $0.9\,v = 0.0009\,v^2$.

a. Write a new differential equation governing the velocity of the jumper (prior to the bungee cord coming into effect).
b. You should notice that this DE is no longer easy to solve. Nonetheless, you are determined to find the velocity of the jumper after 4 seconds by using a numerical technique. Use Euler's method with a step size of 0.5, and estimate the velocity of the jumper after 4 seconds.
c. Find the terminal velocity, if any exists.
d. How do your results compare with those found in requirement one under the linear assumption?
e. What is the velocity of the jumper after 200 feet? (Hint: You can find this from your numerical Table)

Before the bungee cord begins to retard the fall of the jumper, the only forces that act on the jumper are his weight and the force due to wind (air) resistance.

If the force due to the wind resistance is 0.9 times the velocity of the jumper, then use Newton's second law ($\Sigma F = MA$) to write a differential equation that models the fall of the jumper.

$$\sum F = MA$$

$$mx'' = -mg - kx'$$

Since k = .5 since from Hooke's Law ($F = Ks$) 4 = k * 8 so k = 4/8 = 0.5
g = 32.17 ft/s^2 or 9.81 m/s^2
W = weight and W = mg so W/g = m. This implies m = 160/32.17 = 4.9735
We divide the entire equation by m to get

$$x'' = -g - k/mx' - 0.9x' \text{ or}$$
$$x'' = -32.17 - 4.9735x - 0.9x'$$

If we let the starting position be $x(0) = 0$ and $x'(0) = 0$ then we may solve this differential equation,

$$-2.76305555 + -5.30188348i$$

Exercises

9.1 Solve the following higher order differential equations
  a. $y'' - 3y' - 10, y = 0, y(0) = 0, y'(0) = 7$
  b. $y'' + 6y' + 13y, y(0) = 0, y'(0) = 2$
  c. $y'' - 2y' + 10y = 0, y(0) = 2, y(0) = -1$

9.2 Suppose the following initial-value problem modeling the position (in feet) of mass in a spring-mass system at any given time, t, as $y'' + 14y' = 49y = 0$, $y(0) = 0$, $y'(0) = 1$. Solve and graph the solution. What are the positions at $t = 0$ and $t = 1$? In what direction is the mass moving at each of these times?

PROJECTS

9.1 Projectile motion: A spherical projectile weighing 100 lbs is observed to have a limiting velocity of magnitude 400 ft/sec. If the air resistance is given by $v' = -g - (gc/w)v|v|$, *where* $c = 1/1600$ and $w = mg$ is the weight of the projectile where the m is the mass. Model the projectile motion and describe the motion.

9.2 **Skydiving.** A skydiver with equipment weighs 250 lbs. In free fall the skydiver reaches a limiting velocity of 250 ft/sec. After the parachute opens, the skydiver reaches a limiting velocity of 17 ft/sec. Assume the skydiver jumps from 10,000 feet. Use g = 32.2 ft/sec and answer the following: (a) How much time elapses before the free-falling skydiver falls to a speed of 100 ft/sec? (b) When falling at 100 ft/sec, the skydiver pulls the rip cord and assume the chute open immediately. How long does it take for the speed to drop to 25 ft/sec? (c) How long does the entire jump last, and (d) what is the skydiver's speed at landing?

References and Additional Readings

Boyce, W., and DiPrima, R. (2017). *Elementary Differential Equations and Boundary Value Problems*. 11th edition, John Wiley & Sons.

Simmons, G. (1991). *Differential Equations with Applications and Historical Notes*. 2nd edition, McGraw-Hill.

10

System of Linear and Nonlinear ODEs

Systems of Linear Ordinary Differential Equations.

Consider a brine mixture system as shown in Figure 10.1. We want to find the amount of salt mixture in each tank at time *t*, $t > 0$.

A simple example of three stage system is shown in Figure 10.1. Each tank hold their respective volumes V_1, V_2, and V_3 gallons of brine, respectively. Fresh water flows into tank A and is mixed with salt, and flows into tank B and then from tank B into tank C. Let xi(t) be the amount of salt (in pounds) at time *t* for $i = 1, 2,$ and 3. If each flow rate is r gallons per minute, then a simple model yields:

$$x_1'(t) = -k_1 x_1$$

$$x_2'(t) = k_1 x_1 - k_2 x_2$$

$$x3'(t) = k_2 x_2 - k_3 x_3$$

where $k_i = r/V_i$ *for* $i = 1, 2\ 3$.

For this model $V_1 = 20$, $V_2 = 40$, $V_3 = 50$ and r = 4 gal/min.

The initial conditions are $x_1(0) = 15$, $x_2(0) = 0$, $x_3(0) = 0$

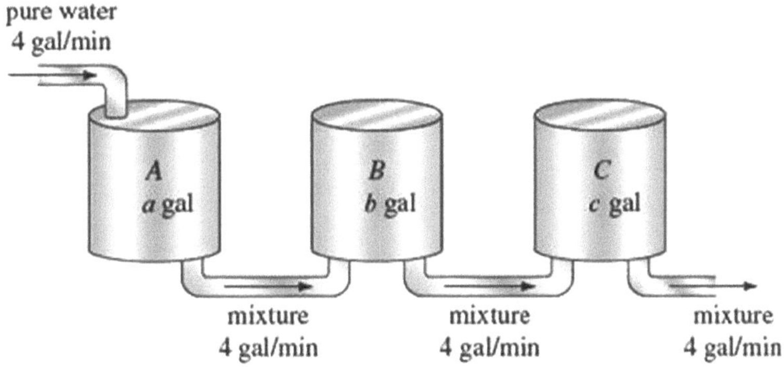

FIGURE 10.1
Three stage system.

DOI: 10.1201/9781003582274-10

Our systems of first ODE is:

$$x_1'(t) = -k_1 x_1 = -4/20\,x_1$$
$$x_2'(t) = k_1 x_1 - k_2 x_2 = 4/20\,x_1 - 4/40\,x_2$$
$$x3'(t) = k_2 x_2 - k_3 x_3 = 4/40\,x_2 - 4/50\,x_3$$

We find the solution as:

$$x_1(t) = 13.125 e^{(-.2t)}$$
$$x_2(t) = -8.75 e^{(-.2t)} + 8.74999 e^{(-.1t)}$$
$$x_3(t) = 8.74999 e^{(-.1t)} - 4.861111 e^{(-0.08*t)}$$

All the salt goes to zero over time, as shown in Figure 10.2.

We will return to this scenario later in the chapter to see how to completely find the solution presented here.

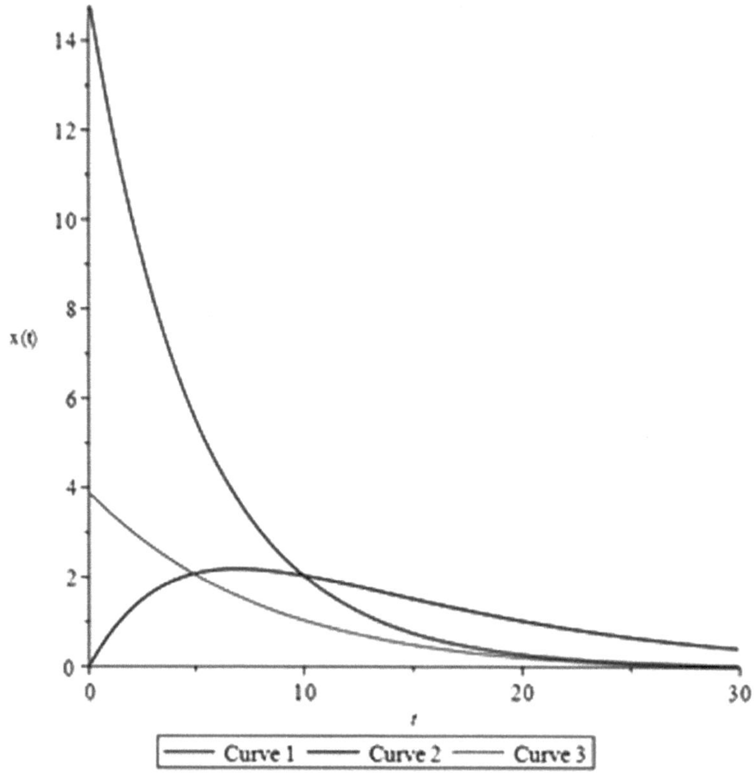

FIGURE 10.2
Three tank system.

Lanchester's Equations of Combat

Consider a battle between a Red Force of size 50 soldiers and a Blue Force of size 100 soldiers.

For example, consider a battle between a Red Force, *R(n)*, and a Blue Force, *B(n)*, as given below:

$$B'(t) = -0.1R(t)$$
$$R'(t) = -0.05B(t)$$
$$B(0) = 100, R(0) = 50$$

The ratio of *B(0)/R(0) = 100/50* = 2.

We are given the attrition coefficients, $k_1 = -0.1$ and $k_2 = -0.050$.

We can quickly obtain the analytical solution.

$$\sqrt{k_1 k_2} = \sqrt{-.1 \cdot -.05} = .0707$$

The eigenvalues are .0707 and −0.0707. We could build the closed form solution with the ratio of the vectors as ±1 and $\frac{\sqrt{k_1 k_2}}{k_1}$. We find $\frac{\sqrt{k_1 k_2}}{k_1}$ = 0.7070. Our general solution would be:

$$\mathrm{B}(t) = -c_1 e^{(-.0707*t)} + c_2 e^{(-.0707)} \text{ and}$$
$$\mathrm{R}(t) = (0.707) c_1 e^{(.0707*t)} + (0.707) c_2 e^{(-.0707t)}$$

With our initial conditions of (100, 50) at period 0, we have the particular solution:

$$\mathrm{B}(t) = 14.64 e^{(.0707*t)} + 85.36 e^{(-.0707t)} \text{ and}$$
$$\mathrm{R}(t) = (0.707)\,(-14.64) e^{(.0707*t)} + (0.707)(85.36) e^{(-.0707t)}$$

We can graph these separately and observe the behavior:

These two graphs (Figure 10.3a and b) of the analytical solution show that the *y*-force (initially at size 50) approaches 0 as the *x*-force (initially at 100) is slightly below 70. Thus, we know the *x*-force or the Blue Force wins.

These two graphs (Figure 10.3a and b) of the analytical solution show that the Red Force (initially at size 50) approaches 0 as the *Blue* Force (initially at 100) is slightly below 70. Thus, we know the *Blue* Force wins.

Since Lanchester's initial combat models in 1914, differential equations have been the methodology to present and solve these combat models. James G. Taylor alluded to difficulties in solving "real" equations and suggested numerical methods in his work (Taylor, 1983). The use of computers to analytically solve or numerically solve combat models is the standard method. We suggest using difference equations, the discrete form of Lanchester equations, in our combat models. We show the discrete forms and their solutions, where applicable. We also show a numerical solution. We compare several of these solutions with

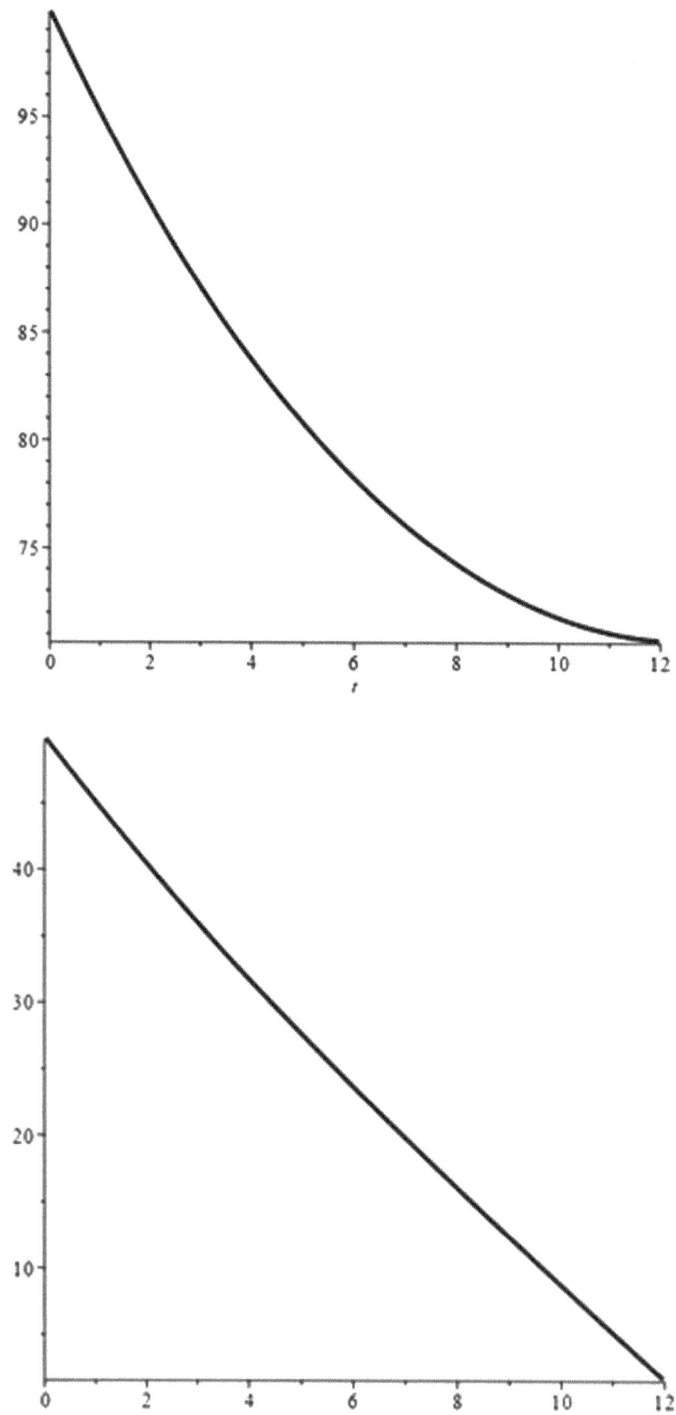

FIGURE 10.3
Solution graphs for (a) Blue Force over time. (b) Red Force over time.

the differential equation form to show how close the discrete form matches. We also suggest the uses of the equations in decision making for our leaders.

Forms of Lanchester Equations

History is filled with examples of the unparalleled heroism and barbarism of war. Specific battles like Bunker Hill, the Alamo, Gettysburg, Little Big Horn, Iwo Jima, and the Battle of the Bulge are a part of our culture and heritage. Campaigns like the Cuban Revolution, Vietnam, and now the conflicts in Afghanistan and Iraq are a part of our personal history. Although combat is continuous, the models of combat usually employ discrete time simulation. For years, Lanchester equations were the norm for computer simulations of combat. The diagram of simple combat as modeled by Lanchester is illustrated in Figure 10.4. We investigate the use of a discrete version of these equations. We will use models of discrete dynamical systems via difference equations to model these conflicts and gain insight into the different methods of "directed fire" conflicts like Nelson's Battle at Trafalgar and the Alamo, and Iwo Jima. We employ *difference equations* that allow for a complete numerical and graphical solution to be analyzed and does not require the mathematical rigor of differential equations. We further investigate the analytical form of the "direct fire" solutions to provide a solution template to be used in modeling efforts.

Lanchester model stated that "under conditions of modern warfare" that combat between two homogeneous forces could be modeled from the state condition of this diagram. We will call this diagram (Figure 10.4) the change diagram.

We will use the paradigm,

$$Future = Present + Change$$

to build our mathematical models. This will be paramount as eventually models will be built that cannot be solved analytically but can be solved by numerical (iteration) methods

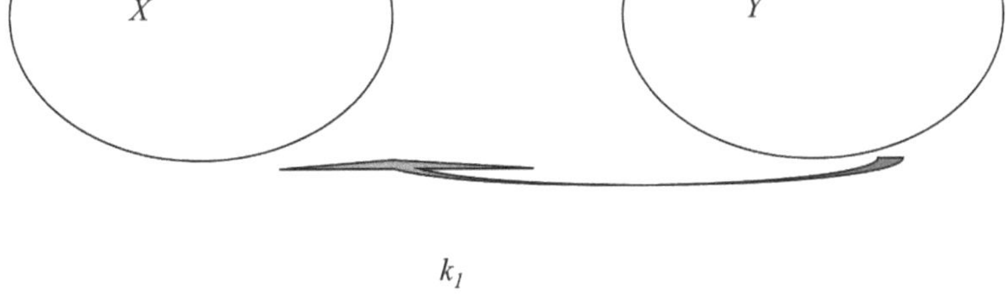

FIGURE 10.4
Change diagram of combat modeled by Lanchester.

We begin by defining the following variables:

$x(n)$ = the number of combatants in the X-force after period n.
$y(n)$ = the number of combatants in the Y-force after period n.

Future is then $x(n + 1)$ and $y(n + 1)$, respectively.
So, we have

$$x(n+1) = x(n) + Change$$
$$y(n+1) = y(n) + Change.$$

Figure 10.4 provides the information on the change diagram that reflects the change. Our dynamical system of equations is shown in Equation 10.1:

$$\begin{aligned} x(n+1) &= x(n) - k_1 y(n) \\ y(n+1) &= y(n) - k_2 x(n) \end{aligned} \tag{10.1}$$

We define our starting conditions as the size of the combatant forces at time period zero: $x(0) = x_0$ and $y(0) = y_0$.
If we move to a continuous form ode we have the following system

$$dx/dt = -k_1 y(t)$$
$$dy/dt = -k_2 x(t)$$

Now, we write these in matrix form shown in Equation 10.2:

$$X'(t) = \begin{bmatrix} 0 & -k_1 \\ -k_2 & 0 \end{bmatrix} \begin{bmatrix} x(t) \\ t(t) \end{bmatrix}, X_0 = \begin{bmatrix} x_0 \\ y_0 \end{bmatrix} \tag{10.2}$$

Eigenvalues and Eigenvectors

Definition of eigenvectors and eigenvalues

Let A be a $\boldsymbol{n \times n}$ matrix. The real number λ is called an eigenvalue of A if there exists a nonzero vector x in $\boldsymbol{R}^n$ such that

$$Ax = \lambda x. \tag{10.3}$$

The nonzero vector x is called an eigenvector of A associated with the eigenvalue λ.

Equation 10.3 is written as $Ax - \lambda x = 0, or (A - \lambda I)x = 0$

where I is a 2×2 identity matrix. The solution to finding λ comes from taking the determinant of the $(A - \lambda I)$ matrix, setting it equal to zero, and solving for λ.

We know that matrix A is $\begin{bmatrix} 0 & -k_1 \\ -k_2 & 0 \end{bmatrix}$.

We set up the form for the use of eigenvalues:

$$\det\begin{bmatrix} 0-\lambda & -k_1 \\ -k_2 & 0-\lambda \end{bmatrix} = 0$$

that yields the characteristic equation

$$\lambda^2 - \mathrm{k}_1\mathrm{k}_2 = 0$$

We solve for λ. Although not intuitively obvious to the casual observer, the two eigenvalues are

$$\lambda_1 = \sqrt{k_1 k_2}$$
$$\lambda_2 = -\sqrt{k_1 k_2}$$

Therefore, we have the eigenvalues from the initial form of the equation. We note that the eigenvalues are a function of the kill rates, k_1 and k_2. If you know the kill rates then you can easily obtain the two eigenvalues.

We also note two other characteristics of the eigenvalues: (1) $\lambda_1 + \lambda_2 = 2$ and (2) $\lambda_1 \geq \lambda_2$. For most of these combat models one eigenvalue will be > 1 and the other eigenvalue will be less than 1 (< 1). The equation explores which forces are being attrited by the larger value of k_1 or k_2 has the eigenvalue, λ >1.

Most literature suggests that the dominant eigenvalue (that eigenvalue is the largest eigenvalue and greater than 1 in this case) will control the system. However, these combat models are observed to be controlled by the smaller eigenvalue.

The general form of the solution is shown in Equation 10.4:

$$\mathrm{X}(\mathrm{k}) = \mathrm{c}_1\mathrm{V}_1\mathrm{e}^{(\mathrm{k1t})} + \mathrm{c}_2\mathrm{V}_2\mathrm{e}^{(\mathrm{k2t})} \tag{10.4}$$

where the vectors V_1 and V_2 are the corresponding eigenvectors.

These eigenvectors, interestingly enough, are in a ratio of the attrition coefficients, k_1 & k_2. The vector for the dominant eigenvalue always has both a positive and a negative component as its eigenvector while the vector for the other smaller of the two eigenvalues always has two positive entries in this same ratio. This is because the equation for finding the eigenvector comes from Equation 10.5:

$$\sqrt{k_1 k_2}\, c_1 - k_1 c_2 = 0 \quad and \quad -\sqrt{k_1 k_2}\, d_1 - k_1 d_2 = 0$$

So,

$$c_1 = k_1, c_2 = \sqrt{k_1 k_2} \quad and \quad d_1 = -k_1, d_2 = \sqrt{k_1 k_2} \tag{10.5}$$

Having simplified formulas for obtaining eigenvalues and eigenvectors allows us to quickly obtain the general form of the analytical solution. We can then use the initial conditions to obtain the particular solution.

Example 10.1: The Alamo

First, consider the situation at the Alamo. According to some historical records, there were approximately 189 Texans barricaded in the Alamo being attacked by 2000 Mexicans in the open fields surrounding the Alamo. We are interested in describing the loss of combatants in each force over the course of the engagement. We will do this by measuring or defining change. We define *T(n)* to be the number of Texans after period n and *M(n)* to be the number of Mexican soldiers after time period n. That is, we want to devise a way to express $\Delta T = T(n+1) - T(n)$ (the loss of Texican combatants over time) and $\Delta M = M(n+1) - M(n)$ (the loss of Mexican combatants over time). The Battle of the Alamo is an example of a "directed fire" battle. The combatants on each side can see their opponents and can direct their fire at them. The Texans hiding behind the barricades were the more difficult target, and we need to have our models reflect this fact. Figure 10.5 illustrates the strategy of this battle.

First, consider ΔM. Upon what does this depend? It depends on the number of bullets being fired by the Texican defenders and how accurately they are being fired at the Mexican army. We can use a proportionality model, $\Delta M \propto (\textit{number of bullets})(\textit{probability of hit})$. The number of bullets capable of being fired depends upon how many men are firing and how rapidly each can fire. Given the weaponry at the time, it might be more effective to have only a portion of the combatants firing with the rest loading for them. This might increase the intensity of fire. There is also an issue of what portion of the force is in a position to fire on the enemy. If the force is in a rectangular formation, with several lines of combatants one behind the other, only the first one or two rows may be capable of firing freely at the enemy. Thus,

$$\Delta M = (\textit{Texicans})(\%\ \textit{firing})\big(\textit{bullets/Texican/min}(\textit{probhit/bullet})(\textit{Mexicans disabled/hit})\big)$$

All of these variables can be combined into a single proportionality constant k. Some of these variables will vary over distance or time. For example, the probability

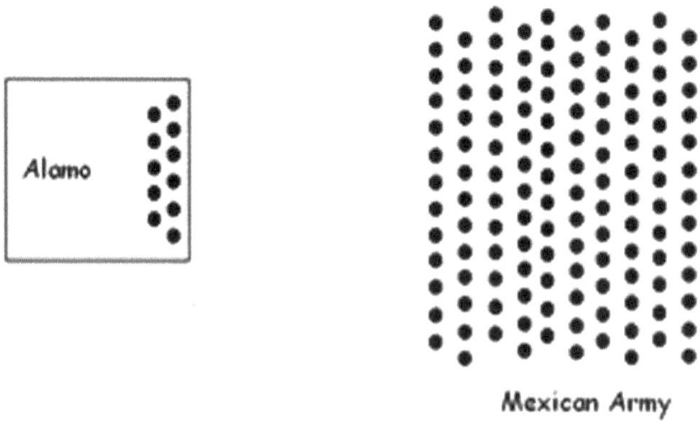

FIGURE 10.5
Mexican Army approaching the Alamo.

of a hit will likely increase as the Mexican army closes in on the Alamo. However, our model assumes each of these except for the number of combatants is constant over the course of the battle. Consequently, we can write $\Delta M = -kT(n)$, where $T(n)$ is the number of Texans remaining in the battle after period n. The negative sign indicates the number of Mexican combatants is decreasing.

Now, consider ΔT. It is similarly composed of terms like the number of Mexicans, percent firing, number of bullets per Mexican combatant per minute, the probability of a hit, and the number of Texans disabled per hit. We would expect that the rate of fire for the Mexican army to be smaller than the Texans, since they will be reloading while marching instead of reloading while standing still. Similarly, the probability of a hit will also be higher for the Texans shooting from a stance behind a wall than for the Mexicans shooting while marching in the open fields. So,

$$dM/dt = -k_1 T(t)$$
$$dT/dt = -k_2 M(t)$$

but the values of k_1 and k_2 will be very different for the two forces. The constants k and c are known as the coefficients of combat effectiveness.

The Battle of the Alamo is actually two battles. The first battle was waged while the Mexicans were in the open field and the effectiveness constant k_2 was very much smaller than k_1 was to the advantage of the Texans. Once the Alamo walls were breached, the values of k_1 and k_2 were vastly altered, and the battle ended in a very short time. We model only the first battle as if it were a fight to the finish.

The differential equation model described is modeled as:

$$\begin{bmatrix} dM/dt \\ dT/dt \end{bmatrix} = \begin{bmatrix} -0.06\ T(t) \\ -.5\ M(t) \end{bmatrix}, \begin{bmatrix} T(0) = 200 \\ M(0) = 1200 \end{bmatrix}$$

The attrition plot is shown in Figure 10.6.

From our equation $\sqrt{k_1 k_2} \cdot T_0 < k_1 \cdot M_0$, we have 0.1732 (200) < 0.5(1200) and we know that the Mexican army wins decisively. In Table 10.1, we obtained the values to achieve parity in each case. We can easily see that many of these values are unrealistic for the event. The Texans were going to lose this battle without outside help.

Rental Cars as a system of linear differential equations

Consider a rental car company with branches at John Wayne Airport, LAX, and Long Beach. Each rents to tourists in the LA area. Depending on what the tourists do, they might rent the car in one location and return to another location. Data was collected over the past year and Table 10.2 yields the results.

This is a three-stage Markov chain that we will solve using Maple (Figures 10.7 and 10.8).

Let $x_1(t)$ = John Wayne airport, $x_2(t)$ = LAX, and $x_3(t)$ = Long Beach

$$x_1'(t) = 0.75 * x_1(t) + 0.2 x_2(t) + 0.40 x_3(t)$$
$$x_2'(t) = 0.05 * x_1(t) + 0.6 x_2(t) + .020 x_3(t)$$
$$x_3'(t) = 0.20 * x_1(t) + 0.2 x_2(t) + 0.40 x_3(t)$$

Initial condition have $x_1(0) = 200$, $x_2(0) = 200$, $x_3(0) = 200$

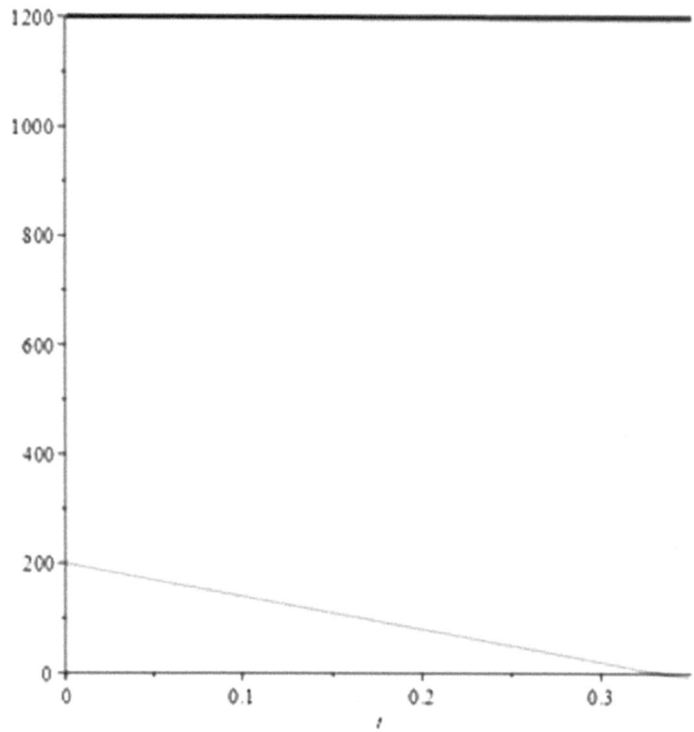

FIGURE 10.6
Attrition for the Battle of the Alamo.

TABLE 10.1
Parity Values

Parity	k_1	k_2	T(0)	M(0)
k_1 variable	2.16 (very unrealistic value)	.06	200	1200
k_2 variable	.5	.01388	200	1200
T(0) variable	.5	.06	200	578
M(0) variable	.5	.06	3464 (unrealistic)	1200

TABLE 10.2
Historical Data

	John Wayne	LAX	Long Beach
John Wayne	0.75	0.05	0.20
LAX	0.20	0.60	0.20
Long Beach	0.40	0.20	0.40

```
> with(DEtools):with(plots):with(linalg):with(LinearAlgebra):
> M:=
> with(DEtools):with(plots):with(linalg):with(LinearAlgebra):
> M:=Matrix(3,3,[-.25,.2,.4,0.05,-.4,.2,.2,.2,-.6]);
> lambda:=eigenvects(M);
> homsol:=matrixDE(M,t);
```

$$M := \begin{bmatrix} -0.25 & 0.2 & 0.4 \\ 0.05 & -0.4 & 0.2 \\ 0.2 & 0.2 & -0.6 \end{bmatrix}$$

$$\lambda := [-0.8000000000, 1., \{[-0.4417613170 \quad -0.3313209880 \quad 0.7730823057]\}], [3.45333 \times 10^{-11}, 1., \{[0.891257721 \quad 0.3119402028 \quad 0.4010659747]\}],$$
$$[-0.4500000000, 1., \{[0.8063033355 \quad -0.8063033366 \quad 1.5 \times 10^{-9}]\}]$$

$$homsol := \left[\begin{bmatrix} 1. & 1.e^{-\frac{4t}{5}} & 1.e^{-\frac{9t}{20}} \\ 0.3500000000 & 0.750000000e^{-\frac{4t}{5}} & -1.000000000e^{-\frac{9t}{20}} \\ 0.4500000000 & -1.750000000e^{-\frac{4t}{5}} & 0. \end{bmatrix}, [0. \quad 0. \quad 0.]\right]$$

```
> M^200;
```

$$\begin{bmatrix} 5.92787938411570 \times 10^{-21} & 5.92787938411570 \times 10^{-21} & -1.77836381523471 \times 10^{-20} \\ 4.44590953808678 \times 10^{-21} & 4.44590953808678 \times 10^{-21} & -1.33377286142603 \times 10^{-20} \\ -1.03737889222025 \times 10^{-20} & -1.03737889222025 \times 10^{-20} & 3.11213667666075 \times 10^{-20} \end{bmatrix}$$

FIGURE 10.7
Markov solution to rental cars at LA airports.

```
> phi:=homsol[1];
> Eigenvalues(M);
```

$$\phi := \begin{bmatrix} 1. & 1.e^{-\frac{4t}{5}} & 1.e^{-\frac{9t}{20}} \\ 0.3500000000 & 0.750000000e^{-\frac{4t}{5}} & -1.000000000e^{-\frac{9t}{20}} \\ 0.4500000000 & -1.750000000e^{-\frac{4t}{5}} & 0. \end{bmatrix}$$

$$\begin{bmatrix} 0. + 0.I \\ -0.450000000000001 + 0.I \\ -0.800000000000000 + 0.I \end{bmatrix}$$

The Complimentary Solution:
The matrix(2,1,[c1,c2]) means that we want a 2 row 1 columns matrix of c1, c2.

```
> Xc:=multiply(phi, matrix(3,1,[c1,c2,c3]));
```

$$Xc := \begin{bmatrix} 1.c1 + 1.e^{-\frac{4t}{5}}c2 + 1.e^{-\frac{9t}{20}}c3 \\ 0.3500000000\,c1 + 0.750000000e^{-\frac{4t}{5}}c2 - 1.000000000e^{-\frac{9t}{20}}c3 \\ 0.4500000000\,c1 - 1.750000000e^{-\frac{4t}{5}}c2 \end{bmatrix}$$

If we only had a homogeneous systesm,w e would solve for the c1 and c2 now, but since this is a nonhomogeneous system we wait until we get **X=Xc+Xp.**

```
> eq4:=solve({200=c1+c2+c3,200=.35*c1+.75*c2-c3,.45*c1-1.75*c2=200},{c1,c2,c3});
```

$$eq4 := \{c1 = 333.3333333, c2 = -28.57142857, c3 = -104.7619048\}$$

FIGURE 10.8
Complimentary solution to rental cars at LA airports.

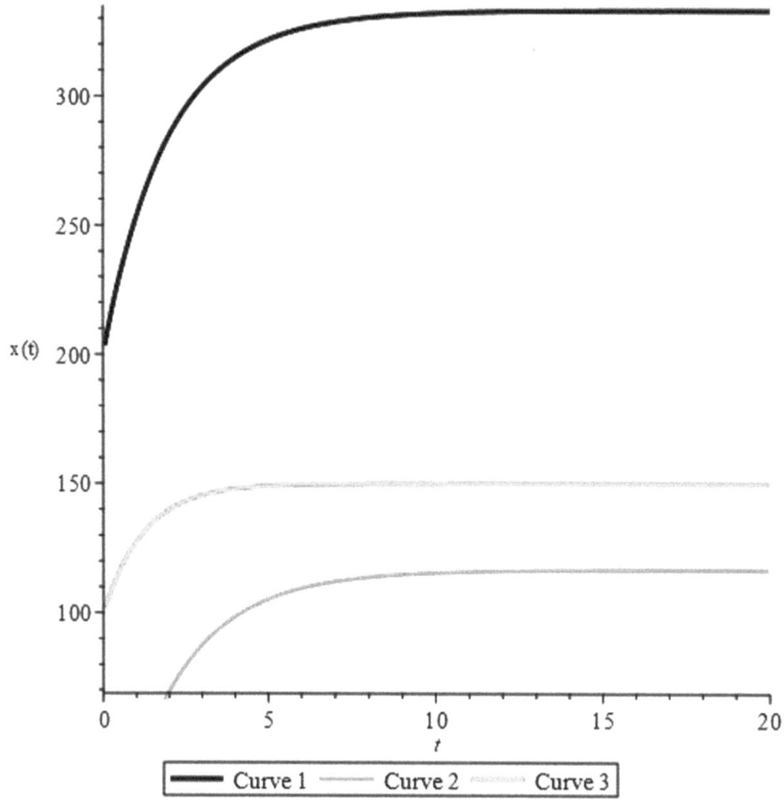

FIGURE 10.9
Rental cars at LA airports.

Steady state values are seen in Figure 10.9 with x1(t) = 333.33, x2(t) 116.64, and x3(t) = 149.94. Rounded we would accept 333, 117, and 150 as our values in the long run.

Review of Matrix Notation, Terminology, and Operations

This section will extend our review of Matrix Algebra. An $m \times n$ matrix **A** is a rectangular array of m rows and n columns.

$$A = \begin{bmatrix} a_{1,1} & a_{1,2} & a_{1,3} & \cdots \\ a_{2,1} & a_{2,2} & a_{2,3} & \cdots \\ a_{3,1} & a_{3,2} & a_{3,3} & \cdots \\ \cdot & \cdot & \cdot & \cdot \\ \cdot & \cdot & \cdot & \cdot \\ \cdot & \cdot & \cdot & \cdot \end{bmatrix}$$

We will use bold capital letters to denote a matrix such as **A**.

If we need to add matrices they need to be of the same dimension, $m \times n$. We then add like entries.

Example 10.2: Addition of Matrices

$$A = \begin{bmatrix} 1 & -4 \\ 5 & 6 \end{bmatrix}, B = \begin{bmatrix} -6 & 7 \\ 8 & 1 \end{bmatrix}, \text{ then } A + B = \begin{bmatrix} 1+-6 & -4+7 \\ 5+8 & 6+1 \end{bmatrix} = \begin{bmatrix} -5 & 3 \\ 13 & 7 \end{bmatrix}$$

For multiplication by a scalar, c, we multiply c by each term.

Example 10.3: Multiplication by Scalar

$$4A = 4\begin{bmatrix} 1 & -4 \\ 5 & 6 \end{bmatrix} = \begin{bmatrix} 4 & -16 \\ 20 & 24 \end{bmatrix}$$

Subtraction is defined as **A – B = A + (–1 * B)**
Common properties are:

$$\mathbf{A} + \mathbf{0} = \mathbf{0} + \mathbf{A} = \mathbf{A}$$
$$\mathbf{A} - \mathbf{A} = \mathbf{0}$$
$$\mathbf{A} + \mathbf{B} = \mathbf{B} + \mathbf{A}$$
$$\mathbf{A} + (\mathbf{B} + \mathbf{C}) = (\mathbf{A} + \mathbf{B}) + \mathbf{C}$$
$$\mathbf{C}(\mathbf{A} + \mathbf{B}) = \mathbf{cA} + \mathbf{cB}$$
$$(\mathbf{c} + \mathbf{d})\mathbf{A} = \mathbf{cA} + \mathbf{dA}$$

Example 10.4: The transpose of a matrix, A^T of a $m \times n$ matrix yields a $n \times m$ matrix.

$$\textbf{If } B = \begin{bmatrix} 1 \\ 2 \\ 3 \end{bmatrix} \textbf{ then } B^T = \begin{bmatrix} 1 & 2 & 3 \end{bmatrix}$$

Matrix multiplication is a bit trickier. If **A** is $m \times n$ then to perform matrix multiplication of A * B then B must be a $n \times p$ matrix.

Example 10.5: Given $A = \begin{bmatrix} 1 & -4 \\ 5 & 6 \end{bmatrix}, B = \begin{bmatrix} -6 & 7 \\ 8 & 1 \end{bmatrix}, C = \begin{bmatrix} 1 & 2 & 3 \\ 4 & 5 & 6 \end{bmatrix}$

Then

$$\mathbf{AB} = \begin{bmatrix} 1 & -4 \\ 5 & 6 \end{bmatrix}\begin{bmatrix} -6 & 7 \\ 8 & 1 \end{bmatrix} = \begin{bmatrix} (1*-6)+(-4*8) & 1*7+(-4*1) \\ 5*-6+6*8 & 5*7+6*1 \end{bmatrix} = \begin{bmatrix} -38 & 3 \\ 18 & 41 \end{bmatrix}$$

$$AC = \begin{bmatrix} -15 & -18 & -21 \\ 29 & 40 & 51 \end{bmatrix}$$

We note that in multiplication, in general, **AB ≠ BA**

We denote the matrix **I** as a square matrix ($m \times m$) of order m whose main diagonal is all **1.**

$$\mathbf{AI} = \mathbf{A} = \mathbf{IA}$$

If **A** is a square matrix, then the inverse of **A** is a square matrix **B** such that **AB=I=BA.** We will also refer to **B** as $\mathbf{A}^{-1}$. In order for $\mathbf{A}^{-1}$ to exist then the determinant of **A**, det(**A**) must be nonzero. If so, the matrix **A** is said to be nonsingular.If the determinant of **A** is zero, then the matrix **A** is singular.

There are many ways to compute the determinant, we assume the student has already been exposed to these methods.

Example 10.6: Given $\mathbf{A} = \begin{bmatrix} 1 & -4 \\ 5 & 6 \end{bmatrix}$**, compute** $\mathbf{A}^{-1}$**.**

First we find the determinant: $|\mathbf{A}| = (1)(6) - (-4)(5) = 26$

$$A^{-1} = \begin{bmatrix} 3/13 & 2/13 \\ -5/26 & 1/26 \end{bmatrix}$$

Example 10.7: Show $AA^{-1} = I$**.**

$$AA^{-1} = \begin{bmatrix} 1\left(\frac{3}{13}\right)+(-4)\left(-\frac{5}{26}\right) & 1\left(\frac{2}{13}\right)+(-4\left(\frac{1}{26}\right) \\ 5\left(\frac{3}{13}\right)+6\left(-\frac{5}{26}\right) & 5\left(\frac{2}{13}\right)+6\left(\frac{1}{26}\right) \end{bmatrix} = \begin{bmatrix} 1 & 0 \\ 0 & 1 \end{bmatrix}$$

If **A** is a matrix valued function, then we might differentiate the matrix-valued function one element at a time.

Example 10.8: Given $x(t) = \begin{bmatrix} t^2 \\ t+3 \\ e^{-2t} \end{bmatrix}$ **and** $A(t) = \begin{bmatrix} \cos(t) & t \\ 2t & \sin(t) \end{bmatrix}$

Then

$$x'^{(t)} = \begin{bmatrix} 2t^2 \\ 1 \\ -2e^{-2t} \end{bmatrix} \text{ and } A'(t) = \begin{bmatrix} -\sin(t) & 1 \\ 2 & \cos(\mathrm{t}) \end{bmatrix}$$

Eigenvalues and Eigenvector

Eigenvalues are found by solving the det|(**A** – λ**I**)| = 0. The eigenvalues will be the coefficient of the exponential in $x(t) = c_1\, e^{(\lambda 1t)} + c_2 e^{\lambda 2t}$. The corresponding eigenvectors, V_1 and V_2 become part of the matrix solution

$$\mathbf{x}(\mathbf{t}) = c_1 V_1 e^{(\lambda 1t)+} c_2 V_2 e^{(\lambda 2t)}$$

Three Tank Brine system

Let's revisit our simple model for three stage system as shown in Figure 10.10.:

$$x_1'(t) = -k_1 x_1$$
$$x_2'(t) = k_1 x_1 - k_2 x_2$$
$$x3'(t) = k_2 x_2 - k_3 x_3$$

where $k_i = r/V_i$ *for* $i = 1,2\ 3$.

For this model $V_1 = 20$, $V_2 = 40$, $V_3 = 50$ and r = 4 gal/min.
The initial conditions are $x_1(0) = 15$, $x_2(0) = 0$, $x_3(0) = 0$
Our system of first ODE is

$$x_1'(t) = -k_1 x_1 = -4/20\,x_1$$
$$x_2'(t) = k_1 x_1 - k_2 x_2 = 4/20\,x_1 - 4/40\,x_2$$
$$x3'(t) = k_2 x_2 - k_3 x_3 = 4/40\,x_2 - 4/50\,x_3$$

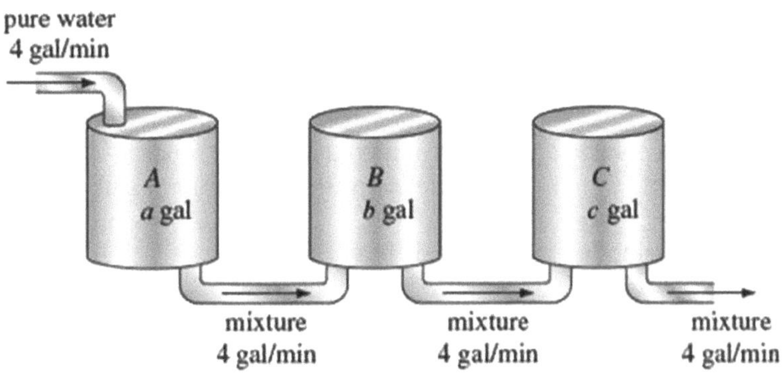

FIGURE 10.10
Three-stage system.

We will put this system in matrix form as-

$$x'(t)=\begin{bmatrix} -0.2 & 0 & 0 \\ 0.2 & -0.1 & 0 \\ 0 & 0.1 & 0.08 \end{bmatrix} x, x(0)=\begin{bmatrix} 15 \\ 0 \\ 0 \end{bmatrix}$$

We require the characteristic equation found by solving the determinant of $A - \lambda I = 0$. The characteristic equation is $(-0.2 - \lambda)(-0.1 - \lambda)(-0.08 - \lambda) = 0$ so our eigenvalues are

$$\lambda_1 = -0.2, \lambda_2 = -0.1, \lambda_3 = -0.08$$

Next, we need the corresponding eigenvectors.

$$[4.2, 2.8, 0]^T, [0, 1.8, 1]^T, [0, 0, 1]^T$$

$$x_1(t) = 4.2c_1e^{(-.2t)}$$

$$x_2(t) = 2.8c_1e^{(-.2t)} + 1.8c_2e^{(-.1t)}$$

$$x_3(t) = c_2e^{(-.1t)} + c3e^{(-0.08*t)}$$

We use the initial conditions to solve for c_1, c_2 and c_3.

$$\text{solve}(\{4.8*\text{c1} = 15, 2.8*\text{c1} + 1.8*\text{c2} = 0, \text{c2} + \text{c3} = 0\}, \{\text{c1}, \text{c2}, \text{c3}\});$$

$$\{\text{c1} = 3.125000000, \text{c2} = -4.861111111, \text{c3} = 4.861111111\}$$

$$x_1(t) = 13.125e^{(-.2t)}$$

$$x_2(t) = -8.75e^{(-.2t)} + 8.74999e^{(-.1t)}$$

$$x_3(t) = 8.74999e^{(-.1t)} - 4.861111e^{(-0.08*t)}$$

As can be seen in Figure 10.11, all the salt quantities go to zero over time.

Systems and Diffusion through a Membrane

Suppose the two salt concentrations of equal volume, *V* are separated by a membrane of permeability *P*. Assume the *P* + *V*, and determine the amount of salt in each concentration at time t if $x_1(0) = 2$, and $x_2(0) = 10$.

We know that diffusion is a system of differential equations

$$x_1'(t) = P/V_1(x_2 - x_1)$$

$$x_2'(t) = P/V_2(x_1 - x_2)$$

with initial amounts $x_1(0) = x_{10}$, and $x_2(0) = x_{20}$.

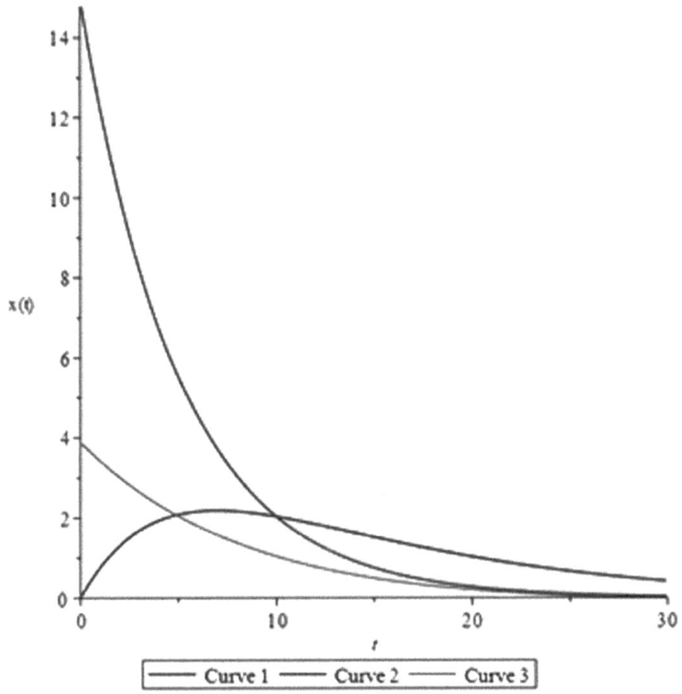

FIGURE 10.11
Solution for the three-tank system.

In the case above,

$$x_1'(t) = (x_2 - x_1)$$
$$x_2'(t) = (x_1 - x_2)$$
$$x_1(0) = 2,\ x_2(0) = 10$$

We can find the solution to the salt concentration problem using Maple (Figures 10.12 and 10.13).

Let's now model through a double wall membrane. Consider a situation where the two salt concentrations are separated by a double-walled membrane where the inner wall has permeability P1 and the outer wall has permeability P2. Further, assume the $0 < P1 < P2$. Suppose the volume of the inner wall is V1 and between the two walls is V2. Let x represent the concentration of the solution within the inner wall and y, the concentration between the two walls. Assume the concentration outside the two walls is a constant, C. The following system of differential equations models this:

$$x'(t) = P1/V1(y - x)$$
$$y'(t) = 1/v2(P2(C - y) + P1(x - y))$$
$$x_1(0) = x_o, \text{ and } y(0) = y_0.$$

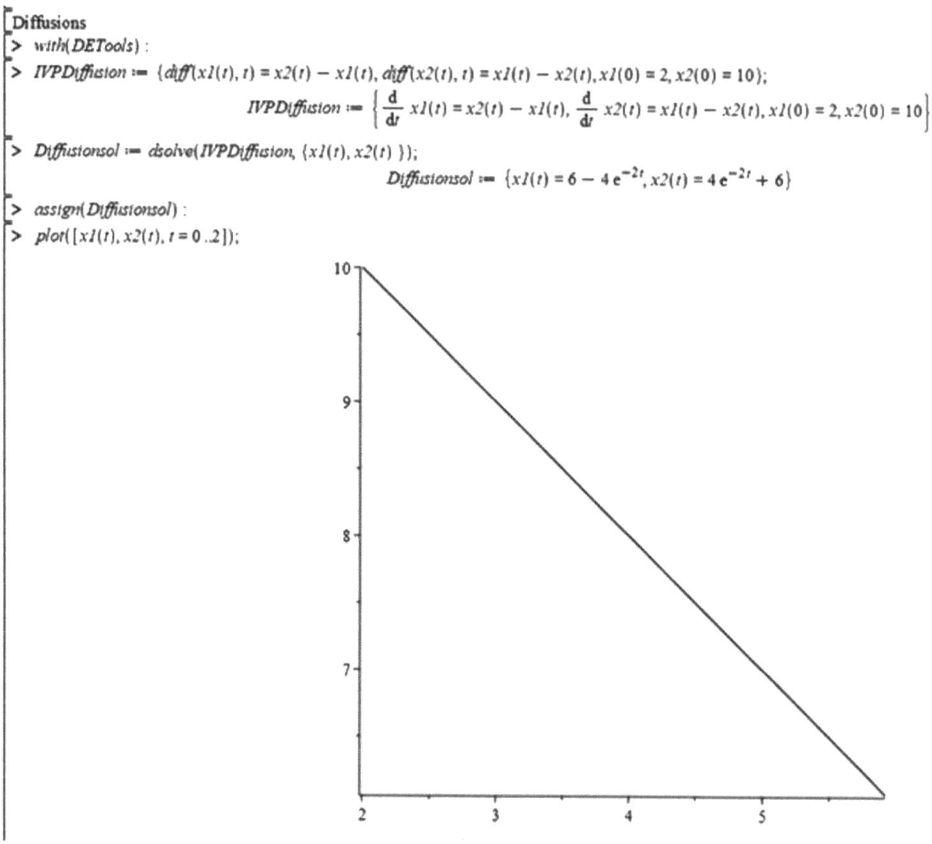

FIGURE 10.12
Diffusion through a membrane.

Further assume that are measurements yield $P1 = 3$, $P2 = 8$, $V1 = 2$, and $V2 = 10$. Our experts think the constant C = 10 and the initial conditions are $x_1(0) = 2$, and $y(0) = 1$.

$$x'(t) = 3/2\,(y - x)$$
$$y'(t) = 1/10\big(8(10 - y) + 3(x - y)\big) = -11y/10 + 3x/10 + 8$$
$$x_1(0) = 2, \text{ and } y(0) = 1.$$

The solution is (Figure 10.14).

$$x(t) = 4.5e^{-2t} - 12.5e^{-3t/5} + 10$$
$$y(t) = -1.5e^{-2t} - 7.5e^{-3t/5} + 10$$

We can now utilize what we know about matrix algebra and Maple to find the solution to this problem. The Maple code is included in Figures 10.15–10.17.

```
> plot({x1(t), x2(t)}, t = 0..2);
```

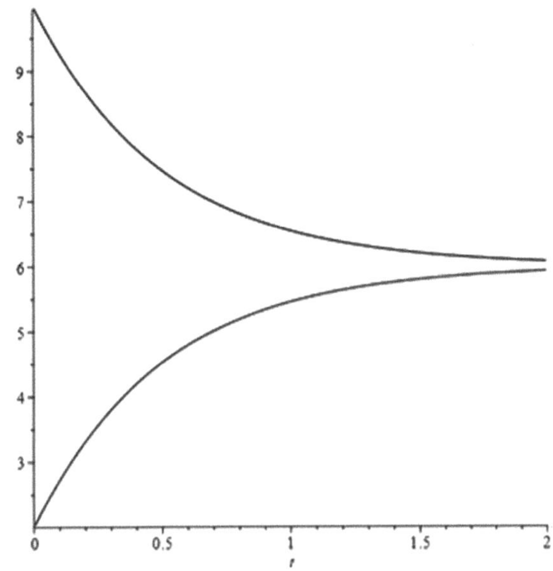

FIGURE 10.13
Double membrane diffusion.

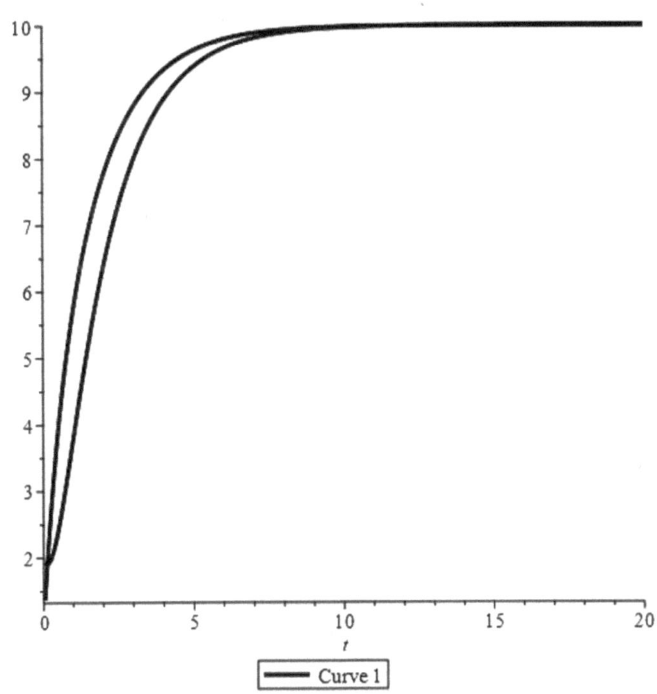

FIGURE 10.14
Systems through double membrane.

```
dx/dt=1.5(y-x), x(0)=2
dy/dt= .8 (10-y)+.3(x-y) = .3 x -1.1y+8, y(0)=1
> with(linalg):
> M:=array([[-1.5,1.5],[.3,-1.1]]);
> lambda:=eigenvects(M);
```

$$M := \begin{bmatrix} -1.5 & 1.5 \\ 0.3 & -1.1 \end{bmatrix}$$

$$\lambda := \left[-0.6000000000, 1, \left\{\left[\, -1.129384879 \quad -0.6776309273 \,\right]\right\}\right], \left[-2.000000000, 1, \left\{\left[\, -0.9486832981 \quad 0.3162277660 \,\right]\right.\right.$$

```
> homsol:=matrixDE(M,t);
```

$$homsol := \left[\begin{bmatrix} 1.\,e^{-2t} & 1.\,e^{-\frac{3t}{5}} \\ -0.333333333\,e^{-2t} & 0.6000000000\,e^{-\frac{3t}{5}} \end{bmatrix}, \begin{bmatrix} 0. & 0. \end{bmatrix}\right]$$

```
> phi:=homsol[1];
```

$$\phi := \begin{bmatrix} 1.\,e^{-2t} & 1.\,e^{-\frac{3t}{5}} \\ -0.333333333\,e^{-2t} & 0.6000000000\,e^{-\frac{3t}{5}} \end{bmatrix}$$

FIGURE 10.15
Building the Maple code for the double membrane problem.

```
> phi_inv:=inverse(phi);
```

$$phi\_inv := \begin{bmatrix} \dfrac{0.6428571431}{e^{-2t}} & -\dfrac{1.071428572}{e^{-2t}} \\ \dfrac{0.3571428569}{e^{-\frac{3t}{5}}} & \dfrac{1.071428572}{e^{-\frac{3t}{5}}} \end{bmatrix}$$

```
> f:=matrix(2,1,[0,8]);
```

$$f := \begin{bmatrix} 0 \\ 8 \end{bmatrix}$$

```
> p1:=multiply(phi_inv,f);
```

$$p1 := \begin{bmatrix} -\dfrac{8.571428576}{e^{-2t}} \\ \dfrac{8.571428576}{e^{-\frac{3t}{5}}} \end{bmatrix}$$

```
> p1_int:=map(int,p1,t);
```

$$p1\_int := \begin{bmatrix} -\dfrac{4.285714288}{e^{-2.t}} \\ \dfrac{14.28571429}{e^{-0.6000000000t}} \end{bmatrix}$$

```
> p2:=multiply(phi,p1_int);
```

$$p2 := \begin{bmatrix} -\dfrac{4.285714288\,e^{-2t}}{e^{-2.t}} + \dfrac{14.28571429\,e^{-\frac{3t}{5}}}{e^{-0.6000000000t}} \\ \dfrac{1.428571428\,e^{-2t}}{e^{-2.t}} + \dfrac{8.571428574\,e^{-\frac{3t}{5}}}{e^{-0.6000000000t}} \end{bmatrix}$$

```
> xp:=simplify(%);
```

$$xp := \begin{bmatrix} 10.00000000 \\ 10.00000000 \end{bmatrix}$$

FIGURE 10.16
Maple code continues for the double membrane problem.

```
> x:=evalm(multiply(phi, matrix(2,1,[c1,c2]))+xp);
```

$$x := \begin{bmatrix} 1.\,e^{-2t} c1 + 1.\,e^{-\frac{3t}{5}} c2 + 10.00000000 \\ -0.333333333\, e^{-2t} c1 + 0.6000000000\, e^{-\frac{3t}{5}} c2 + 10.00000000 \end{bmatrix}$$

Solve for c1, c2

```
> solve({2=c1+c2+10,10-(1/3)*c1+.6*c2=1},{c1,c2});
```

$$\{c1 = 4.500000000, c2 = -12.50000000\}$$

```
> x1:=subs({c1 = 4.500000000, c2 = -12.50000000},x[1,1]);
```

$$x1 := 4.500000000\, e^{-2t} - 12.50000000\, e^{-\frac{3t}{5}} + 10.00000000$$

```
> x2:=subs({c1 = 4.500000000, c2 = -12.50000000},x[2,1]);
```

$$x2 := -1.499999998\, e^{-2t} - 7.500000000\, e^{-\frac{3t}{5}} + 10.00000000$$

FIGURE 10.17
Maple solution for the concentration levels in the double membrane problem.

Fick's law can be written as an equation that can be used to calculate the rate of diffusion. Rate = P × A × ((C1 − C2)T), where P = A permeability constant that is a quantitative measure of the rate at which a particular molecule can cross a particular membrane A = surface area, C1 − C2 = the difference in concentration between two areas.

What are Fick's Laws of Diffusion

Fick's laws of diffusion are mathematical statements describing how particles under random thermal motion tend to spread from a region of higher concentration to a region of lower concentration to equalize concentration on both the regions. The laws also describe the relationship between the rate of diffusion and the three factors that affect diffusion.

They state that "the rate of diffusion is directly proportional to both the surface area and concentration difference and is inversely proportional to the thickness of the membrane."

A diffusion process that obeys Fick's laws is called normal or Fickian diffusion; otherwise, it is called anomalous diffusion or non-Fickian diffusion.

Thus, Fick's laws can be mathematically written as:

Rate of Diffusion ∝ Surface area × Concentration gradient/Thickness of membrane

Example 10.9

A sample of alveolar epithelium tissue from a mammal is 1.5 m thick and has a surface area of 3 m^2. The concentration of oxygen in the alveolus is 1.8×10^{-16} mol m^{-3} and the concentration of oxygen in the blood is 7.5×10^{-17} mol m^{-3}. The permeability constant for oxygen across the alveolar epithelium is 0.012 molecule s^{-1}.

Calculate the rate of diffusion across the section of the alveolar epithelium.

Step 1: Substitute numbers into the equation

$$\text{Rate} = P \times A \times \left(\left(C_1 - C_2\right) T\right)$$
$$\text{Rate} = 0.012 \times 3 \times \left(\left(1.8 \times 10^{-16} - 7.5 \times 10^{-17}\right) 1.5\right)$$

Step 2: Complete the calculation

$$\text{Rate} = 0.012 \times 3 \times \left(1.05 \times 10^{-16}\ 1.5\right)$$
$$\text{Rate} = 0.012 \times 3 \times 7 \times 10^{-17}$$
$$\text{Rate} = 2.52 \times 10^{-18}\ \text{molecules m}^{-2}\ \text{s}^{-1}$$

Exercises

Find and classify all the rest points. Then sketch a few trajectories to indicate the motion.

10.1 $dx/dt = x$, $dy/dt = y$
10.2 $dx/dt = -x$, $dy/dt = 2y$
10.3 $dx/dt = y$, $dy/dt = -2x$
10.4 $dx/dt = -x + 1$, $dy/dt = -2y$

PROJECTS

10.1 Given the following system of Linear first ODEs of species cooperation (symbiosis):

$$dx_1/dt = -0.5x_1 + x_2$$
$$dx_2/dt = 0.25x_1 - 0.5x_2$$

and $x_1(0) = 200$ and $x_2(0) = 500$.

a. Perform Euler's method with step-size $h = 0.1$ to obtain graphs of numerical solutions for $x_1(t)$ and $x_2(t)$ versus t and for x_1 versus x_2. You can put both $x_1(t)$ and $x_2(t)$ versus t on one axis if you want.
b. From the graphs discuss the long-term behavior of the system (discuss stability).
c. Analytically using Eigenvalues and Eigenvectors solve the system of DEs to determine the population of each species for $t > 0$.
d. Determine if there is a steady state solution for this system.
e. Obtain real plots of $x_1(t)$ and $x_2(t)$ versus t and for $x_1(t)$ versus $x_2(t)$. Compare to the numerical plots. Briefly discuss.

10.2 Given a competitive hunter model defined by the system:

$$dx/dt = 15x - x^2 - 2xy = x\left(15 - x - 2y\right)$$
$$dy/dt = 12y - y^2 - 1.5xy = y\left(12 - y - 1.5x\right)$$

a. Perform a graphical analysis of this competitive hunter model in the *x-y* plane.
b. Identify all equilibrium points and classify their stability.
c. Find the numerical solutions using Euler's method with step size $h = 0.05$. Try it from two separate initial conditions: first, use $x(0) = 5$ and $y(0) = 4$, then use $x(0) = 3$, $y(0) = 9$. Obtain graphs of $x(t)$, $y(t)$ individually (or on the same axis) and then a plot of x versus y using your numerical approximations. Compare to your phase portrait analysis.

10.3 Since bass and trout both live in the same lake and eat the same food sources, they are competing for survival. The rates of growth for bass (dB/dt) and for trout (dT/dt) are estimated by the following equations:

$$dB/dt = (10 - B - T)B$$
$$dT/dt = (15 - B - 3T)T$$

Coefficients and values are in thousands.

a. Obtain a "qualitative" graphical solution of this system. Find all equilibrium points of the system and classify each as unstable, stable, or asymptotically stable.
b. If the initial conditions are $B(0) = 5$ and $T(0) = 2$, determine the long-term behavior of the system from your graph in part (a). Sketch it out.
c. Using Euler's method, h = 0.1 and the same initial conditions as above, obtain estimates for *B* and *T*. Using these estimates determine a more accurate graph by plotting *B* versus *T* for the solution from t = 0 to t = 7.
d. Euler's method: $x_{n+1} = x_n + h\, f(x_n, y_n)$ & $y_{n+1} = y_n + h\, g(x_n, y_n)$.
e. Compare the graph in part (c) to the possible solutions found in (a) and (b). Briefly comment.

10.4 Three Brine system

We looked at the simple example of three stage system as shown in Figure 10.18. Each tanks hold their respective volumes V_1, V_2, and V_3 gallons of brine, respectively. Fresh water flows into tank A and is mixed with salt, and flows in tank B and then from tank B into tank C. Let xi(t) be the amount of salt (in pounds) at time *t* for *i* = *1*, *2* and *3*. If each flow rate is r gallons per minute. In this case, increase the amounts to 6 gal/min form the 4 gal/min.

a. Model this situation
b. Solve the model
c. graph the solution, and describe the long term behavior.

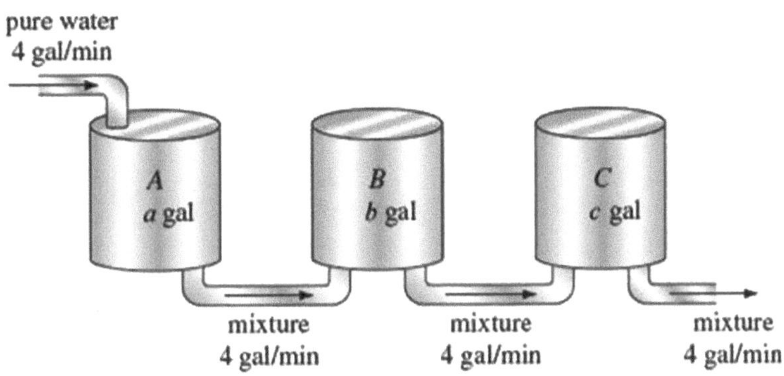

FIGURE 10.18
Three-stage system.

References and Additional Readings

Edwards, C., and Penney, D. (2008). *Linear Algebra and Differential Equations*. 1st edition, Pearson. ISBN: 978-0136054252.

Goode, S., and Annin, S. (2017). *Differential Equations and Linear Algebra*. 4th edition, Pearson.

Taylor, J. (1983). A Lanchester-type aggregated-force model of conventional ground combat. *Naval Research Logistics Quarterly*, 30(1), 113–131.

11

Numerical Solutions to Systems of ODEs

Consider a friend who recently retired and bought farmland in South Carolina. He desires to have a fishing pond and his favorite fish to catch are bass and trout. He finds he has a fair-sized freshwater pond on his land but it contains no fish. He takes a water sample to the local fish and game authority and they analyze the water and conclude that the water can sustain a fish population. He visits the local fish hatchery where they provide him the growth rates of bass and trout in isolation, call these values r and s, respectively. The experts tell him that bass and trout have the same food sources in the water and will compete for the oxygen in the water as well as food for survival. The experts estimate the interaction rates between the bass and trout for survival and call these rates m and n, respectively. We desire to build a mathematical model to help our friend determine if the pond can sustain both species of fish. This leads to a competitive hunter system of differential equations that we will examine later. This will be a nonlinear system of ODEs. This type of ODE is not easily solved analytically nor may not be solvable analytically so we turn to numerical methods.

Competition between Species

Imagine a small to medium fish pond supporting both trout and bass. Let T(t) denote the population of trout at time t and B(t) denote the population of bass at time t. We want to know if both can coexist in the pond. Although population growth depends on many factors we will limit ourselves to basic isolated growth and the interaction with the other competing species for the scarce life-support resources.

We assume that the species grows in isolation. The levels of the population of the trout and the bass, B(t) and T(t), depend on many variables such as their initial numbers, the amount of competition, the existence of predators, their individual species' birth and death rates, and so forth. In isolation, we assume the following proportionality models (following the same arguments as the basic populations models that we have discussed before) to be true where the environment can support an unlimited number of trout and/or bass. Later, we might refine this model to incorporate the limited growth assumptions of the logistics model:

$$\frac{dB}{dt} = mB$$

$$\frac{dT}{dt} = aT$$

DOI: 10.1201/9781003582274-11

Next, we modify the proceeding differential equations to take into account the competition of the trout and bass for living space, oxygen, and food supply. The effect is that the interaction decreases the growth of the species. The interaction terms for competition led to a decay rate that we call n for bass and b for trout. This leads to the following simplified model:

$$\frac{dB}{dt} = mB - nBT$$

$$\frac{dT}{dt} = aT - bBT$$

If we have the initial stockage level, B_0 and T_0, we determine how the species coexist over time.

If the model is not reasonable we might try logistic growth instead of isolated growth. Logistic growth in isolation was discussed in first ODE models as a refinement.

$B(t)$ = number of Bass fish after time t

$T(t)$ = number of Trout after t time

Rate of Change of growth = rate in isolation + rate in competition for resources

$$dB/dt = 0.7B - 0.02B * T$$

$$dT/dt = 0.5 - 0.01B * T$$

Solve dB/dt = 0 and dT/dt = 0 to find the equilibrium points:

$$dB/dt = 0 = B(0.7 - 0.02T) = 0 \text{ so } B = 0 \text{ or } T = 35$$

$$dT/dt = 0 = T(0.5 - 0.01B) = 0 \text{ so } T = 0 \text{ or } B = 50$$

The equilibrium values are (0, 0) and (50, 35).

```
> with(plots):with(DEtools):
> eqn1:=diff(B(t),t)=.7*B(t)-.02*B(t)*T(t);
```

$$\text{eqn1} := \frac{d}{dt}B(t) = 0.7B(t) - 0.2B(t)T(t)$$

```
> eqn2:=diff(T(t),t)=.5*T(t)-.01*B(t)*T(t);
```

$$\text{eqn2} := \frac{d}{dt}T(t) = 0.5T(t) - 0.1B(t)T(t)$$

```
> DEplot([eqn1,eqn2], [B(t),T(t)], t=0..20, B=0..75,T=0..75);
```

As shown in Figure 11.1, we see both equilibrium values are not stable. Depending on the starting value one species will dominate over time.

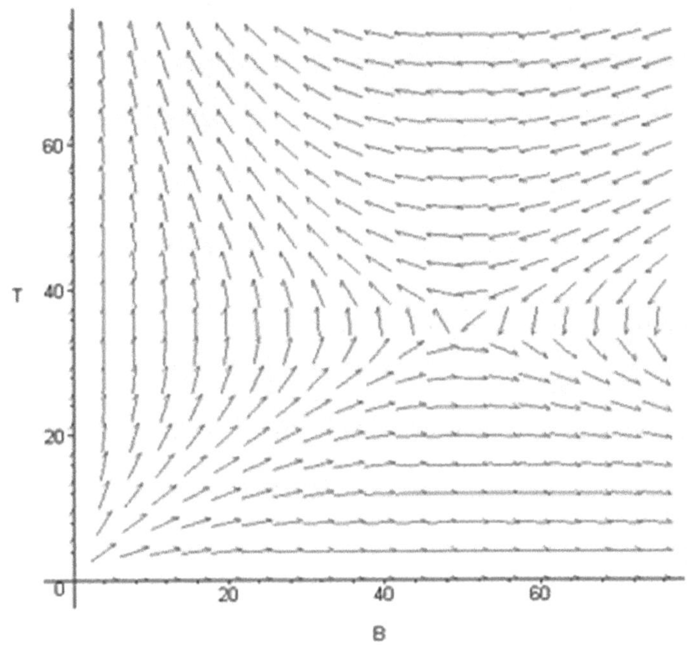

FIGURE 11.1
Phase portrait of competitive hunter model.

Thus, the phase portrait is useful to give us a sense of the possible solutions.

In Chapter 6, we discussed the use of numerical solutions (Euler, Improved Euler, and Runge-Kutta methods) to first ODEs. In this chapter, we extend the use of numerical solutions to systems of differential equations. We show only Euler's and Runge-Kutta methods. Our goal here is to provide a solution method for many models of systems of ODEs that do not have closed form analytical solutions.

Throughout most of the chapter, we have investigated and modeled autonomous systems of first ODEs. A more general form of systems of two ordinary first ODEs is given by Equation 11.1:

$$\frac{dx}{dt} = f(t,x,y)$$
$$\frac{dy}{dt} = g(t,x,y) \tag{11.1}$$

If the variable t appears explicitly in one of the functions f or g the system is not autonomous. In this section, we present numerical techniques for approximating solutions for x(t) and y(t) subject to initial conditions $x(t_0) = x_0$ and $y(t_0) = y_0$.

We will give the algorithm for each and show the Maple commands to execute a numerical solution. We also show how to obtain both the phase portraits and the plots of approximate numerical solutions.

Euler's Method with Systems

Consider the iterative formula for Euler's method for systems as

$$x(n)=x(n-1)+f(t(n-1),x(n-1),y(n-1))\Delta t$$
$$y(n)=y(n-1)+g(t(n-1),x(n-1),y(n-1))\Delta t$$

We illustrate a few iterations for the following initial value problem with a step size of $\Delta t = 0.1$:

$$x' = 3x-2y, x(0)=3$$
$$y' = 5x-4y, y(0)=6$$
$$x(0)=3, y(0)=6 \textit{ Given}$$
$$x(1)=3+(0.1)\cdot(3\cdot 3-2\cdot 6)=2.7$$
$$y(1)=6+(0.1)\cdot(5\cdot 3-4\cdot 6)=5.1$$

and

$$x(2)=2.7+(0.1)\cdot(3\cdot(2.7)-2\cdot(5.1))=2.49$$
$$y(2)=5.1+(0.1)\cdot(5\cdot(2.7)-4\cdot(5.1))=4.41$$

and so forth.

In Maple, we enter the system and initial conditions and then use the dsolve with classical numerical methods. Here is the command sequence to obtain the Euler estimates for our example (Figure 11.2).

```
>ode1:=diff(x(t),t)=3x(t)-2y(t);
```

$$ode1 := \frac{d}{dt}x(t)=3x(t)-2y(t)$$

```
>ode2:=diff(y(t),t)=5x(t)-4y(t);
```

$$ode2 := \frac{d}{dt}y(t)=5x(t)-4y(t)$$

```
>inits:=x(0)=3,y(0)=6;
```

```
inits:=x(0)=3,y(0)=6
```

eulersol := [[t x(t) y(t)], [

0.	3.	6.
0.1	2.70000000000000016	5.09999999999999964
0.2	2.49000000000000022	4.41000000000000014
0.3	2.35500000000000042	3.89100000000000044
0.4	2.28330000000000056	3.51210000000000022
0.5	2.26587000000000050	3.24891000000000042
0.6	2.29584900000000046	3.08228100000000050
0.7	2.36814750000000052	2.99729310000000070
0.8	2.47913313000000058	2.98244961000000064
0.9	2.62638314700000075	3.02903633100000080
1.	2.80849082490000113	3.13061337210000090
1.1	3.02491539795000097	3.28261343571000142
1.2	3.27586733019300080	3.48202576040100098
1.3	3.56222237717070068	3.72714912133710108
1.4	3.88545926605448954	4.01740066138760987
1.5	4.24761691359331550	4.35317002985981106
1.6	4.65126798169934742	4.73571047471254403
1.7	5.09950628126664274	5.16706027567719950
1.8	5.59594611051119450	5.64998930603964044
1.9	6.14473208245662317	6.18796663887938080
2.	6.75055837941773440	6.78514602455594052
2.1	7.41869668833186857	7.44636680444243292
2.2	8.15503233394294114	8.17716842683139332

]

FIGURE 11.2
Maple solution for system of equations.

```
>eulersol:=dsolve({ode1,ode2,inits}, numeric, method=classical[foeuler],
  output=array)([0,.1,.2,.3,.4,.5,.6,.7,.8,.9,1,1.1,1.2,1.3,1.4,1.5,.6,1.7,1.8,
  1.9,2,2.1,2.2]), stepwise = 0.1);
```

The power of Euler's method is twofold. First, it is easy to use and second, as a numerical method it can be used to estimate a solution to a system of differential equations that does not have a closed form solution.

Predator-Prey Model 1

Assume we have the following predator-prey system that does not have a closed form analytical solution:

$$\frac{dx}{dt} = 3x - xy$$

$$\frac{dy}{dt} = xy - 2y$$

$$x(0) = 1,\ y(0) = 2$$

$$t_0 = 1, \Delta t = .1$$

We will obtain an estimate of the solution using Euler's method (Figure 11.3).

eulersol := [[t $x(t)$ $y(t)$]], [

t	$x(t)$	$y(t)$
0.	3.	6.
0.1	2.70000000000000016	5.09999999999999964
0.2	2.49000000000000022	4.41000000000000014
0.3	2.35500000000000042	3.89100000000000044
0.4	2.28330000000000056	3.51210000000000022
0.5	2.26587000000000050	3.24891000000000042
0.6	2.29584900000000046	3.08228100000000050
0.7	2.36814750000000052	2.99729310000000070
0.8	2.47913313000000058	2.98244961000000064
0.9	2.62638314700000075	3.02903633100000080
1.	2.80849082490000113	3.13061337210000090
1.1	3.02491539795000097	3.28261343571000142
1.2	3.27586733019300080	3.48202576040100098
1.3	3.56222237717070068	3.72714912133710108
1.4	3.88545926605448954	4.01740066138760987
1.5	4.24761691359331550	4.35317002985981106
1.6	4.65126798169934742	4.73571047471254403
1.7	5.09950628126664274	5.16706027567719950
1.8	5.59594611051119450	5.64998930603964044
1.9	6.14473208245662317	6.18796663887938080
2.	6.75055837941773440	6.78514602455594052
2.1	7.41869668833186857	7.44636680444243292
2.2	8.15503233394294114	8.17716842683139332

]

FIGURE 11.3
Euler solution for predator and prey model.

```
> with(linalg):with(DEtools):

> ode1:=diff(x(t),t)=3*x(t)-x(t)*y(t);
```

$$ode1 := \frac{d}{dt}x(t) = 3x(t) - x(t)y(t)$$

```
> ode2:=diff(y(t),t)=x(t)*y(t)-2*y(t);
```

$$ode2 := \frac{d}{dt}y(t) = x(t)y(t) - 2y(t)$$

```
> inits:=x(0)=1,y(0)=2;
```

$$inits := x(0) = 1,\ y(0) = 2$$

```
>
> eulersol:=dsolve({ode1,ode2,inits},numeric, method=classical
[foreuler], output=array([0,.1,.2,.3,.4,.5,.6,.7,.8,.9,1,1.1,
1.2,1.3,1.4,1.5,1.6,1.7,1.8,1.9,2,2.1,2.2]),stepsize=0.1);
```

We will use Maple to plot the solution to the predator and prey problem (Figure 11.4).

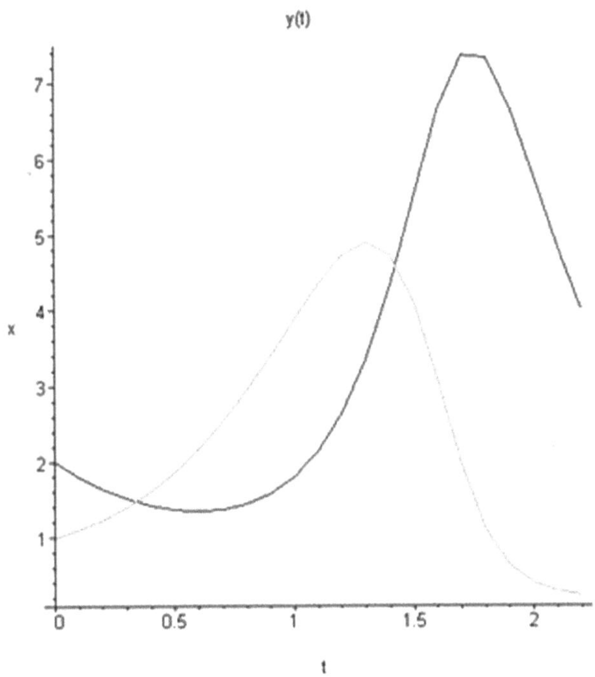

FIGURE 11.4
Predator-prey model.

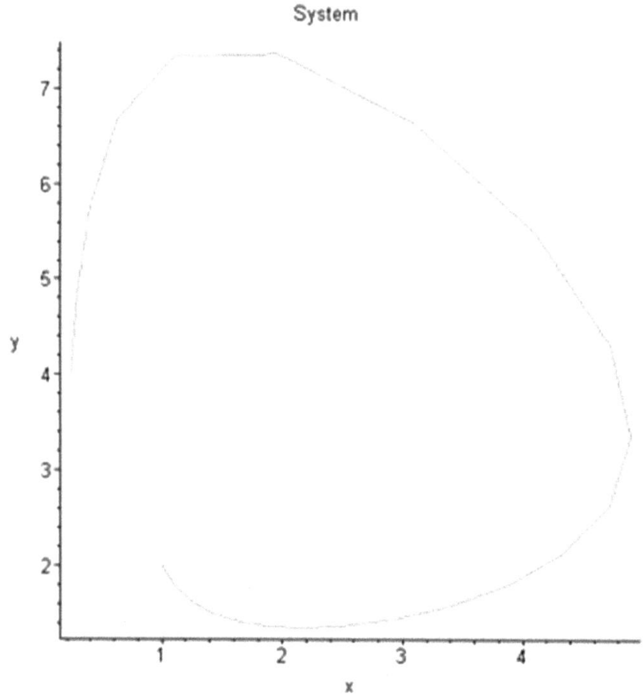

FIGURE 11.5
Closed loop system solution.

```
> with(plots):

> plot1:=odeplot(eulersol,[t,x(t)],0..10, color=green,title=`x(t)`):

> plot2:=odeplot(eulersol,[t,y(t)],0..10,color=blue,title=`y(t)`):

> display(plot1,plot2);
```

We experiment and find that when we plot x(t) versus y(t) we have approximately a closed loop (Figure 11.5).

```
> odeplot(eulersol,[x(t),y(t)],0..22, color=green,title=`System`);
```

Another method for numerical estimates is the Runge-Kutta applied to systems. We illustrate with the same predator-prey example and develop a solution (Figure 11.6) and plot our solution (Figures 11.7 and 11.8).

```
> restart;
> with(linalg):with(DEtools):
```

Warning, the protected names norm and trace have been redefined and unprotected.

rk4sol :=

t	$x(t)$	$y(t)$
0.	1.	2.
0.1	1.11554071453956705	1.81968188493959970
0.2	1.26463746535620780	1.67761981669985926
0.3	1.45146389024689194	1.57278931221810914
0.4	1.68031037053259168	1.50542579227745321
0.5	1.95456986108324960	1.47762563817635396
0.6	2.27500034092209802	1.49412980191491540
0.7	2.63687047094120208	1.56336399876409616
0.8	3.02561346371586382	1.69866644464148696
0.9	3.41107833928229720	1.91916110241906601
1.	3.74212348522216676	2.24852435676575446
1.1	3.94706090545572064	2.70772001337807611
1.2	3.95001172191016892	3.29658238745704324
1.3	3.70934802583802936	3.96638771088000476
1.4	3.25874674267819132	4.60727591934640213
1.5	2.70448435236999174	5.08419319694337712
1.6	2.16641586882149006	5.30768472071020270
1.7	1.71962027608752543	5.27274016506954890
1.8	1.38441109626295588	5.03717315896791806
1.9	1.14891557150367586	4.67751425643222784
2.	0.991726731312949417	4.25984946993614156
2.1	0.893335577155144112	3.83076393278872685
2.2	0.839403883501402824	3.41909584046305914

FIGURE 11.6
RK4 solution for predator–prey model.

Warning, the previous binding of the name adjoint has been removed and it now has an assigned value (Figure 11.6).

```
> ode1:=diff(x(t),t)=3*x(t)-x(t)*y(t);
```

$$ode1 := \frac{d}{dt}x(t) = 3x(t) - x(t)y(t)$$

$$ode2 := \frac{d}{dt}y(t) = x(t)y(t) - 2y(t)$$

```
> inits:=x(0)=1,y(0)=2;
```

$$inits := x(0) = 1,\ y(0) = 2$$

```
>

> rk4sol:=dsolve({ode1,ode2,inits},numeric, method=classical[rk4], out
put=array([0,.1,.2,.3,.4,.5,.6,.7,.8,.9,1,1.1,1.2,1.3,1.4,1.5,1.6,1.7,
1.8,1.9,2,2.1,2.2]),stepsize=0.1);

> with(plots):
```

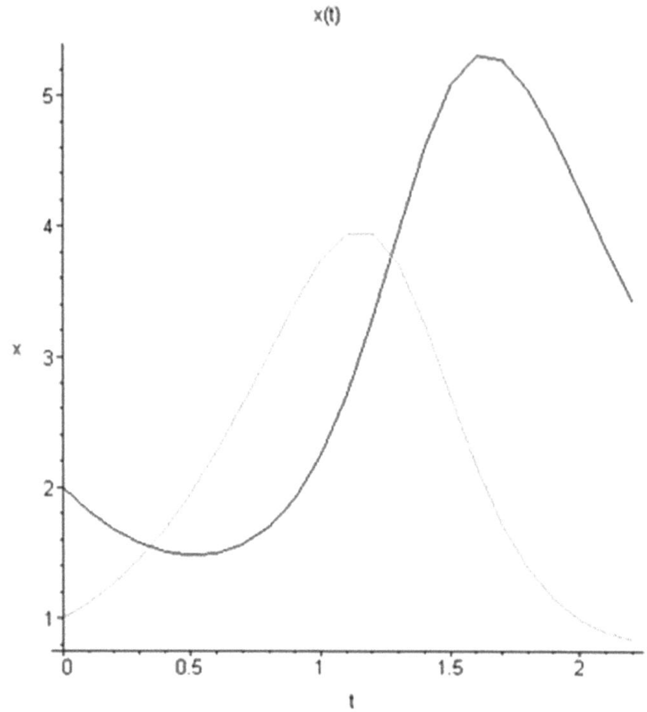

FIGURE 11.7
Runge-Kute solution plot for the predator and prey problem.

Warning, the name changecoords have been redefined (Figure 11.7)

```
> plot1:=odeplot(rk4sol, [t,x(t)],0..10, color=green,title=`x(t)`):

> plot2:=odeplot(rk4sol, [t,y(t)],0..10,color=blue,title=`y(t)`):

> display(plot1,plot2);
```

We can plot the solution (Figure 11.8) using:

```
> odeplot(rk4sol, [x(t),y(t)],0..22, color=green,title=`System`);
```

Foxes versus Rabbits as a Predator-Prey Revisited

We examined predator-prey models in Chapter 4 as a dynamical system. In this section, we revisit these as differential equations. We repeat our (admittedly simplistic) assumptions from Chapter 4:

- The predator species is totally dependent on the prey species as its only food supply.
- The prey species has an unlimited food supply and no threat to its growth other than the specific predator.

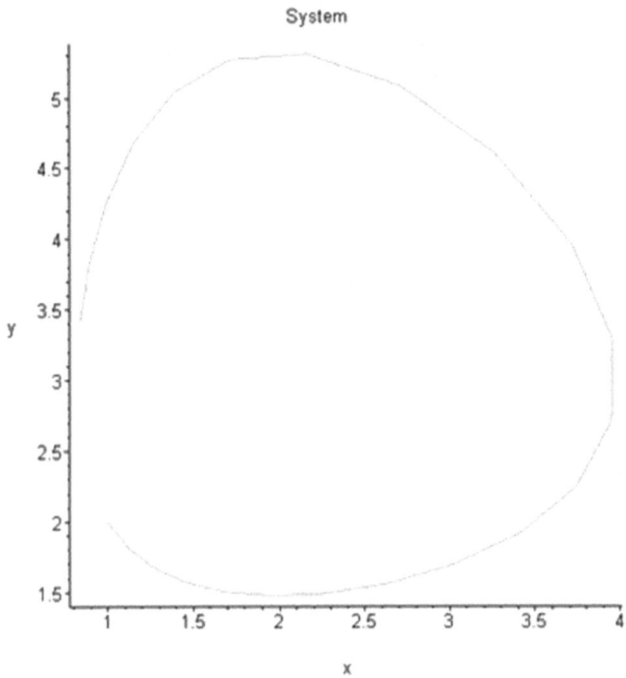

FIGURE 11.8
RF4 plot of solution.

If there were no predators, the second assumption would imply that the prey species grows exponentially, i.e., if $x = x(t)$ is the size of the prey population at time *t*, then we would have

$$\frac{dx}{dt} = ax.$$

This represents exponential growth when $a > 0$ and decay when $a < 0$.

But there *are* predators, which must account for a negative component in the prey growth rate.

Suppose we write $y = y(t)$ for the size of the predator population at time *t*. Here are the crucial assumptions for completing the model:

- The rate at which predators encounter prey is jointly proportional to the sizes of the two populations.
- A fixed proportion of encounters leads to the death of the prey.

These assumptions lead to the conclusion that the negative component of the prey growth rate is proportional to the product *xy* of the population sizes, i.e.,

$$\frac{dx}{dt} = ax - bxy.$$

Now we consider the predator population. If there were no food supply, the population would die out at a rate proportional to its size, i.e., we would find

$$\frac{dy}{dt} = -cy.$$

(Keep in mind that the "natural growth rate" is a composite of birth and death rates, both presumably proportional to population size. In the absence of food, there is no energy supply to support the birth rate.) But there is a food supply: the prey. And what's bad for hares is good for lynx. That is, the energy to support the growth of the predator population is proportional to the deaths of prey, so

$$\frac{dy}{dt} = -cy + pxy.$$

This discussion leads to the Lotka-Volterra predator-prey model:

$$\frac{dx}{dt} = ax - bxy$$
$$\frac{dy}{dt} = -cy + pxy$$

where a, b, c, and p are positive constants. We assume the *{a,b,c,p}={0.1, 0.005/60, 0.04, 0.00004}* (Figure 11.9)

```
with (DETools): with (plots)
Predator-Prey Model
```

$$>\texttt{hare1:=diff(h(t),t)=0.1h(t)-}\frac{0.005}{60}\texttt{h(t)f(t);}$$

$$\texttt{hare1:=diff(h(t),t)=0.1h(t)-}\frac{0.005}{60}\texttt{h(t)f(t)}$$

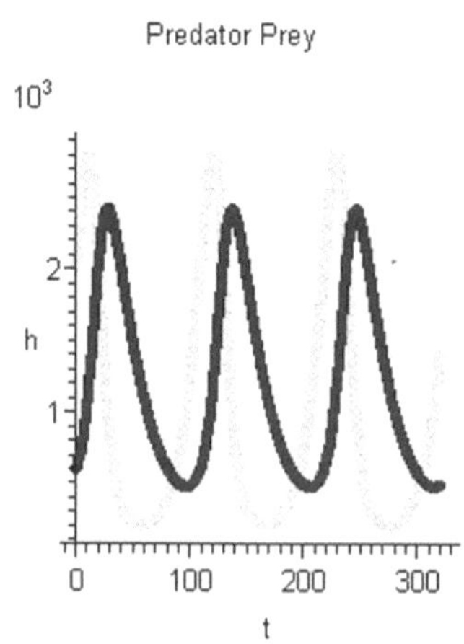

FIGURE 11.9
Predator-prey plot.

```
>fox1:=diff(f(t),t)=-0.04f(t)+0.00004h(t)f(t);
  fox1:=diff(f(t),t)=-0.04f(t)+0.00004h(t)f(t)
```

```
>Model:={hare1,fox1}
```

$$\text{Model} := \left\{ \frac{d}{dt} f(t) = -0.04 f(t) + 0.00004 h(t) f(t) \right\}$$

$$\frac{d}{dt} h(t) = 0.1 h(t) - 0.0000833 h(t) f(t)$$

```
>vars:={h(t),f(t)};
  vars:={h(t),f(t)}

>init1:=[h(0)=2000, f(o)=600];
  init1:=[h(0)=2000, f(o)=600]

>domain t=0...320;

>L:=DEplot(Model1, vars, domain, [init1], stepsixe=0.5, scene=[t,f],
 arrows=NONE, linecolor=blue)

>H:=DEplot(Model1, vars, domain, [init1], stepsixe=0.5, scene=[t,f],
 arrows=NONE);

>display({L, H}, title="Predator prey");

>
display ({L, H}, title = 'Predator Prey');

>init2:=[h(0)=2000,f(o)=1200];
  init2:=[h(0)=2000,f(o)=1200]

>init3:=[h(0)=2000,f(o)=3000];
  init3:=[h(0)=2000,f(o)=3000]

>DEplot(Model1, vars, domain, [init1, init2, init3], stepsixe=0.5,
 scene=[h,f], title='Predator Prey', arrows=NONE);
```

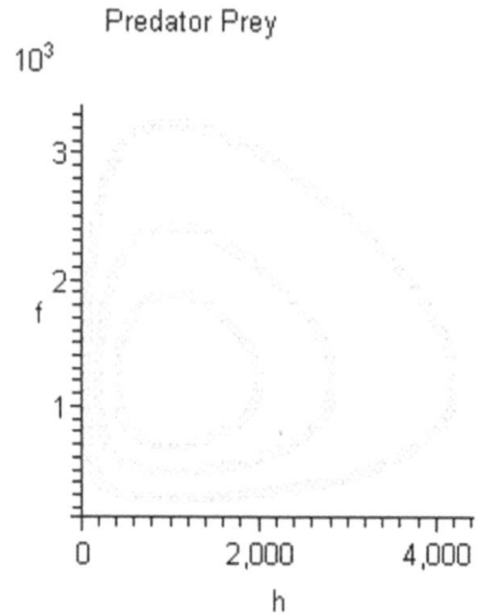

FIGURE 11.10
Predator-prey solution.

We can find the solution (Figure 11.10) using:

```
> equi:=solve({rhs(hare1),rhs(fox1)}, vars);

        equil:={f(t)=0, h(t)=0}, {h(t)=1000, f(t)=1200.}
```

Model Interpretation

We see that our predator-prey model is in equilibrium as we move around the equilibrium value (1000,1200). The point (0,0) is not stable. The ecological system appears stable and does need human intervention at this time.

We illustrate Euler's method as a system in Excel.

We show a few lines of Excel formulas (Figure 11.11) using h = 1 as the step size.

We provide a plot of the rabbits versus foxes from Excel (Figure 11.12).

Again, we see that the system moves around the equilibrium value of (1000, 1200) and away from (0, 0).

t	R	F	dr	df	NewR	Newf
0	2000	=600	=$A$5*D9-$B$5*D9*	=-$C$5*E9+$D$5*D9'	=D9+$G$5*F9	=E9+$G$5*G9
=C9+$G$5	=H9	=I9	=$A$5*D10-$B$5*D1	=-$C$5*E10+$D$5*D:	=D10+$G$5*F10	=E10+$G$5*G10

FIGURE 11.11
Excel code for rabbit and fox model.

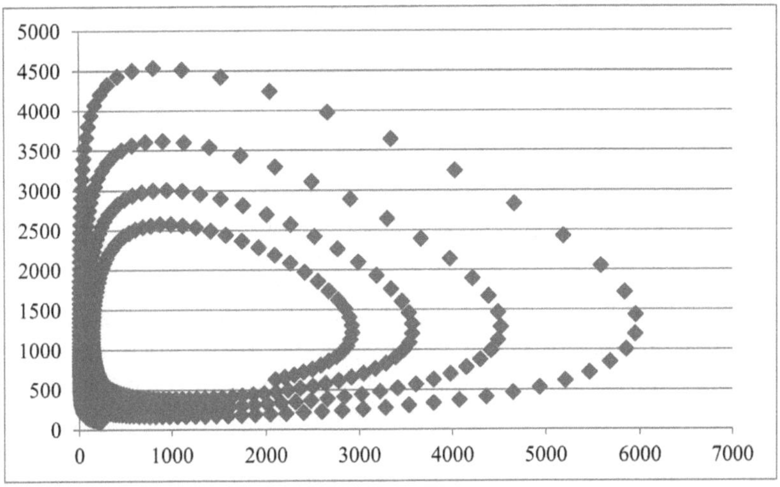

FIGURE 11.12
Excel plot for rabbit–fox model.

Continuous SIR Models of Epidemics

Consider a disease that is spreading throughout the United States such as the new flu. The CDC is interested in knowing and experimenting with a model for this new disease prior to it actually becoming a "real" epidemic. Let us consider the population being divided into three categories: susceptible, infected, and removed. We make the following assumptions for our model:

- No one enters or leaves the community and there is no contact outside the community.
- Each person is either susceptible, S (able to catch this new flu); infected, I (currently has the flu and can spread the flu); or removed, R (already had the flu and will not get it again that includes death).
- Initially, every person is either S or I.
- Once someone gets the flu this year they cannot get again.
- The average length of the disease is two weeks over which the person is deemed infected and can spread the disease.
- Our time period for the model will be per week.

The model we will consider is the SIR model. Let's assume the following definition for our variables.

$S(n)$ = number in the population susceptible after period n.

$I(n)$ = number infected after period n.

$R(n)$ = number removed after period n.

Let's start our modeling process with *R(n)*. Our assumption for the length of time someone has the flu is two weeks. Thus, half the infected people will be removed each week,

$$\frac{dR}{dt} = 0.5 * I(t)$$

The value, 0.5, is called the removal rate per week. It represents the proportion of the infected persons who are removed from infection each week. If real data is available, then we could do "data analysis" in order to obtain the removal rate.

I(t) will have terms that both increase and decrease its amount over time. It is decreased by the number that is removed each week, *0.5*I(tn)*. It is increased by the number of susceptible that come into contact with an infected person and catch the disease, *aS(t)I(t)*. We define the rate, *a*, as the rate at which the disease is spread or the transmission coefficient. We realize this is a probabilistic coefficient. We will assume, initially, that this rate is a constant value that can be found from initial conditions.

Let's illustrate as follows. Assume we have a population of 1,000 students in the dorms. Our nurse found 3 students reporting to the infirmary initially. The next week, 5 students came into the infirmary with flu-like symptoms. *I(0) = 3, S(0) = 997.* In week 1, the number of newly infected is 30.

$$5 = \mathrm{a}\mathrm{I}(\mathrm{n})\mathrm{S}(\mathrm{n}) = \mathrm{a}(3) * (995)$$
$$\mathrm{a} = 0.00167$$

Let's consider *S(t)*. This number is decreased only by the number that becomes infected. We may use the same rate, *a*, as before to obtain the model:

$$\frac{dS}{dt} = -0.00167 \cdot S(t) \cdot I(t)$$

Our coupled SIR model is shown in the systems of differential equations below:

$$\frac{dR}{dt} = 0.5I(t)$$
$$\frac{dI}{dt} = -0.5I(t) + 0.00167I(t)S(t)$$
$$\frac{dS}{dt} = -0.00167S(t)I(t)$$
$$I(0) = 3, S(0) = 997, R(0) = 0$$

The SIR model above can be solved iteratively and viewed graphically. Let's iterate the solution and obtain the graph to observe the behavior to obtain some insights.

```
>removed:=diff(r(t),t)=0.5Inf(t);
```

$$\mathrm{removed} := \frac{\mathrm{d}}{\mathrm{dt}}\mathrm{r}(\mathrm{t}) = 0.5\mathrm{Inf}(\mathrm{t})$$

```
>infect:=diff(Inf(t),t)=-0.5Inf(t)+0.00167s(t)Inf(t);
```

$$infect := \frac{d}{dt}Inf(t) = -0.5Inf(t)+0.00167s(t)Inf(t)$$

```
>succept:=diff(s(t),t)=-0.00167s(t)Inf(t);
```

$$succept := \frac{d}{dt}s(t)-0.00167s(t)Inf(t)$$

```
>Model1:={removed, infec, succept};
```

$$Model1 := \left\{\frac{d}{dt}Inf(t) = -0.5Inf(t)+0.00167s(t)Inf(t)\right\}$$

$$\frac{d}{dt}r(t)=0.5Inf(t),\ \frac{d}{dt}s(t)=-0.00167s(t)Inf(t)$$

```
>vars:={r(t), Inf(t), s(t)};
 ars:={r(t), Inf(t), s(t)}
>init1:=[s(0)=997, r(0)=0, Inf(0)=3];
 init1:=[s(0)=997, r(0)=0, Inf(0)=3]
>domain:=[t=0..25;
 domain:=[t=0..25
>L:=DEplot(Model1, vars, domain, [init1], stepsixe=0.5, scene=[t,s],
 arrows=NONE, linecolor=blue)
>H:=DEplot(Model1, vars, domain, [init1], stepsixe=0.5, scene=[t,Inf],
   arrows=NONE, linecolor=red)
>G:=DEplot(Model1, vars, domain, [init1], stepsixe=0.5, scene=[t,r],
 arrows=NONE)
>display:=({L, H, G}, title='SIR')
```

In this example (Figure 11.13), we see that the maximum number of inflected persons occurs at about day 7. Everyone survives and not everyone gets the flu. Let's see what happens in another case example.

SIR with Hong Kong Flu

During the winter of 1968–1969, the United States was swept by a virulent new strain of influenza, named *Hong Kong flu* for its place of discovery. At that time, no flu vaccine was available, so many more people were infected than would be the case today. We will study the spread of the disease through a single urban population, that of New York City.

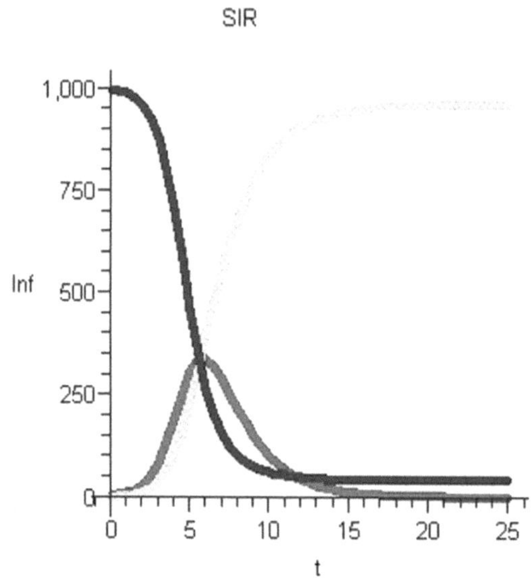

FIGURE 11.13
SIR model.

The data displayed in Figure 11.11 are weekly totals of "excess" pneumonia-influenza deaths, that is, the number of such deaths above the average numbers to be expected from other sources. The graph displays the same data. (Source: Centers for Disease Control.)

Relatively few flu sufferers die from the disease or its complications, even without a vaccine. However, we may reasonably assume that the number of excess deaths in a week was proportional to the number of new cases of flu in some earlier week, say, three weeks earlier. Thus the Table in Figure 11.14 reflects (proportionally) the rise and subsequent decline in the number of new cases of Hong Kong flu. We will model the spread of such a disease so that we can predict what might happen with similar epidemics in the future.

Week	Flu-related deaths	Week	Flu-related deaths
1	14	8	108
2	28	9	68
3	50	10	77
4	66	11	33
5	156	12	65
6	190	13	24
7	156		

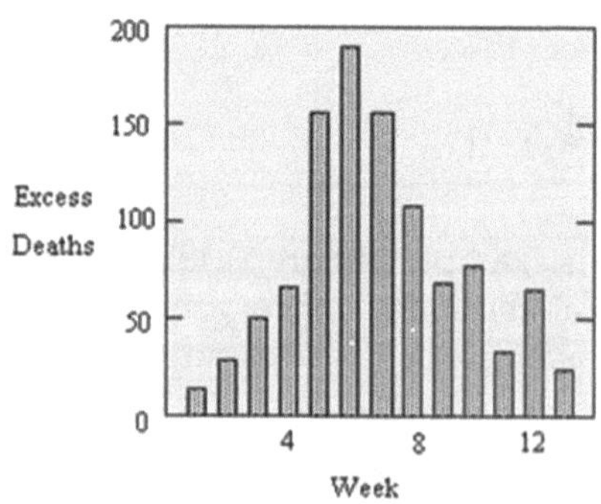

FIGURE 11.14
Flu-related deaths.

At any given time during a flu epidemic, we want to know the number of people who are infected. We also want to know the number who have been infected and have recovered, because these people now have an immunity to the disease. (As a matter of convenience, we include in the recovered group the relative handful who do not recover but die – they too can no longer contract the disease.) If we ignore movement into and out of the infected area, then the remainder of the population is still susceptible to the disease. Thus, at any time, the fixed total population (approximately 7,900,000 in the case of New York City in the late 1960s) may be divided into three distinct groups:

The Model

The first set of dependent variables counts *people* in each of the groups, each as a function of time:

$S = S(t)$ is the number of susceptible individuals,

$I = I(t)$ is the number of infected individuals, and

$R = R(t)$ is the number of recovered individuals.

The second set of dependent variables represents the *fraction* of the total population in each of the three categories. So, if **N** is the total population (7,900,000 in our example), we have

$s(t) = S(t)/N$, the susceptible fraction of the population,

$i(t) = I(t)/N$, the infected fraction of the population, and

$r(t) = R(t)/N$, the recovered fraction of the population.

It may seem more natural to work with population counts, but some of our calculations will be simpler if we use the fractions instead. The two sets of dependent variables are proportional to each other, so either set will give us the same information about the progress of the epidemic.

Finally, we complete our model by giving each differential equation an initial condition. For this particular virus – Hong Kong flu in New York City in the late 1960s – hardly anyone was immune at the beginning of the epidemic, so almost everyone was susceptible. We will assume that there was a trace level of infection in the population, say, 10 people. Thus, our initial values for the population variables are

$$S(0) = 7{,}900{,}000$$
$$I(0) = 10$$
$$R(0) = 0$$

In terms of the scaled variables, these initial conditions are

$$S(0) = 7{,}900{,}000$$
$$I(0) = 10$$
$$R(0) = 0$$

$$\frac{ds}{dt} = -bs(t)i(t), \quad s(0) = 1$$

$$\frac{di}{dt} = bs(t)i(t) - ki(t), \quad i(0) = 1.27 \times 10^{-6}$$

$$\frac{dr}{dt} = ki(t), \quad i(0) = 0$$

This model is identical to the one we just developed. So what is different? We must keep in mind the number lost due to deaths from the flu. We have some data so we can estimate the parameters we need. We don't know the values for the parameters **b** and **k** yet, but we can estimate them, and then adjust them as necessary to fit the excess death data. We have already estimated the average period of infectiousness at three days, so that would suggest **k = 1/3**. If we guess that each infected would make a possibly infecting contact every two days, then **b** would be **1/2**. We emphasize that this is just a guess. The following plot shows the solution curves for these choices of **b** and **k**.

The following is the Maple Code

```
>removed:=diff(r(t),t)=0.3333Inf(t);
```

$$\text{removed} := \frac{d}{dt}r(t) = 0.3333\text{Inf}(t)$$

```
>infect:=diff(Inf(t),t)=-0.3333Inf(t)+0.05s(t)Inf(t);
```

$$\text{infect} := \frac{d}{dt}\text{Inf}(t) = -0.3333\text{Inf}(t) + 0.05s(t)\text{Inf}(t)$$

```
>succept:=diff(s(t),t)=-0.05s(t)Inf(t);
```

$$\text{succept} := \frac{d}{dt}s(t) - 0.05s(t)\text{Inf}(t)$$

```
>Model1:={removed, infec, succept};
```

$$\textbf{Model1} := \left\{\frac{d}{dt}\text{Inf}(t) = -0.3333\text{Inf}(t) + 0.05s(t)\text{Inf}(t)\right\}$$

$$\frac{d}{dt}r(t) = 0.333\text{Inf}(t), \quad \frac{d}{dt}s(t) = -0.05s(t)\text{Inf}(t)$$

```
>vars:={r(t), Inf(t), s(t)};
  ars:={r(t), Inf(t), s(t)}

>init1:=[s(0)=1, r(0)=0, Inf(0)=1.27·10^-63];
  init1:=[s(0)=1, r(0)=0, Inf(0)=0.00000127]

>domain:=[t=0..150;
  domain:=[t=0..150
```

```
>L:=DEplot(Model1, vars, domain, [init1], stepsixe=0.5,
    scene=[t,s], arrows=NONE, linecolor=blue)

>H:=DEplot(Model1, vars, domain, [init1], stepsixe=0.5, scene=[t,Inf],
    arrows=NONE, linecolor=red)

>G:=DEplot(Model1, vars, domain, [init1], stepsixe=0.5, scene=[t,r],
    arrows=NONE)

>display:=({L, H, G}, title='SIR')
```

Let's look at these results (Figure 11.15). The model shows the number of susceptible decreasing as a percentage from 100% to just less than 50%. We also see the number removed goes from 0% to close to 60%. Finally, we see the number infected peaking around 75 to a value of about 5%. We started with 7,900,000 people. So about 395,000 have the flu at the maximum time period. The area under the curve would represent the total number that had been infected in the epidemic. So what about our losses due to deaths? The average deaths per week was 79.6 or about 0.001% of the population.

Now let's compare our model with the data. Recall that these were the numbers of deaths each week that could be attributed to the flu epidemic. If we assume that the fraction of deaths among infected individuals is constant, then the number of deaths per week should be roughly proportional to the number of infected people in some earlier week. We repeat the graph of the data, along with the graph of **i(t)** with **k = 1/3** and **b = 6/10** and we examine the graphs. The shape of the model appears reasonable (Figure 11.16).

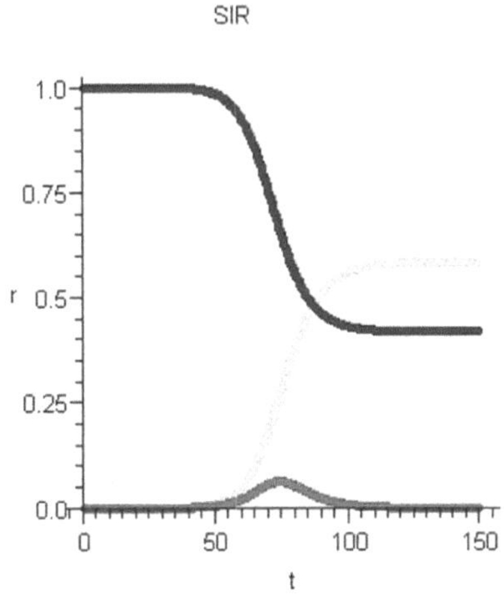

FIGURE 11.15
SIR model in Maple.

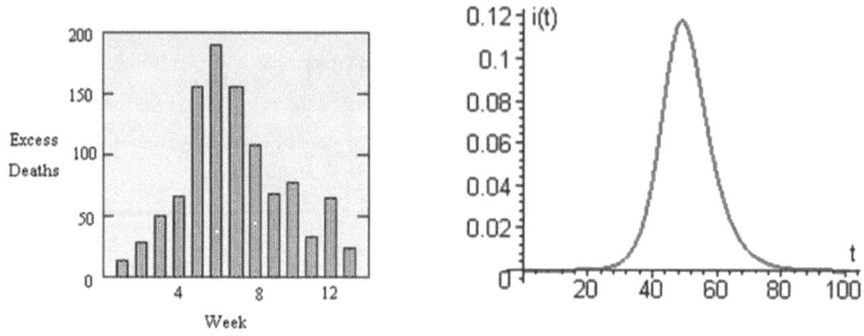

FIGURE 11.16
Excess deaths results.

Lanchester's Equations of Combat: Models of Combat Iwo Jima

At Iwo Jima in WWII, the Japanese had 21,500 soldiers and the US had 73,000 soldiers (Figure 11.17). They engaged in conventional warfare, but the Japanese were fighting from reinforced entrenchments. The kill rate for the Japanese against the US was 0.0544 while that of the US side against the Japanese was 0.0106 (based on data after the battle). If these are correct, which side should win? How many should remain on the winning side when the other side has only 1500 remaining? Give a brief explanation of the kill rates. (Historical note: The battle ended with 1500 Japanese survivors and 44314 US survivors and took approximately 33–34 days.)

Fredrick W. Lanchester developed equations that have been used to model combat for almost 100 years. He developed the following model that is called the square law model for modern combat.

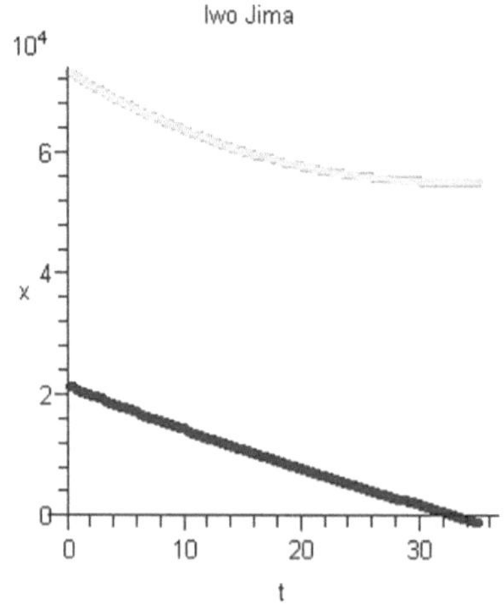

FIGURE 11.17
Iwo Jima attrition.

Square Law Modern Combat

$$\frac{dx}{dt} = -a \cdot y(t)$$
$$\frac{dy}{dt} = -b \cdot x(t)$$

where a and b represent the kill rates against the x and y force, respectively by their opponents.

The following is the Maple code for the Lanchester equations of combat.

```
>b :=0.0544: a := 0.0106:
>forcex := diff(x(t), t) = -ay(t);
```

$$\text{forcex:=}\frac{d}{dt}x(t) = -0.0106y(t)$$

```
>forcex := diff(y(t), t) = -by(t);
```

$$\text{forcex:=}\frac{d}{dt}x(t) = -0.0544y(t)$$

```
>Model1:={forcex, forcey};
```

$$\text{Model1:=}\left\{\frac{d}{dt}x(t) = -0.0106y(t), \frac{d}{dt}x(t) = -0.0544y(t)\right\}$$

```
>vars:={x)t),y(t)};
  vars:={x)t),y(t)}

>intel1:=[x(0)=21500, y(0)=73500];
  intel1:=[x(0)=21500, y(0)=73500]

>domain:=t=0.35;
  domain:=t=0.35

>L:=DEplot(Model1, vars, domain, [init1], stepsixe=0.5, scene=[t,x],
    arrows=NONE, linecolor=blue);
>H:=DEplot(Model1, vars, domain, [init1], stepsixe=0.5, scene=[t,y],
    arrows=NONE, linecolor=red);

>display:=({L, H}, title='Iwo Jima');
```

How did we do? Actually not well. We had a 91% error for the Japanese and a 24% error for the US force. What could account for this?

History shows two facts that were not modeled correctly for the square law. First, the Japanese were embedded in a hillside like guerilla warfare. They saw the US forces from attacking whereas the US probably could not see the Japanese soldiers well. Additionally, the US force landed amphibiously over a two-week period. They were not there all at once. You will be asked to consider this in the exercise set to see if you can do better in modeling this historical event/

You might want to consider this model form:

Brackney's Mixed law (also called the Parabolic Law was developed in 1959) is used to represent Guerilla warfare:

$$\frac{dx}{dt} = -a \cdot y(t)$$

$$\frac{dy}{dt} = -b \cdot x(t) \cdot y(t)$$

where a and b *are* kill rates, x represents the conventional force, and y is the guerilla force.

Numerical Methods Using Python

Predator-prey in Python

```
import matplotlib.cm as cm
    import matplotlib.pyplot as plt
    import numpy as np

    alpha = 1


    #abscisses x
    x = np.linspace(0.1,5,20)
    #ordonnées y
    y = np.linspace(0.1,5,20)
    #matrices de x et y
    mat_x,  mat_y = np.meshgrid(x,y)

    #abscisses a
    a = np.linspace(0.1,1,20)
    #ordonnées b
    b = np.linspace(0.1,1,20)
    #matrices de a et b
    mat_a,  mat_b = np.meshgrid(a,b)
```

```
#abscisses c
c = np.linspace(1,2.5,20)
#ordonnées d
d = np.linspace(0.1,1,20)
#matrices de c et d
mat_c,  mat_d = np.meshgrid(c,d)

#abscisses e
e = np.linspace(1,2.5,20)
#ordonnées f
f = np.linspace(1,2.5,20)
#matrices de e et f
mat_e, mat_f = np.meshgrid(e,f)

#abscisses g
g = np.linspace(0.1, 1, 20)
#ordonnées h
h = np.linspace(1, 2.5, 20)
#matrices de g et h
mat_g, mat_h = np.meshgrid(g, h)

def H(x,y):
    return (x-np.log(x) + y - np.log(y))
#1_ère composante de F
def F_1(x,y):
    return x*(1-y)
#2__ème composante de F
def F_2(x,y):
    return alpha*y*(x-1)
#Traçage pour des abscisses différents pour mettre en évidence
l'aspect périodique des solutions
plt.ylabel("Predators")
plt.xlabel("Prey")
plt.contour(mat_x,mat_y,H(mat_x,mat_y))
plt.quiver(mat_x,mat_y,F_1(mat_x,mat_y),F_2(mat_x,mat_y))
plt.show()

#the next 3 plots could be considered as a zoom in to observe
better the vector field
plt.contour(mat_a,mat_b,H(mat_a,mat_b))
plt.quiver(mat_a,mat_b,F_1(mat_a,mat_b),F_2(mat_a,mat_b))
plt.xlabel("Prey")
plt.ylabel("Predators")
plt.show()

plt.contour(mat_c,mat_d,H(mat_c,mat_d))
plt.quiver(mat_c,mat_d,F_1(mat_c,mat_d),F_2(mat_c,mat_d))
plt.xlabel("Prey")
plt.ylabel("Predators")
plt.show()
```

```
plt.contour(mat_e,mat_f,H(mat_e,mat_f))
plt.quiver(mat_e,mat_f,F_1(mat_e,mat_f),F_2(mat_e,mat_f))
plt.xlabel("Prey")
plt.ylabel("Predators")

plt.contour(mat_g,mat_h,H(mat_g,mat_h))
plt.quiver(mat_g,mat_h,F_1(mat_g,mat_h),F_2(mat_g,mat_h))
plt.xlabel("Prey")
plt.ylabel("Predators")


plt.show()
```

We will now provide the Python code for the last several problems using both Euler and RK4.

First is the code for the Fox versus Rabbit system of equations (Figure 11.19).

```
%matplotlib notebook
import numpy as np
import pandas as pd
import matplotlib.pyplot as plt
```

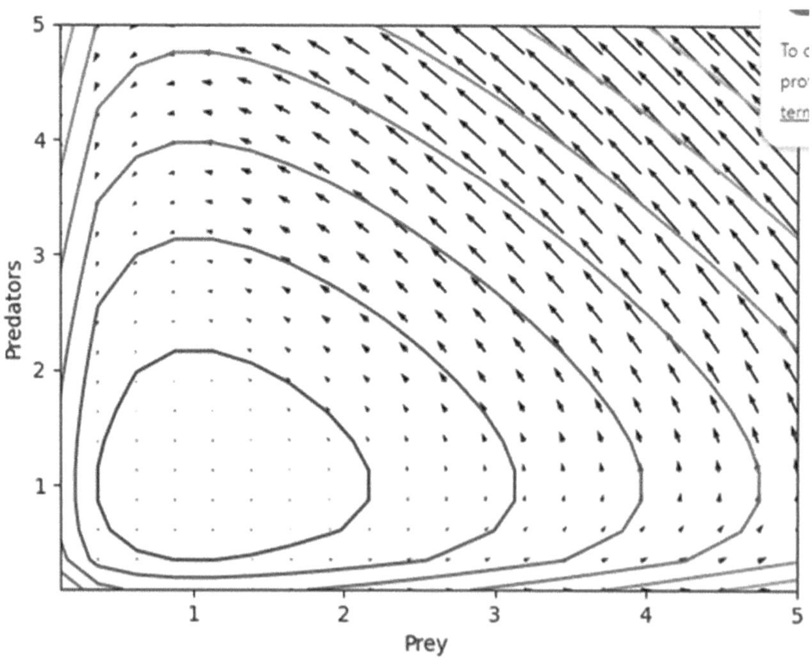

FIGURE 11.18
Predator-prey model.

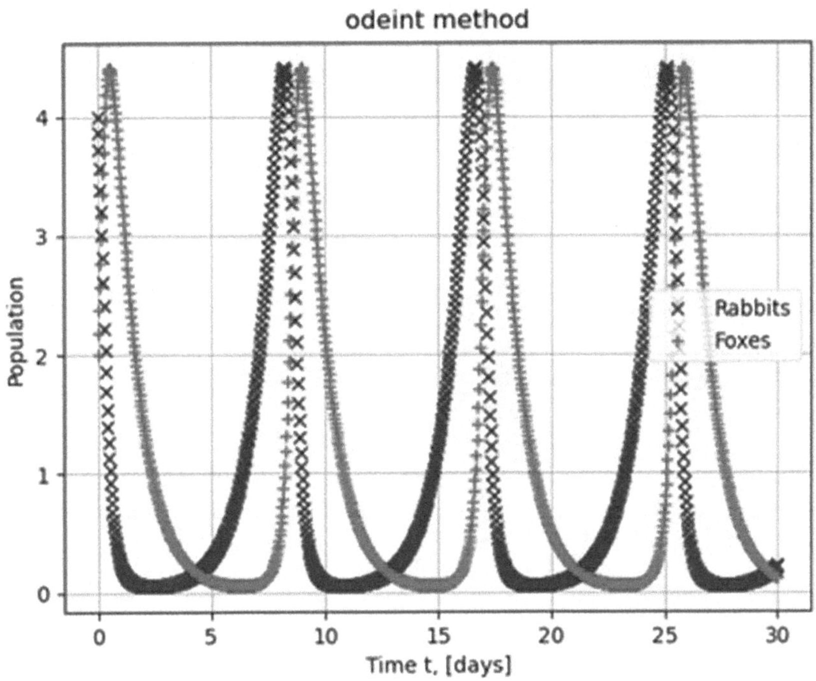

FIGURE 11.19
Rabbits versus foxes with different coefficients.

```
from scipy import integrate
import ipywidgets as ipw
alpha = 1. #mortality rate due to predators
beta = 1.
delta = 1.
gamma = 1.
x0 = 4.
y0 = 2.

def derivative(X, t, alpha, beta, delta, gamma):
    x, y = X
    dotx = x * (alpha - beta * y)
    doty = y * (-delta + gamma * x)
    return np.array([dotx, doty])
Nt = 1000
tmax = 30.
t = np.linspace(0.,tmax, Nt)
X0 = [x0, y0]
res = integrate.odeint(derivative, X0, t, args = (alpha, beta, delta,
gamma))
x, y = res.T
(2, 1000)
```

```
plt.Figure()
plt.grid()
plt.title("odeint method")
plt.plot(t, x, 'xb', label = 'Rabbits')
plt.plot(t, y, '+r', label = "Foxes")
plt.xlabel('Time t, [days]')
plt.ylabel('Population')
plt.legend()

plt.show()


import random
import matplotlib.cm as cm
betas = np.arange(0.9, 1.4, 0.1)


nums=np.random.random((10,len(betas)))
colors = cm.rainbow(np.linspace(0, 1, nums.shape[0]))  # generate the
colors for each data set


fig, ax = plt.subplots(2,1)


for beta, i in zip(betas, range(len(betas))):
    res = integrate.odeint(derivative, X0, t, args = (alpha,beta,
delta, gamma))
    ax[0].plot(t, res[:,0], color = colors[i],  linestyle = '-',
label = r"$\beta = $" + "{0:.2f}".format(beta))
    ax[1].plot(t, res[:,1], color = colors[i], linestyle = '-',
label = r" $\beta = $" + "{0:.2f}".format(beta))
    ax[0].legend()
    ax[1].legend()


ax[0].grid()
ax[1].grid()
ax[0].set_xlabel('Time t, [days]')
ax[0].set_ylabel('Rabbits')
ax[1].set_xlabel('Time t, [days]')
ax[1].set_ylabel('Foxes');
plt.Figure()
IC = np.linspace(1.0, 6.0, 21) # initial conditions for Rabbits
population (prey)
for Rabbits in IC:
    X0 = [Rabbits, 1.0]
    Xs = integrate.odeint(derivative, X0, t, args = (alpha, beta,
delta, gamma))
    plt.plot(Xs[:,0], Xs[:,1], "-", label = "$x_0 =$"+str(X0[0]))
plt.xlabel("Rabbits")
plt.ylabel("Foxes")
plt.legend()
plt.title("Rabbits vs Foxes");
```

Euler Numerical Method for Fox Versus Rabbits Problem in Python (Figure 11.20)

```
def Euler(func, X0, t, alpha, beta, delta, gamma):
    """
    Euler solver.
    """
    dt = t[1] - t[0]
    nt = len(t)
    X  = np.zeros([nt, len(X0)])
    X[0] = X0
    for i in range(nt-1):
        X[i+1] = X[i] + func(X[i], t[i], alpha,  beta, delta,
gamma) * dt
        return X
Xe = Euler(derivative, X0, t, alpha, beta, delta, gamma)
plt.Figure()
plt.title("Euler method")
plt.plot(t, Xe[:, 0], 'xb', label = 'Rabbits')
plt.plot(t, Xe[:, 1], '+r', label = "Foxes")
plt.grid()
```

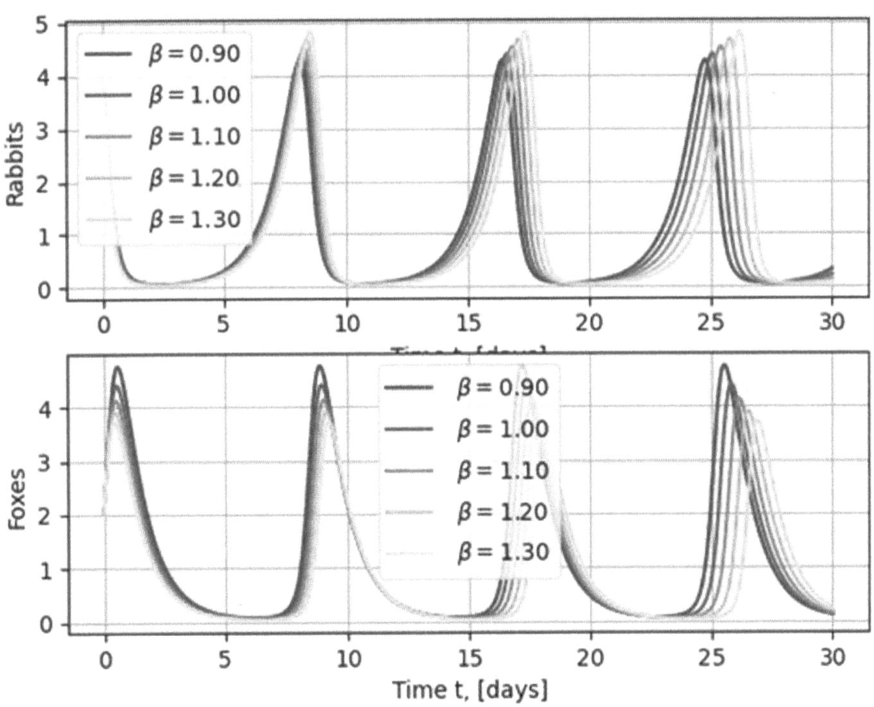

FIGURE 11.20
Phase portrait for the fox versus rabbity problem.

```
plt.xlabel("Time, $t$ [s]")
plt.ylabel('Population')
plt.ylim([0.,3.])
plt.legend(loc = "best")

plt.show()
```

RK4 Numerical Method for Fox Versus Rabbits (Figure 11.21)

```
def RK4(func, X0, t, alpha,  beta, delta, gamma):
    """
    Runge Kutta 4 solver.
    """
    dt = t[1] - t[0]
    nt = len(t)
    X  = np.zeros([nt, len(X0)])
    X[0] = X0
    for i in range(nt-1):
        k1 = func(X[i], t[i], alpha,  beta, delta, gamma)
```

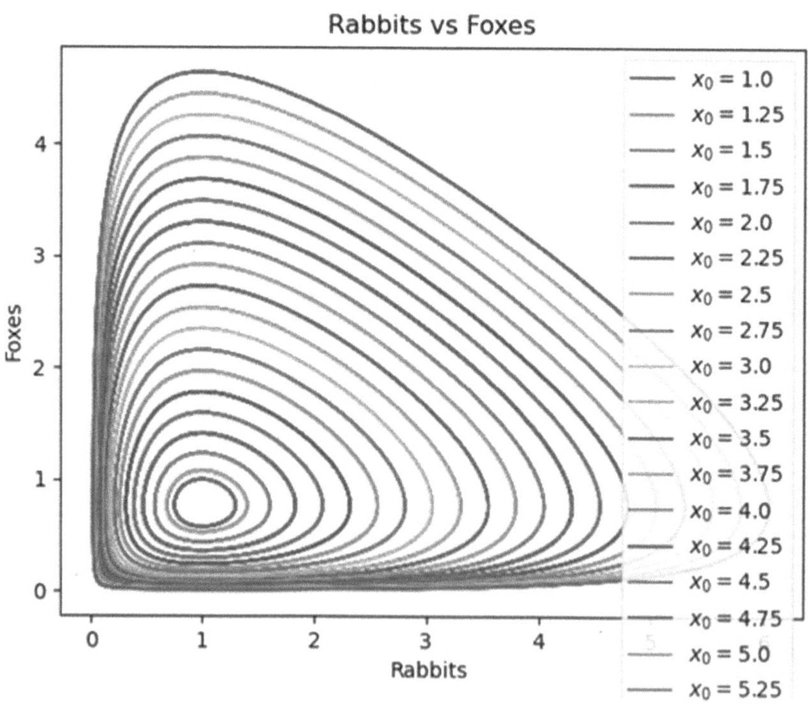

FIGURE 11.21
RK4 numerical method for the fox versus rabbit problem.

```
        k2 = func(X[i] + dt/2. * k1, t[i] + dt/2., alpha,  beta,
delta, gamma)
        k3 = func(X[i] + dt/2. * k2, t[i] + dt/2., alpha,  beta,
delta, gamma)
        k4 = func(X[i] + dt * k3, t[i] + dt, alpha,  beta, delta,
gamma)
        X[i+1] = X[i] + dt / 6. * (k1 + 2. * k2 + 2. * k3 + k4)
    return X


    Xrk4 = RK4(derivative, X0, t, alpha,  beta, delta, gamma)
    plt.Figure()
    plt.title("RK4 method")
    plt.plot(t, Xrk4[:, 0], 'xb', label = 'Rabbits')
    plt.plot(t, Xrk4[:, 1], '+r', label = "Foxes")
    plt.grid()
    plt.xlabel("Time, $t$ [s]")
    plt.ylabel('Population')
    plt.legend(loc = "best")


    plt.show();


    plt.Figure()
    plt.plot(Xrk4[:, 0], Xrk4[:, 1], "-")
    plt.xlabel("Rabbits")
    plt.ylabel("Foxes")
    plt.grid()
    plt.title("Phase plane : Rabbits vs Foxes (RK4)");
    %matplotlib notebook
    import numpy as np
    import pandas as pd
    import matplotlib.pyplot as plt
    from scipy import integrate
    import ipywidgets as ipw
    alpha = 1. #mortality rate due to predators
    beta = 1.
    delta = 1.
    gamma = 1.
    x0 = 4.
    y0 = 2.


    def derivative(X, t, alpha, beta, delta, gamma):
        x, y = X
        dotx = x * (alpha - beta * y)
        doty = y * (-delta + gamma * x)
        return np.array([dotx, doty])
    Nt = 1000
    tmax = 30.
    t = np.linspace(0.,tmax, Nt)
    X0 = [x0, y0]
```

```
    res = integrate.odeint(derivative, X0, t, args = (alpha, beta,
delta, gamma))
    x, y = res.T
    (2, 1000)
    plt.Figure()
    plt.grid()
    plt.title("odeint method")
    plt.plot(t, x, 'xb', label = 'Rabbits')
    plt.plot(t, y, '+r', label = "Foxes")
    plt.xlabel('Time t, [days]')
    plt.ylabel('Population')
    plt.legend()

    plt.show()

    import random
    import matplotlib.cm as cm
    betas = np.arange(0.9, 1.4, 0.1)

    nums=np.random.random((10,len(betas)))
    colors = cm.rainbow(np.linspace(0, 1, nums.shape[0]))  # generate
the colors for each data set

    fig, ax = plt.subplots(2,1)

    for beta, i in zip(betas, range(len(betas))):
        res = integrate.odeint(derivative, X0, t, args = (alpha,beta,
delta, gamma))
        ax[0].plot(t, res[:,0], color = colors[i],  linestyle = '-',
label = r"$\beta = $" + "{0:.2f}".format(beta))
        ax[1].plot(t, res[:,1], color = colors[i], linestyle = '-',
label = r" $\beta = $" + "{0:.2f}".format(beta))
        ax[0].legend()
        ax[1].legend()

    ax[0].grid()
    ax[1].grid()
    ax[0].set_xlabel('Time t, [days]')
    ax[0].set_ylabel('Rabbits')
    ax[1].set_xlabel('Time t, [days]')
    ax[1].set_ylabel('Foxes');
    plt.Figure()
    IC = np.linspace(1.0, 6.0, 21) # initial conditions for Rabbits
population (prey)
    for Rabbits in IC:
        X0 = [Rabbits, 1.0]
        Xs = integrate.odeint(derivative, X0, t, args = (alpha, beta,
delta, gamma))
        plt.plot(Xs[:,0], Xs[:,1], "-", label = "$x_0 =$"+str(X0[0]))
```

```
    plt.xlabel("Rabbits")
    plt.ylabel("Foxes")
    plt.legend()
    plt.title("Rabbits vs Foxes");
```

The Euler method for foxes versus rabbits (Figure 11.22)

```
def Euler(func, X0, t, alpha, beta, delta, gamma):
        """
        Euler solver.
        """
        dt = t[1] - t[0]
        nt = len(t)
        X  = np.zeros([nt, len(X0)])
        X[0] = X0
        for i in range(nt-1):
            X[i+1] = X[i] + func(X[i], t[i], alpha,  beta, delta,
gamma) * dt
        return X
```

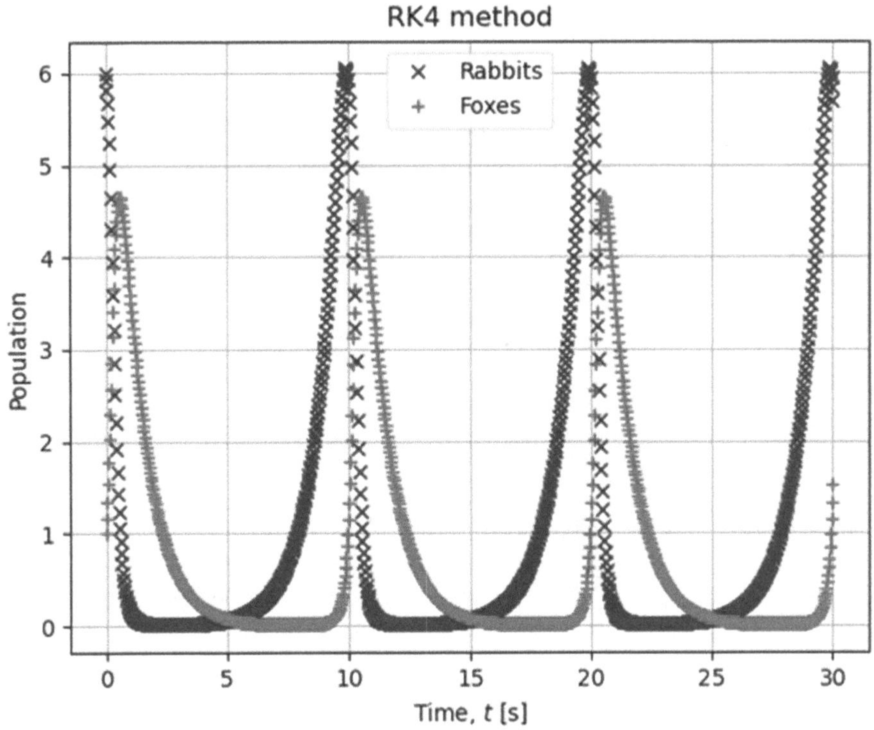

FIGURE 11.22
RK4 numerical method for fox versus rabbits problem.

```
    Xe = Euler(derivative, X0, t, alpha, beta, delta, gamma)
    plt.Figure()
    plt.title("Euler method")
    plt.plot(t, Xe[:, 0], 'xb', label = 'Rabbits')
    plt.plot(t, Xe[:, 1], '+r', label = "Foxes")
    plt.grid()
    plt.xlabel("Time, $t$ [s]")
    plt.ylabel('Population')
    plt.ylim([0.,3.])
    plt.legend(loc = "best")

    plt.show()
```

The RK4 Numerical Method for Fox Versus Rabbits (Figure 11.23)

```
def RK4(func, X0, t, alpha,  beta, delta, gamma):
    """
    Runge Kutta 4 solver.
    """
    dt = t[1] - t[0]
    nt = len(t)
    X  = np.zeros([nt, len(X0)])
    X[0] = X0
    for i in range(nt-1):
        k1 = func(X[i], t[i], alpha,  beta, delta, gamma)
        k2 = func(X[i] + dt/2. * k1, t[i] + dt/2., alpha,  beta,
delta, gamma)
        k3 = func(X[i] + dt/2. * k2, t[i] + dt/2., alpha,  beta,
delta, gamma)
        k4 = func(X[i] + dt    * k3, t[i] + dt, alpha,  beta,
delta, gamma)
        X[i+1] = X[i] + dt / 6. * (k1 + 2. * k2 + 2. * k3 + k4)
    return X

    Xrk4 = RK4(derivative, X0, t, alpha,  beta, delta, gamma)
    plt.Figure()
    plt.title("RK4 method")
    plt.plot(t, Xrk4[:, 0], 'xb', label = 'Rabbits')
    plt.plot(t, Xrk4[:, 1], '+r', label = "Foxes")
    plt.grid()
    plt.xlabel("Time, $t$ [s]")
    plt.ylabel('Population')
    plt.legend(loc = "best")

    plt.show();

    plt.Figure()
    plt.plot(Xrk4[:, 0], Xrk4[:, 1], "-")
```

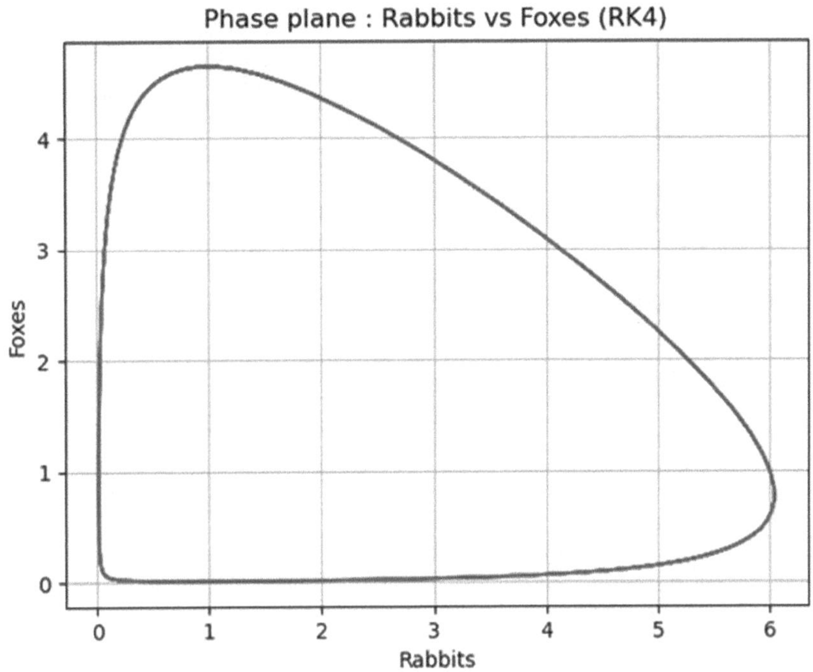

FIGURE 11.23
RK4 phase plane for fox versus rabbits problem.

```
plt.xlabel("Rabbits")
plt.ylabel("Foxes")
plt.grid()
plt.title("Phase plane : Rabbits vs Foxes (RK4)");
```

Solution graph in Python

Exercises

11.1 Solve the following problems using Maple.
   a. $x' = 3x - 2y$, $y' = 5x - 4y$, *with* $x(0) = 3$, $y(0) = 6$
   b. $x' = 3x + 4y$, $y' = 3x + 2y$, $x(0) = 1$, $y(0) = 1$
   c. $x' = 2x - 3y + 5$, $y' = 3x - y + 7$, $x(0) = 1$, $y(0) = 1$

11.2 Given the following system of linear first ODEs of species cooperation (symbiosis):

$$dx_1/dt = -0.5x_1 + x_2$$

$$dx_2/dt = 0.25x_1 - 0.5x_2$$

and $x_1(0) = 200$ and $x_2(0) = 500$.

a. Perform Euler's method with step-size h = 0.1 to obtain graphs of numerical solutions for $x_1(t)$ and $x_2(t)$ versus t and for x_1 versus x_2. You can put both $x_1(t)$ and $x_2(t)$ versus t on one axis if you want.
b. From the graphs discuss the long-term behavior of the system (discuss stability).
c. Analytically using Eigenvalues and Eigenvectors solve the system of DEs to determine the population of each species for $t > 0$.
d. Determine if there is a steady state solution for this system.
e. Obtain real plots of $x_1(t)$ and $x_2(t)$ versus t and for $x_1(t)$ versus $x_2(t)$. Compare to the numerical plots. Briefly discuss.

11.3 Given a competitive hunter model defined by the system:

$$dx/dt = 15x - x^2 - 2xy = x(15 - x - 2y)$$

$$dy/dt = 12y - y^2 - 1.5xy = y(12 - y - 1.5x)$$

a. Perform a graphical analysis of this competitive hunter model in the *x-y* plane.
b. Identify all equilibrium points and classify their stability.
c. Find the numerical solutions using Euler's method with step size h = 0.05. Try it from two separate initial conditions: first, use $x(0) = 5$ and $y(0) = 4$, then use $x(0) = 3$, $y(0) = 9$. Obtain graphs of $x(t)$, $y(t)$ individually (or on the same axis) and then a plot of x versus y using your numerical approximations. Compare to your phase portrait analysis.

11.4 Since bass and trout both live in the same lake and eat the same food sources, they are competing for survival. The rates of growth for bass (dB/dt) and for trout (dT/dt) are estimated by the following equations:

$$dB/dt = (10 - B - T)B$$

$$dT/dt = (15 - B - 3T)T$$

Coefficients and values are in thousands.

a. Obtain a "qualitative" graphical solution of this system. Find all equilibrium points of the system and classify each as unstable, stable, or asymptotically stable.
b. If the initial conditions are $B(0) = 5$ and $T(0) = 2$, determine the long-term behavior of the system from your graph in part (a). Sketch it out.
c. Using Euler's method, h = 0.1 and the same initial conditions as above, obtain estimates for B and T. Using these estimates determine a more accurate graph by plotting B versus T for the solution from t = 0 to t = 7.
d. Compare the graph in part (c) to the possible solutions found in (a) and (b). Briefly comment.

11.5 Find the equilibrium values for the predator-prey model presented.
11.6 Find the equilibrium values for the SIR model presented
11.7 Find the equilibrium values for the combat model presented
11.8 In the predator-prey model determine the outcomes with the following sets of parameters.

a. Initial foxes are 200 and initial rabbits are 400.
b. Initial foxes are 2000 and initial rabbits are 10,000
c. Birth rate of rabbits increases to 0.1

11.9 In the SIR model determine the outcome with the following parameters changed.

a. Initially 5 are sick and 10 the next week.
b. The flu last one week.
c. The flu last four weeks.
d. There are 4,000 students in the dorm and 5 are initially infected and 30 more the next week.

PROJECT

11.1 The Battle of Iwo Jima: The validity of Lanchester's equation can be demonstrated in an actual situation where US forces captured the island of Iwo Jima. Information required for the verification and "what if" analysis is the number of friendly troops put ashore each day and the number of friendly causalities for each day's engagement, the knowledge that the enemy troops were not reinforced or withdrawn, the number of enemy troops at the start of the battle and the number at the end of the engagement, and the length of the engagement. The enemy was well entrenched in the rocks on the island. The US forces were attacking the enemy's prepared defenses. In an idealized situation, the US forces would be considered to follow a modified Lanchester's square law with replacement troops landing each day while the enemy could be considered to follow the standard square law.

Since the enemy is entrenched and looking down on the US troops attacking it is easier to hit and kill US forces. The P(hit US troops with an enemy weapon) = 0.54 and the P(kill US troops | a hit) = 0.1. We assume these events are independent and that their product represents the kill coefficient of the aggregated Japanese forces against US forces. The P(hit Japanese troops with a US weapon) = 0.12 and we will assume P(Kill Japanese troops | a hit) is also 0.1. We assume these events are independent and that their product represents the kill coefficient of the aggregated US forces against the Japanese forces.

i. Determine the kill rates for the US and Japanese forces. Explain from your knowledge of probability and statistics why they might be reasonable.
ii. Determine who wins a fight to the finish.
iii. Parity: Is parity possible in this problem? Can we find it easy? Is it possible or easier to find parity after all the US troops have landed and now assume that the battle is new? Under this scenario at what kill ratio could the enemy have reached parity? Is that value feasible? Explain.
iv. The real battle ended with 1500 Japanese survivors and 44314 US survivors and took approximately 33–34 days. Relate your results with these real results. If different, why do you think these results are different?
v. Reflect on your use of Lanchester equations to adequately explain the results of the Battle of Iwo Jima Data.

Enemy's initial strength was 21,000 troops in fortified positions on the Island.

Friendly troop strength was modified by following on landings (Table 11.1):

TABLE 11.1

Follow on Troop Landings on Iwo Jima

Day	Number of Troops
1	30,000
2	1,200
3	6,735
4	3,626
5	5,158
6	13,227
7	3,054
8	3,359
9	3,180
10	1,456
11	250
Thereafter	0
Total Troops	71,245

References and Additional Readings

Abell, M. L., and Braselton, J. P. (2000). *Differential Equation with Maple V*. 2nd edition. Academic Press.

Barrow, D., Belmonte, A., Boggess, A., Bryant, J., Kiffe, T., Morgan, J., Rahe, M., Smith, K., and Stecher, M. (1998). *Solving Differential equations with Maple V Release 4*. Pacific Grove, CA: Brooks-Cole Publisher.

Giordano, F. R., and Maurice, D. W. (1991). *Differential Equations: A Modeling Approach*. Reading, MA: Addison-Wesley Publishers.

12

Modeling Using Laplace Transforms

The Laplace transform is a mathematical technique used to simplify the process of solving linear differential equations. The technique converts calculus functions from the time domain into the complex frequency domain, it transforms differential equations into algebraic equations, which are easier to solve. Once the algebraic solution is found, the inverse Laplace transform is applied to convert the result back to the time domain. This method is widely used in engineering and physics to analyze systems involving signals, circuits, and control processes.

Introduction to Laplace Transform

Let's consider a problem developed at the United States Military Academy (see The Cannon problem by Arney et al.). We solved the first order cannon problem in Chapter 5. We now turn our attention to the second order model. In this model, we initially assume gravity is 32 ft/s^2.

$$\text{For } 0 < t < 1/5:$$

$$\sum F = ma$$

$$F_1 - F_2 = 150a$$

$$4500 - 600\sqrt{2}\,x' = 150x''$$

$$30 - 4\sqrt{2}\,x' = x'$$

$$For\, t > 1/5$$

$$\sum F = ma$$

$$-F_2 = 150a$$

$$-600\sqrt{2}\ x' = 150x''$$

$$-4\sqrt{2}x' = x'$$

DOI: 10.1201/9781003582274-12

Therefore, we have the following:

$$x'' + 4\sqrt{2}x' = \begin{cases} 30 & 0 \le t < 1/5 \\ 0 & t \ge 1/5 \end{cases}, \ x(0) = 0, x'(0) = 0$$

We will use Laplace transforms to solve this due to piece-wise f(t) functions. The solution is

$$x(t) = \begin{cases} 0.94(e^{-4\sqrt{2}t} - 1 + 5.3t & 0 \le t < 1/5 \\ 1.1 + .94e^{-4\sqrt{2}t} - .94e^{-4\sqrt{2}\left(t-\frac{1}{5}\right)} & t \ge 1/5 \end{cases}$$

The plot, shown in Figure 12.1, shows the piecewise solution to our problem. We see from Figure 12.1 that the limit as t → infinity is about 1.06.
Let's look at bungee jumping with Laplace transforms:
Bungee jumping: We have a second ODE for Bungee jumping as

$$mx''(t) + cx'(t) + kx(t) = 0 \text{ with initial condition } x(0) = \alpha, x'(0) = \beta$$

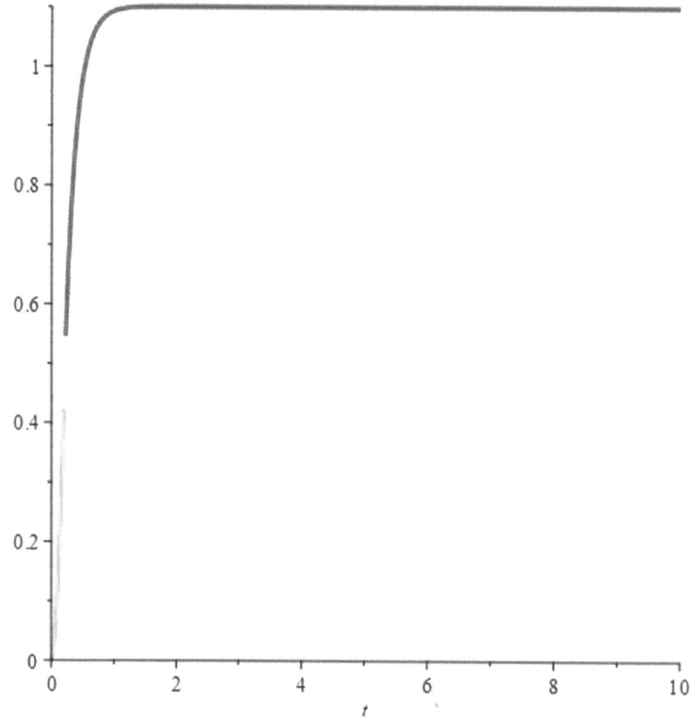

FIGURE 12.1
Plot of the piecewise solution.

When the forcing function f(t) is a continuous function, we might use Laplace transforms to solve the problem. An object with mass m = 4 is attached to the end of a spring with a spring constant of k = 1. Determine the motion of the mass if $c = 2$, $x(0) = -1$, $x'(0) = 1$ and the forcing function, $f(t)$, is given by $f(t) = 0$ *then* $f(t) = 1$.

$$x''(t)+2x'(t)+(1)x(t)=0 \text{ with initial condition } x(0)=-1, x'(0)=1$$

a. We take the Laplace (L) transforms of both sides and obtain

$$4L\{x''\}+2L\{x'\}+L\{x\}=L\{0\} \text{ so}$$

$$4\left[s^2X(s)-sx(0)-x'(0)\right]+2\{sX(s)-x(0)+X(s)=0$$

$$\left(4s^2+2s+1\right)X(s)=-4s+2$$

We solve for $X(s) = (-4s + 2)/4\ s^2 + 2s + 1$.
We apply the inverse Laplace transform to obtain

$$x(t)=-e^{4}\left[\cos\left(\frac{\sqrt{3}}{4}t\right)-\sqrt{3}\sin\frac{\sqrt{3}}{4}t\right]$$

We plot the solution as shown in Figures 12.2 and 12.3.

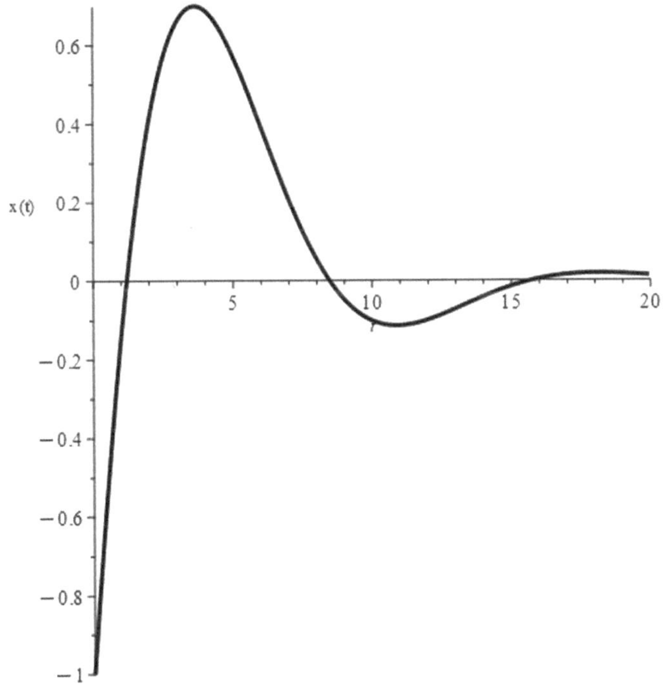

FIGURE 12.2
Plot of our spring-mass system for f(t) = 0.

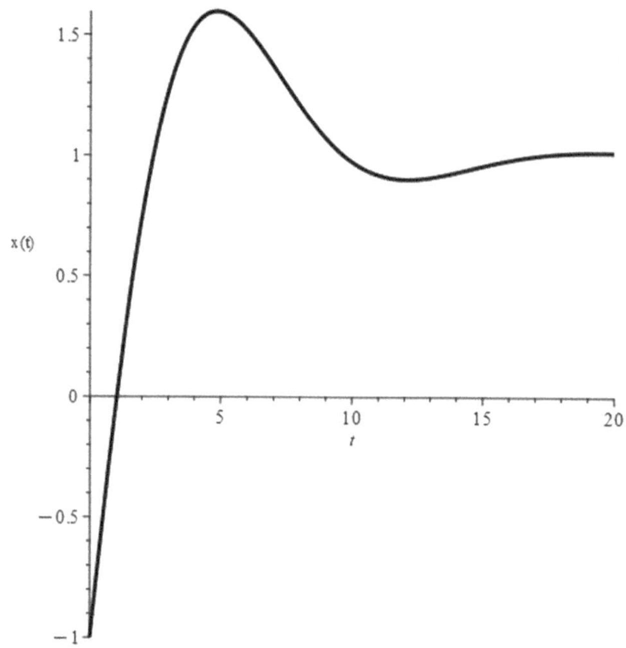

FIGURE 12.3
Plot of solution when the forcing function *f(t)* = *1*.

Using Laplace Transforms and Inverse Laplace Transforms

Given a function, *f(t)*, defined for all $t \geq 0$, the Laplace transform of *f* is the function, *F* defines as

$$F(s) = L\{f(t)\} = \int_0^\infty e^{-st} f(t)\,dt$$

For all values of *x* for which the improper integral converges.

Example 12.1: If f(t) = 1 then from the definition we replace f(t) with 1

$$F(s) = L\{f(t)\} = \int_0^\infty e^{-st}\,dt = -1/s\ e^{(-st)}\Big]_0^\infty = \lim_{c\to\infty}\left[\frac{-1}{s}e^{-bs} + \frac{1}{s}\right] = 1/s,\ \text{for } s > 0$$

Example 12.2: If f(t) = e^{at} for $t \geq 0$

$$F(s) = L\{f(t)\} = \int_0^\infty e^{-st}e^{at}\,dt = \left[\frac{-e^{-(s-a)t}}{s-a}\right]_0^\infty = \frac{1}{s-a}$$

$f(t) = \mathcal{L}^{-1}\{F(s)\}$	$F(s) = \mathcal{L}\{f(t)\}$
1. 1	$\frac{1}{s}$
2. e^{at}	$\frac{1}{s-a}$
3. $t^n, \quad n = 1, 2, 3, \ldots$	$\frac{n!}{s^{n+1}}$
4. $t^p, p > -1$	$\frac{\Gamma(p+1)}{s^{p+1}}$
5. $\sqrt{t}$	$\frac{\sqrt{\pi}}{2s^{\frac{3}{2}}}$
6. $t^{n-\frac{1}{2}}, \quad n = 1, 2, 3, \ldots$	$\frac{1 \cdot 3 \cdot 5 \cdots (2n-1)\sqrt{\pi}}{2^n s^{n+\frac{1}{2}}}$
7. $\sin(at)$	$\frac{a}{s^2 + a^2}$
8. $\cos(at)$	$\frac{s}{s^2 + a^2}$
9. $t\sin(at)$	$\frac{2as}{(s^2 + a^2)^2}$
10. $t\cos(at)$	$\frac{s^2 - a^2}{(s^2 + a^2)^2}$
11. $\sin(at) - at\cos(at)$	$\frac{2a^3}{(s^2 + a^2)^2}$
12. $\sin(at) + at\cos(at)$	$\frac{2as^2}{(s^2 + a^2)^2}$
13. $\cos(at) - at\sin(at)$	$\frac{s(s^2 - a^2)}{(s^2 + a^2)^2}$
14. $\cos(at) + at\sin(at)$	$\frac{s(s^2 + 3a^2)}{(s^2 + a^2)^2}$
15. $\sin(at + b)$	$\frac{s\sin(b) + a\cos(b)}{s^2 + a^2}$
16. $\cos(at + b)$	$\frac{s\cos(b) - a\sin(b)}{s^2 + a^2}$

FIGURE 12.4
Laplace transforms for some common functions.

Figure 12.4 provides some Laplace transforms for a few common functions.

For an inverse we merely find the transform in the right column and see the left function that created it.

For example, consider the Laplace transform $F(s) = 4/s + 1/(s - 8) + 3/(s - 4)$ then

$$F(t) = 4(1)e^{8t} + 3e^{4t}$$

Models Using Laplace Transforms

Example 12.3: L-R-C Circuit Revisited

We use Kirchoff laws for a simple circuit (Figure 12.5).

$$LQ''(t)+RQ'(t)+1/CQ(t)=E(t) \text{ with } Q(0)=Q_o \text{ and } Q'(0)=I_o$$

Laplace transforms are most useful when the E(t) function is piecewise defined.

Suppose that we consider the circuit in Figure 12.3 with capacitor C, resistor R, and a voltage supply E(t) defined as

$$E(t)=\begin{cases} E_0\,0 & \le t<1 \\ 0 & t\ge 1 \end{cases}.$$

If L = 0, then find Q(t) and I(t) when Q(0) = 0, C = 10 − 2 farads, R = 100 Ω and E_0 = 100 V.

Our first order IVP is

$$100Q'(t)+100Q(t)=\begin{cases} 100, & 0\le t, \ <1 \\ 0, & t\ge 1 \end{cases}$$

and Q(0) = 0.

Since E(t) is a piecewise continuous function we can define E(t) in terms of a unit step function:

$$E(t)=100\left[u(t-0)-u(t-1)\right]=100\left[u(t)-u(t-1)\right]=100\left[1-u(t-1)\right]$$

Now we take the Laplace transform of each side

$$100SL\{Q(t)\}-Q(0)+100L\{Q(t)\}=100L\{1-u(t-1)\}$$
$$(s+1)L\{Q(t)\}=1/s-e^{-s/}s$$
$$L\{Q(t)\}=1/(s(s+1)-e^{-s}/)s(s+1)$$

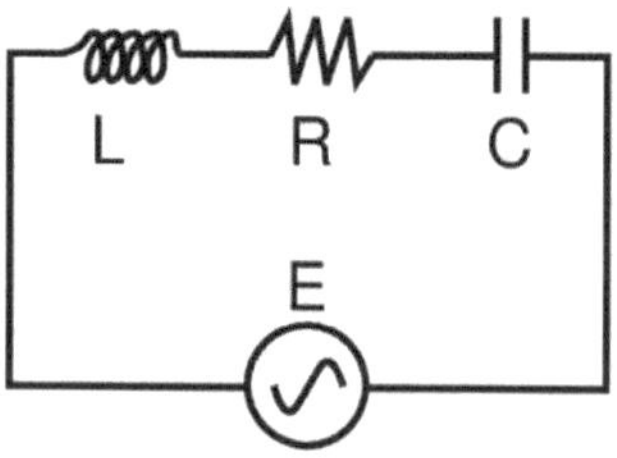

FIGURE 12.5
Simple L-R-C circuit for Example 12.3.

We take L^{-1} of $1/(s(s + 1))$ and L^{-1} of $e^{-s}/(s(s + 1))$.

$$L^{-1}(1/(s(s+1)) = 1 - e^{-t}$$

$$L^{-1}(e^{-s}/(s(s+1)) = [1 - e^{-(t-1)}]u(t-1)$$

$$Q(t) = \begin{cases} e^{-t}, 0 \le t < 1 \\ e^{-t}(e-1), t \ge 1 \end{cases} = I(t)$$

Example 12.4: Population Model

Let's model the population on a medium-sized island. Let *x(t)* be the population at time *t*. The growth rate, *k*, is a positive constant.

$$x'(t) + kx(t) = 1000(1 + a * \sin(t)), x(0) = xo.$$

Let's assume $k = 3$, $x_0 = 2000$ and $a = 0.25$.

$$x'(t) + 3x(t) = 1000(1 + 0.25 * \sin(t)), x(0) = 2000.$$

$$x'(t) = -3 * x(t) + 1000(1 + 0.25 * \sin(t)), x(0) = 2000.$$

We see the solution plot for Example 12.4 in Figure 12.6.
Let's illustrate the Laplace transform method in Maple (Figure 12.7).

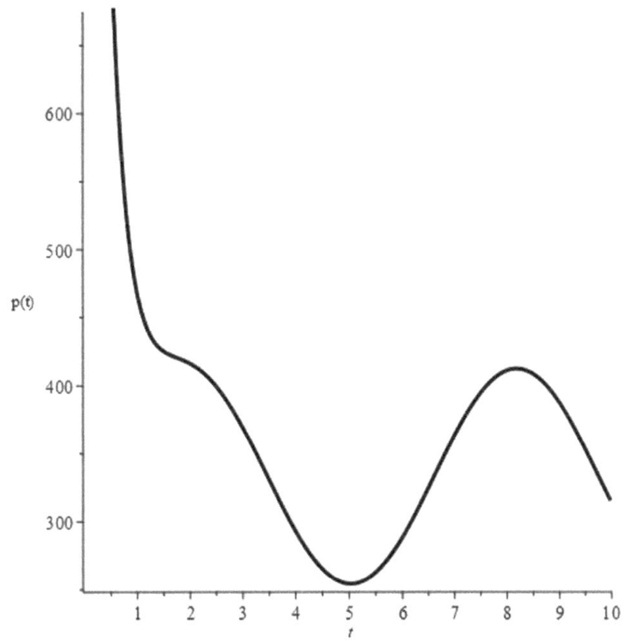

FIGURE 12.6
Population model Example 12.4.

```
> popode := diff(x(t), t) =-3 x(t) + 1000(1 + 0.25 sin( t));
```

$$popode := \frac{d}{dt}\,x(t) = -3\,x(t) + 1000 + 250.00\,\sin(t)$$

```
> lapopode := laplace(popode, t, s);
```

$$lapopode := s\,L(x(t), t, s) - 1.\,x(0) = -3.\,L(x(t), t, s) + \frac{1000.}{s} + \frac{250.}{s^2 + 1.}$$

```
> lapopode2 := subs(x(0) = 2000 , laplace(x(t), t, s) = U(s), lapopode);
```

$$lapopode2 := s\,U(s) - 2000. = -3.\,U(s) + \frac{1000.}{s} + \frac{250.}{s^2 + 1.}$$

```
> Usol := solve(lapopode2, U(s));
```

$$Usol := \frac{250.\,(8.\,s^3 + 4.\,s^2 + 9.\,s + 4.)}{s\,(s^2 + 1.)\,(3. + s)}$$

```
> usol := invlaplace( Usol, s, t);
```

$$usol := 1691.666667\,e^{-3.\,t} + 333.3333333 - 25.\cos(t) + 75.\sin(t)$$

```
> evalf(dsolve( {popode, x(0) = 2000}, x(t), method = laplace));
```

$$x(t) = 1691.666667\,e^{-3.\,t} + 333.3333333 - 25.\cos(t) + 75.\sin(t)$$

FIGURE 12.7
Maple solution for Example 12.4.

Example 12.5: The Cannon Problem Revisited

We write the right-hand side of the equation using the step function

$$U(t) = 1, t > 0 \text{ and } U(t) = 0, t < 0.$$

Therefore,

$$x'' + 4\sqrt{2}x' = 30 - 30U(t - 1/5).$$

We convert to the Laplace space to form an algebraic equation:

$$s^2X(s) - s\,x(0) - x(o) + 4\sqrt{2}s\,X(s) + 4\sqrt{2}\,x(0) = \frac{30}{s} - \frac{30}{s}e^{\left(-\frac{1}{5}s\right)}$$

$$\left(s^2 + 4\sqrt{2}s\right)X(s) = \frac{30}{s} - \frac{30}{s}e^{-s/5}$$

$$X(s) = \frac{30}{s^2\left(s + 4\sqrt{2}\right)} - \frac{30e^{-s/5}}{s^2\left(s + 4\sqrt{2}s\right)}$$

We use partial fraction decomposition to separate the first term and we obtain:

$$\frac{30}{s^2\left(s + 4\sqrt{2}\right)} = \frac{0.94}{s} + \frac{5.3}{s^2} + \frac{0.94}{s + 4\sqrt{2}}.$$

The solution produces the following equation in Laplace space:

$$X(s) = \frac{0.94}{s} + \frac{5.3}{s^2} + \frac{0.94}{s + 4\sqrt{2}} - e^{-\frac{s}{5}}\left(\frac{0.94}{s} + \frac{5.3}{s^2} + \frac{0.94}{s + 4\sqrt{2}}\right).$$

We convert back this equation space by taking the inverse Laplace transform:

$$x(t) = -0.94 + 5.3\,t + 0.94e^{-4\sqrt{2}t} - U\left(t - \frac{1}{5}\right)\left[-.94 + 5.3\left(t - \frac{1}{5}\right) + 0.94e^{-4\sqrt{2}\left(t-\frac{1}{5}\right)}\right]$$

We write the solution as:

$$x(t) = \begin{cases} .94(e^{-4\sqrt{2}t} - 1 + 5.3t & 0 \le t < 1/5 \\ 1.1 + .94e^{-4\sqrt{2}t} - .94e^{-4\sqrt{2}\left(t-\frac{1}{5}\right)} & t \ge 1/5 \end{cases}$$

The plot is shown in Figure 12.1 and we see from Figure 12.1 that the limit as $t \to$ infinity is about 1.06. Therefore, we never exceed 5 feet.

Example 12.6: RC Circuit

Consider an RC circuit where the resistance R = 10 Ω, and the capacitance C = 0.01 F. The circuit is powered by a voltage source E(t) which is a step function: E(t) = 5 V for $t \ge 0$, and the initial charge on the capacitor is Q(0) = 0.

We want to find the charge Q(t) on the capacitor for $t \ge 0$.

Step 1: Formulate the Differential Equation

Using Kirchhoff's law, the governing equation for the circuit is:

$$R\frac{dQ}{dt} + \frac{1}{C}Q(t) = E(t).$$

Substitute the given values for R, C, and E(t):

$$10\frac{dQ}{dt} + \frac{1}{0.01}Q(t) = 5.$$

Simplifying:

$$10\frac{dQ}{dt} + 100Q(t) = 5$$

Step 2: Take the Laplace Transform

Taking the Laplace (L) transform of both sides, we use the following transforms:

$$-\mathcal{L}\{dQ/dt\} = sQ(s) - Q(0),$$

$$-\mathcal{L}\{Q(t)\} = Q(s),$$

$$-\mathcal{L}\{5\} = 5/s.$$

Thus, the transformed equation becomes:

$$10(sQ(s) - 0) + 100Q(s) = 5/s.$$

Solving for Q(s):

$$Q(s) = \frac{5}{s(10(s+10))}$$

Step 3: Simplify and Use Partial Fractions

Simplify the expression and apply partial fractions:

$$Q(s) = \frac{A}{s} + \frac{B}{s+10}$$

$$A = 0.5, B = -0.5$$

$$Q(s) = \frac{0.5}{s} - \frac{0.5}{s+10}$$

Step 4: Take the Inverse Laplace Transform to find *Q*(t)

Thus, the charge on the capacitor for $t \geq 0$ is given by:

$$Q(t) = 0.5(1 - e^{-10t}) \text{ Coulombs.}$$

Example 12.7: Mass-Spring-Damper System

Consider a mass-spring-damper system where the mass m = 2 kg, the damping coefficient c = 4 Ns/m, and the spring constant k = 16 N/m. The external forcing function is f(t) = 10 sin(2t), and the initial conditions are x(0) = 0 and x′(0) = 0.

We want to find the displacement x(t) for $t \geq 0$.

Step 1: Formulate the Differential Equation

The equation of motion for the mass-spring-damper system is:

$$mx''(t) + cx'(t) + kx(t) = f(t).$$

Substituting the given values:

$$2x''(t) + 4x'(t) + 16x(t) = 10\sin(2t).$$

Step 2: Take the inverse Laplace Transform

Taking the Laplace transform of both sides gives:

$$\mathcal{L}\{10 \sin(2t)\} = \frac{20}{s^2+4}.$$

The transformed equation becomes:

$$2s^2X(s) + 4sX(s) + 16X(s) = \frac{20}{s^2+4}$$

Step 3: Solve for X(s)

$$X(s) = \frac{20}{(s^2+4)(2s^2+4s+16)}$$

Use partial fraction decomposition to split the expression, we get:

$$X(s) = \frac{2}{(s^2+4)} - \frac{s+1}{(s^2+2s+8)}$$

Step 4: Apply Inverse Laplace Transform

Applying the inverse Laplace transform, we obtain the solution for the displacement of the mass for $t \geq 0$.

$$x(t) = \sin(2t) - e^{-t}\cos(2t).$$

Exercises

Evaluate the following problems:

12.1 $\mathcal{L}\{e^{-3t}\}$

12.2 $\mathcal{L}\{t\}$

12.3 $\mathcal{L}\{\sin 2t\}$

12.4 $\mathcal{L}\{3t - 5\sin 2t\}$

Evaluate the following inverse transforms for problems.

12.5 $\mathcal{L}^{-1}\left\{\frac{1}{s^5}\right\}$

12.6 $\mathcal{L}^{-1}\left\{\frac{3s+5}{s^2+3}\right\}$

12.7 $\mathcal{L}^{-1}\left\{\frac{1}{(s-1)(s+2)}\right\}$

12.8 $\mathcal{L}^{-1}\left\{\frac{1}{(s-1)(s+2)(s+4)}\right\}$

12.9 $\mathcal{L}^{-1}\left\{\frac{s-4}{(s-1)(s+2)}\right\}$

12.10 Find $\mathcal{L}(f(t))$ for $f(t) = \begin{cases} 0 & 0 \leq t < 3 \\ 2 & t \geq 3 \end{cases}$

12.11 Find $\mathcal{L}(f(t))$ for $f(t) = \begin{cases} -1 & 0 \leq t < 1 \\ 1 & t \geq 1 \end{cases}$

12.12 Evaluate $\mathcal{L}\{(t-2)^3\mathcal{U}(t-2)\}$

12.13 Evaluate $\mathcal{L}\{\mathcal{U}(t-5)\}$

12.14 Determine the current *i(t)* in a single loop L-R_C circuit when L = 0.1 henry, R = 20 ohms,
C = 10^{-3} farads, and *i(0) = 0*. The voltage is given in Figure 12.8.

12.15 Suppose our population model is given by:

$$x'(t) + kx(t) = 1000(1 + a * \sin(t)), x(0) = xo.$$

Let's assume k = 3.5, x_0 = 2500, and *a = 0.25*. Solve and interpret.

12.16 Suppose our population model is given by:

$$x'(t) + kx(t) = 1000(1 + a * \sin(t)), x(0) = xo.$$

Let's assume k = 4, x_0 = 3500, and *a = 0.35*. Solve and interpret.

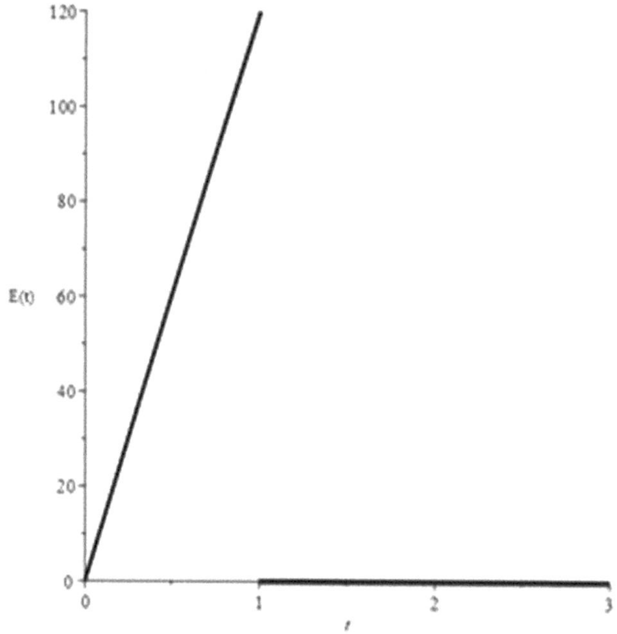

FIGURE 12.8
Graph for Exercise 12.14.

12.17 Let's revisit the cannon problem from Chapter 11. We now turn our attention to the second order model. In this model, we initially assume gravity is 32 ft/s². Solve and interpret.

Assume that for $0 < t < 1/5$:

$$\sum F = ma$$

$$F_1 - F_2 = 175a$$

$$5250 - 700\sqrt{2}x' = 175x''$$

$$30 - 4\sqrt{2}\,x' = x'$$

For $t > 1/5$

$$\sum F = ma$$

$$-F_2 = 175a$$

$$-700x' = 175x''$$

$$-4x' = x'$$

PROJECTS

12.1 Consider the following scenario: we have two tanks as shown in Figure 12.9. Tank 1 initially contains 80 liters of water and 100 g of salt. Tank 2 initially has 20 liters of water and 50 g of salt. Water containing a mix of 50 g/L of salt is poured into tank 1 at a rate of 3L/m while a mixture going into tank 2 contains a mixture of 35 g/L of salt at a rate of 3.5 L/m. The mixture flows from tank 1 to tank 2 at a rate of 2 L/m. The mixture drains from tank 2 at a rate of 6 L/m while some of the mixture flows back into tank 1 at a rate of 2 L/m. The remainder leaves the tank. Let Q_1 and Q_2 be the amount of salt in each tank at time t. Model and solve.

12.2 Cannon 2

A projectile is fired over horizontal ground at an initial speed of 1200 ft/s at an angle of elevation of 30 degrees. Assume there is no air resistance.

a. Write the differential equation model for the projectile's motion. b. Where and when does the projectile land? c. What is the highest point on the trajectory of the round? d. What is the maximum range of a round fired at that initial speed?

b. Write out the differential equation of motion of a projectile moving at high speeds (assume the resistive force is equal to the square of the velocity). Why is this equation more difficult to solve?

c. A projectile is fired over horizontal ground at an initial speed of 1200 ft/s at an angle of elevation of 30 degrees. (Same data as exercise 1.) For this model, assume the air resistance is proportional to the velocity with a proportionality constant of k = 0.03. a. Write the differential equation model for the projectile's motion. b. Where and when does the projectile land? c. What is the highest point on the trajectory of the round?

d. Analyze the firing motion of a Parrott Gun with a total weight of 4200 lbs, a retarding force of just 300 times the velocity of the gun (the gun is located in sandy soil), and a blast force from the projectile launch of 4600 lbs lasting for 0.15 seconds.

e. Use a spring constant of k = 750 lb/ft, assembly weight of 3200 lbs, a damping mechanism of 500 times the velocity of the assembly, and a blast force of 8,000 lbs lasting 0.25 seconds to model a cannon system with a differential equation. a. Find the equation of motion. b. Graph the equation of motion for the time period $0 < t < 3$ seconds.

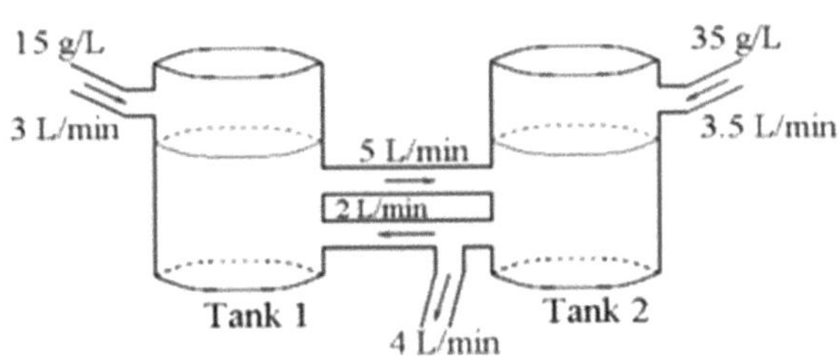

FIGURE 12.9
Tanks for project 12.1.

f. At what time does the breech block first return to the in-battery position? What is its instantaneous velocity at this time? d. At what time does the breech block reach its maximum displacement? What is the maximum displacement?

References and Additional Readings

Arney, D., and Clark, C. The King of Battle: The Cannon.
Zill, D. (2017). *A First Course in Differential Equations with Modeling Applications*. Cengage Publishing.

13

Answers to Selected Exercises

Chapter 1 Exercises

1.1 Clear your worksheet and Declare a Constant, a, to be [[1, 2, 3], [4, 5, 6], [7, 8, 0]]. Also Declare a Constant, b, to be [[1, 3, 1], [2, –1, 1], [3, 2, 2]]. Using Maple create the matrices for the following expressions and simplify:

   i. a + b
   ii. a*b
   iii. a^2
   iv. a^(–1)
   v. b^(–1)

Figure 13.1 provides the solution to Exercise 1.1.

Chapter 2 Exercises

The questions presented in this chapter were designed as general thought exercises and there is not one correct answer, However, we will provide an example of some considerations we developed for Exercise 2.2.

2.2 In modeling the size of any prehistoric creature, what information would you like to be able to obtain? What additional assumptions might be required?

When modeling the size of a prehistoric creature, the goal is typically to estimate its mass, length, or volume based on available fossil evidence. To construct an accurate mathematical model, I would seek the following types of information:

1. **Fossilized Bone Measurements**
   - Lengths and diameters of major bones (e.g., femur, humerus, skull)
   - Overall skeletal proportions
2. **Comparison to Living Analogues**
   - Size and scaling relationships in modern descendants or similar species (e.g., birds for theropods, reptiles for dinosaurs)

DOI: 10.1201/9781003582274-13

```
> with(LinearAlgebra) :
> A := Matrix(3, 3, [[1, 2, 3], [4, 5, 6], [7, 8, 0]] );
```

$$A := \begin{bmatrix} 1 & 2 & 3 \\ 4 & 5 & 6 \\ 7 & 8 & 0 \end{bmatrix}$$

```
> B := Matrix(3, 3, [[1, 3, 1], [2,-1, 1], [3, 2, 2]] );
```

$$B := \begin{bmatrix} 1 & 3 & 1 \\ 2 & -1 & 1 \\ 3 & 2 & 2 \end{bmatrix}$$

```
> A + B
```

$$\begin{bmatrix} 2 & 5 & 4 \\ 6 & 4 & 7 \\ 10 & 10 & 2 \end{bmatrix}$$

```
> MatrixMatrixMultiply(A, B);
```

$$\begin{bmatrix} 14 & 7 & 9 \\ 32 & 19 & 21 \\ 23 & 13 & 15 \end{bmatrix}$$

```
> A^2;
```

$$\begin{bmatrix} 30 & 36 & 15 \\ 66 & 81 & 42 \\ 39 & 54 & 69 \end{bmatrix}$$

```
> MatrixInverse(A);
```

$$\begin{bmatrix} -\frac{16}{9} & \frac{8}{9} & -\frac{1}{9} \\ \frac{14}{9} & -\frac{7}{9} & \frac{2}{9} \\ -\frac{1}{9} & \frac{2}{9} & -\frac{1}{9} \end{bmatrix}$$

```
> MatrixInverse(B);
```

FIGURE 13.1
Solution for question 1.1.

3. **Soft Tissue Estimates**
   - Thickness and distribution of muscle, fat, and skin
4. **Posture and Locomotion**
   - Quadrupedal vs. bipedal stance

Additional Assumptions Required

1. **Allometric Scaling Assumptions**
   - If complete fossils are missing, size can be estimated using known scaling laws (e.g., mass ∝ limb circumference^3/2)
2. **Body Composition and Density**
   - Estimating volume from skeletal structure and assuming a reasonable mass-to-volume ratio
3. **Scaling Laws for Growth**
   - Assuming similar growth patterns to modern analogs
   - Logarithmic or power-law relationships between skeletal dimensions and body mass

Chapter 3 Exercises

3.1 Find the equilibrium value, if any exists, for each of the following dynamical systems:

a. $u(n) = .5\ u(n-1) + 3 \quad X = 0.5 * X + 3, 0.5X = 3, X = 6$

b. $u(n) = 5\ u(n-1) + 3\ x = 5x + 3, -4x = 3 = x = -3/4$

c. $u(n) = 0.75\ u(n-1) + 16\ x = .75x + 16, .25\ x = 16 = x = 64$

d. $u(n) = 0.5\ u(n-1)\ x = 0$

3.3 W(n) = weight in lb after period n in weeks

$W(n+1) = W(n) - 150/3600\ W(n) + 16500/3600$

$W(0) = 105$ lb

$EV = 150/3600\ X = 4.583$ so $X = 109.9992$ lb Equilibrium weight

It takes slightly more than 158 weeks

3.5 Suppose a patient eliminates 40% of a particular drug every 12 hours. The patient takes a mg pill every 12 hours each day. How many mg should the pill be for the drug to stabilize at 45 mg?

D(n) = amount of drug in system after n 12 hr periods

$D(n+1) = D(n) - 0.4 * D(n) + b$

EV will equal 45 mg

$0.4\ D(n) + b = 45$

27 mg every 12 hours by iteration

3.7 $a(n+1) = 1.7\ a(n) - 0.14\ a(n)^2$

$X = 0$

3.9 $a(n+1) = 0.2\ a(n) - 0.2\ a(n)^3$

$X = 0$

3.11 Time period to stabilize at 1000

3.13 EV = 0.2 Stable

3.15 EV = 100,000

3.17 EV = 100,000

3.19 EV = 10

3.21 EV = 257.6923

3.23 At the beginning of every 24-hour period Oliver takes a 10 mg pill of digoxin and 1 mg is immediately absorbed into the blood. Every 24 hours the kidneys filter out 30% of the digoxin in the blood. We want to develop a discrete dynamical system that helps us to model the amount of digoxin in the blood on any given day (24-hour period).

a. Define your variables in words.

b. Make a flow diagram for digoxin in the blood.

c. Write a discrete dynamical system that models the amount of digoxin in the blood. Also, state the initial condition.

d. How much digoxin is in the blood after 10 days?

D(n) = amount of pill left after n periods
$D(n + 1) = D(n) - 1 - .3D(n) + 10$
$0.3 X = 9$
$X = 27$ mg EV
Model $D(k) = -27 (.7^k) + 27$
$D(10) = 26.237$ mg

3.25 Working full time but fed up with gas prices and excited about their bad press, you decide to buy a Toyota Prius. You get a gift from graduation for a significant down payment. You check with your new bank, and you can get a loan for $16,785 at 4.55% compounded monthly for five years. They tell you that your monthly payments will be $313.31.

a. Define your variables in words.

b. Make a flow diagram for the amount owed to the bank.

c. Write a discrete dynamical system that models the amount owed to the bank.

d. What will your final monthly payment be? (Month 60)

M(n) = amount of money after n period
$M(n + 1) = (1 + .0455/12) M(n) - 31.331$, $M(0) = 16{,}785$
Amount to pay each month is not 313.31 but $712.8597

3.27 $a(n + 1) = 1.3\, a(n) + 20$ $X = 20/-3 = -6.66666$

3.29 EV = 10

Chapter 4 Exercises

4.1 The ratio is 1.5 but the number of periods required to get there fluctuates.

4.3

B(n) = number of trucks in Boston after period n
LA(n) = number of trucks in LA after period n
C(n) = number of trucks in Chicago after period n
B(n + 1) = B(n) – .5 B(n) +. 5 C(n)
LA(n + 1) = LA(n) – .5 LA(n) + .5 C(n)
C(n + 1) = C(n) –. C(n) + .5 B(n) +. 5C(n)
B(0) = 150, LA(0) = 75, C(0) = 75
100 trucks at each location.

n	B(n)	LA(n)	C(n)	'otal Trucks
0	150	75	75	300
1	112.5	75	112.5	300
2	112.5	93.75	93.75	300
3	103.125	93.75	103.125	300
4	103.125	98.4375	98.4375	300
5	100.781	98.4375	100.781	300
6	100.781	99.6094	99.6094	300
7	100.195	99.6094	100.195	300
8	100.195	99.9023	99.9023	300
9	100.049	99.9023	100.049	300
10	100.049	99.9756	99.9756	300
11	100.012	99.9756	100.012	300
12	100.012	99.9939	99.9939	300
13	100.003	99.9939	100.003	300
14	100.003	99.9985	99.9985	300
15	100.001	99.9985	100.001	300
16	100.001	99.9996	99.9996	300
17	100	99.9996	100	300
18	100	99.9999	99.9999	300

FIGURE 13.2
Excel solution for Exercise 4.3.

4.5 60% at A and 40% at B in the long run

TABLE 13.1

Solution for Exercise 4.5

n	A	B
0	1	0
1	0.8	0.2
2	0.7	0.3
3	0.65	0.35
4	0.625	0.375
5	0.6125	0.3875
6	0.60625	0.39375
7	0.60313	0.39688
8	0.60156	0.39844
9	0.60078	0.39922
10	0.60039	0.39961
11	0.6002	0.3998
12	0.6001	0.3999
13	0.60005	0.39995
14	0.60002	0.39998
15	0.60001	0.39999
16	0.60001	0.39999

4.7 (17.0588, 25.8823, 13,.9705) Stable

Chapter 5 Exercises

5.6 y(t) = 2 * exp(t * (t + 2)/2)

5.7 y(t) = –1 + 3 * exp(t)

5.8 v(t) = 2 – 2 * exp(–16 * t). Terminal velocity is where v′(t) = 0 at v(t) = 2

5.9 P(t) = 10/(2 + 3 * exp(–15 * t)) and P(10) = 5

5.13 y(t) = 3/2 – exp(–2 * t)/2

5.15 y(t) = 3/2 – exp(–2 * t)/2

5.17 y(t) = sin(4 * t) – 4 * cos(4 * t) + (9 * exp(–t))/2

Chapter 6 Exercises

6.1 h = .1 y(3) 30.8988 $y_{actual}(3) = 38.1707$, % error –23.55

h = .01 y(3) 35.57693 $y_{actual}(3) = 38.1707$, % error –7.29164

h = .01 y(3) 33.3583 $y_{actual}(3) = 38.1707$, % error –10.258

6.3 Consider v′ = 32 – 1.6 v, v(0) = 0. Using Euler's method in Maple or any other program and a step size of 0.05 estimate v(2). What is the terminal velocity? Keep stepping out until you approximate the terminal velocity to 3 decimal places of accuracy. Terminal velocity = 2.

```
> #Numerical Methods
> dsolve({eqn1, ics}, numeric, method = classical[rk4], output = array([0, .05, .1, .15, .2, .3, .4, .5, .6, .7, .8, .9, 1.0, 1.5, 2, 2.5, 3., 3.5]),
    stepsize = 0.05);
```

t	v(t)
0.	0.
0.05	1.09653333333333
0.1	1.59187399111111
0.15	1.81563587758459
0.2	1.91671658043421
0.3	1.98300493518300
0.4	1.99653193601271
0.5	1.99929229644315
0.6	1.99985558388593
0.7	1.99997053001387
0.8	1.99999398626609
0.9	1.99999877281939
1.0	1.99999974957784
1.5	1.99999999991139
2.	1.99999999999997
2.5	2.
3.	2.
3.5	2.

FIGURE 13.3
Solution for Exercise 6.3.

6.5 Consider v′ = 32 – 1.6 v, v(0) = 0. Using Heunform in Maple or any other program and a step size of 0.05 estimate v(2). What is the terminal velocity? Keep stepping out until you approximate the terminal velocity to 3 decimal places of accuracy.

```
> #Numerical Methods
> dsolve({eqn1, ics}, numeric, method = classical[heunform], output = array([0, .05, .1, .15, .2, .3, .4, .5, .6, .7, .8, .9, 1.0, 1.5, 2, 2.5, 3.,
    3.5]), stepsize = 0.05);
```

t	$y(t)$
0.	0.
0.05	0.960000000000000
0.1	1.45920000000000
0.15	1.71878400000000
0.2	1.85376768000000
0.3	1.96045878067200
0.4	1.98930805429371
0.5	1.99710889788102
0.6	1.99921824598703
0.7	1.99978861371489
0.8	1.99994284114851
0.9	1.99998454424656
1.0	1.99999582076427
1.5	1.99999999395870
2.	1.99999999999127
2.5	1.99999999999999
3.	2.00000000000000
3.5	2.00000000000000

```
>
```

FIGURE 13.4
Solution for Exercise 6.5.

Chapter 7 Exercises

Find the general solution to the following Differential Equations,

7.1 $y'' - 2y' + 10y = 0$ *General solution is* $y(t) = e^t * (c_1 \sin(3t) + c_2 \cos(3t))$

7.3 $y'' + 5y' + 6y = 0$ $y(t) = c_1e^{(-2t)} + c2e^{(-3t)}$

7.5 $y'' + 14y' + 49y = 0$ $y(t) = c_1e^{-7t} + c_2te^{-7}$

Find the general and particular solutions to the following Differential Equations and graph the solution.

7.7 $y'' - 3y' - 10y = 0, y(0) = 0, y'(0) = 7$

General solution is $y(t) = c_1e^{-2t} + c_2e^{-5t}$. *Particular solution is founding solving* $c_1 + c_2 = 0$ *and* $-2c_1 - 5c_2 = 7$; $y(t) = 2.33333\, e^{-2t} - 2.33333\, e^{-5t}$

7.9 $y'' - 2y' + 10y = 0, y(0) = 2, y'(0) = -1$

General solution is $y(t) = e^{t}*(c_1\cos(3t) + c_2\sin(3t))$ *Particular solution is founding solving* $c_1 = 2$ *and* $3c_1\sin(3*0) - 3c_2\cos(3*0) = -1$; $y(t) = e^{t}*(2\cos(3t) + -3\sin(3t)$

7.11 $y(t) = \sin(t)c_2 + \cos(t)c_1 + \frac{1}{2} - \frac{\cos(2t)}{6}$

Chapter 8 Exercises

Solve the following higher order differential equations using Euler's method,

8.1 $y'' - 3y' - 10y = 0, y(0) = 0, y'(0) = 7$

```
dsol1 := dsolve({deq1, ic1}, numeric, method = classical[foreuler], output = array([0, .1, .2, .3, .4, .5, .6, .7, .8, .9, 1.0, 1.1,
    1.4, 1.5, 1.6, 1.7, 1.8, 1.9, 2.0]), stepsize = .1);
```

dsol1 :=

t	$y(t)$	$\frac{d}{dt}y(t)$
0.	0.	7.
0.1	0.700000000000000	9.10000000000000
0.2	1.61000000000000	12.5300000000000
0.3	2.86300000000000	17.8990000000000
0.4	4.65290000000000	26.1317000000000
0.5	7.26607000000000	38.6241100000000
0.6	11.1284810000000	57.4774130000000
0.7	16.8762223000000	85.8491179000000
0.8	25.4611340900000	128.480075570000
0.9	38.3091416470000	192.485232331000
1.0	57.5576648801000	288.539943677300
1.1	86.4116592478301	432.659591660590
1.2	129.677618413889	648.869128406597
1.3	194.564531254549	973.207485342466
1.4	291.885279788795	1459.73426219975
1.5	437.858706008770	2189.53982064847
1.6	656.812688073618	3284.26047285179
1.7	985.238735358796	4926.35130278094
1.8	1477.87386563689	7389.49542897401
1.9	2216.82340853429	11084.2179233031
2.0	3325.24520086460	16626.3067088283

FIGURE 13.5
Solution for Exercise 8.1.

8.3 $y'' - 2y' + 10y = 0, y(0) = 2, y(0) = -1$

```
> dsol2 := dsolve({deq2, ic1}, numeric, method = classical[foreuler], output = array([0, .1, .2, .3, .4, .5, .6, .7, .8, .9, 1.0, 1.1, 1.2, 1.3,
    1.4, 1.5, 1.6, 1.7, 1.8, 1.9, 2.0]), stepsize = .1);
```

dsol2 :=

t	$y(t)$	$\frac{d}{dt}y(t)$
0.	2.	−1.
0.1	1.90000000000000	1.20000000000000
0.2	2.02000000000000	2.86000000000000
0.3	2.30600000000000	4.30800000000000
0.4	2.73680000000000	5.75240000000000
0.5	3.31204000000000	7.33872000000000
0.6	4.04591200000000	9.18301600000000
0.7	4.96421360000000	11.3923248000000
0.8	6.10344608000000	14.0780734400000
0.9	7.51125342400000	17.3659048320000
1.0	9.24784390720000	21.4039772896000
1.1	11.3882416361600	26.3710257388800
1.2	14.0253442100480	32.4850622272640
1.3	17.2738504327744	40.0133939918592
1.4	21.2751898319603	49.2845656262618
1.5	26.2036463945865	60.7028423329697
1.6	32.2739306278835	74.7659202609623
1.7	39.7505226539797	92.0866668366533
1.8	48.9591893376450	113.419856123302
1.9	60.3011749499752	139.695074236287
2.0	74.2706823736039	172.057234339005

FIGURE 13.6
Solution for Exercise 8.3.

Chapter 9 Exercises

9.1 Determine the qualitative solution

a. $\frac{dy}{dx} = y$ Rest point = 0

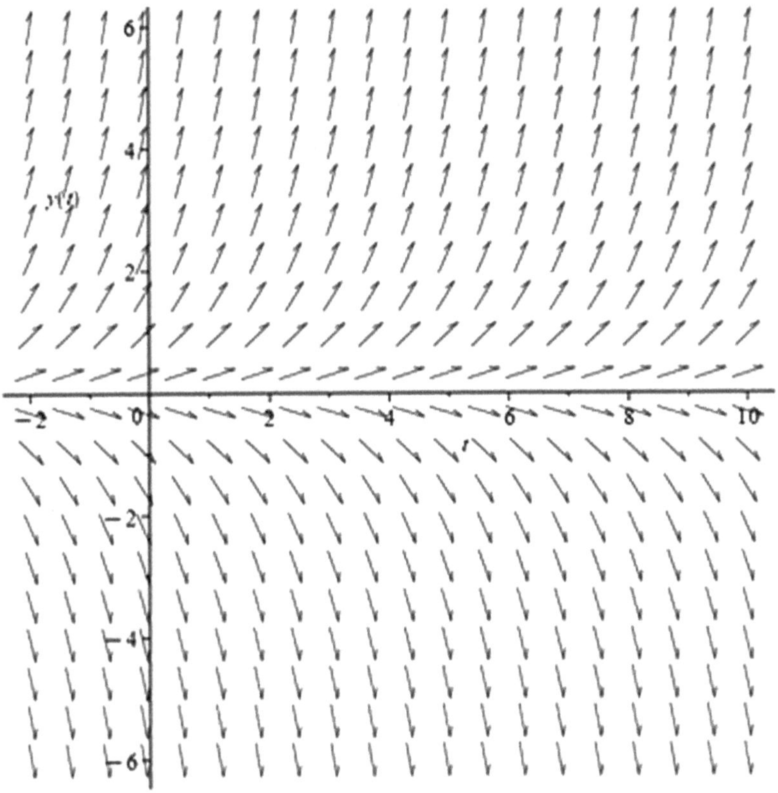

FIGURE 13.7
Solution for Exercise 9.1a.

b. $\frac{dy}{dx} = y - 1$ Rest Point = 1

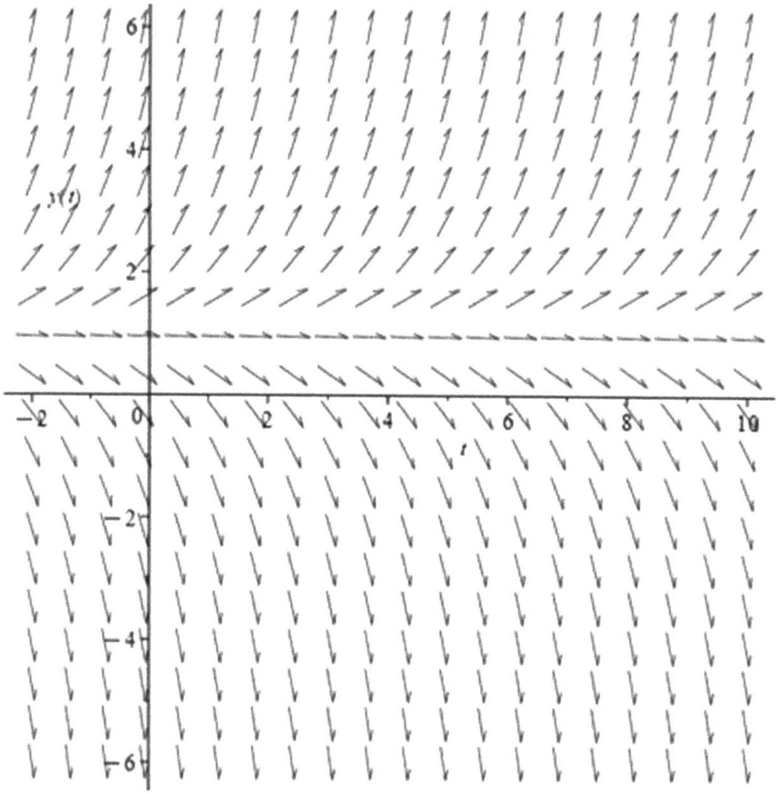

FIGURE 13.8
Solution for Exercise 9.1b.

c. $\frac{dy}{dx} = 7 + y$ Rest point = –7

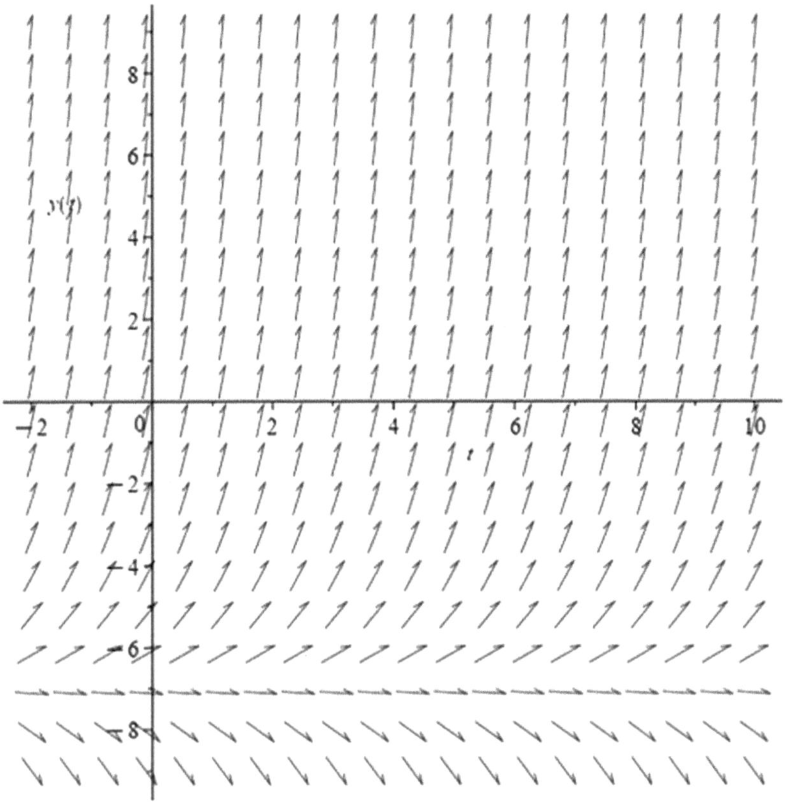

FIGURE 13.9
Solution for Exercise 9.1c.

d. $\frac{dy}{dx} = 7 - y$ Rest point = 7

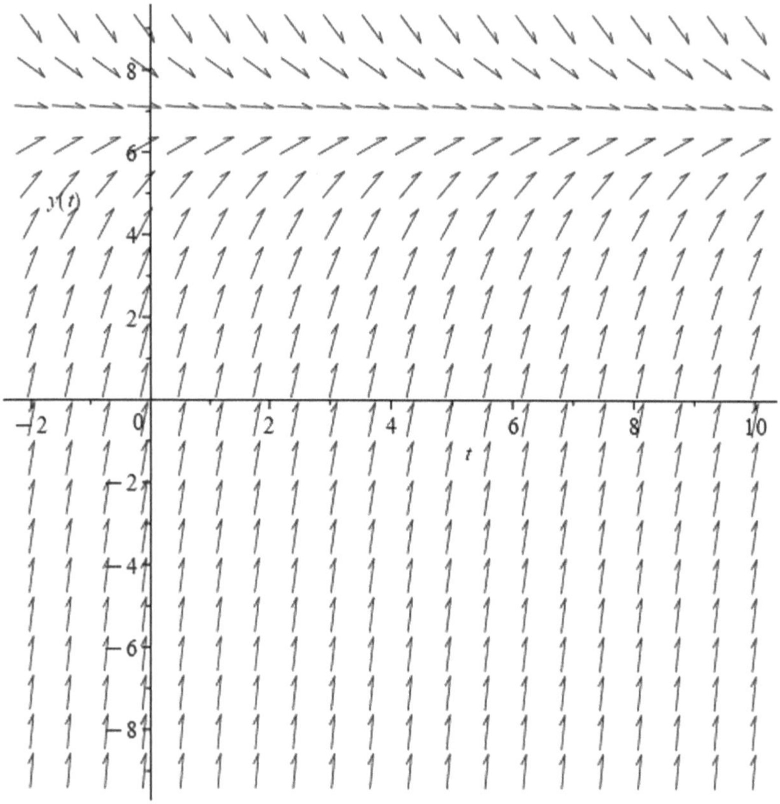

FIGURE 13.10
Solution for Exercise 9.1d.

e. $\dfrac{dy}{dx} = (y-1)y$ Rest points are 0 and 1.

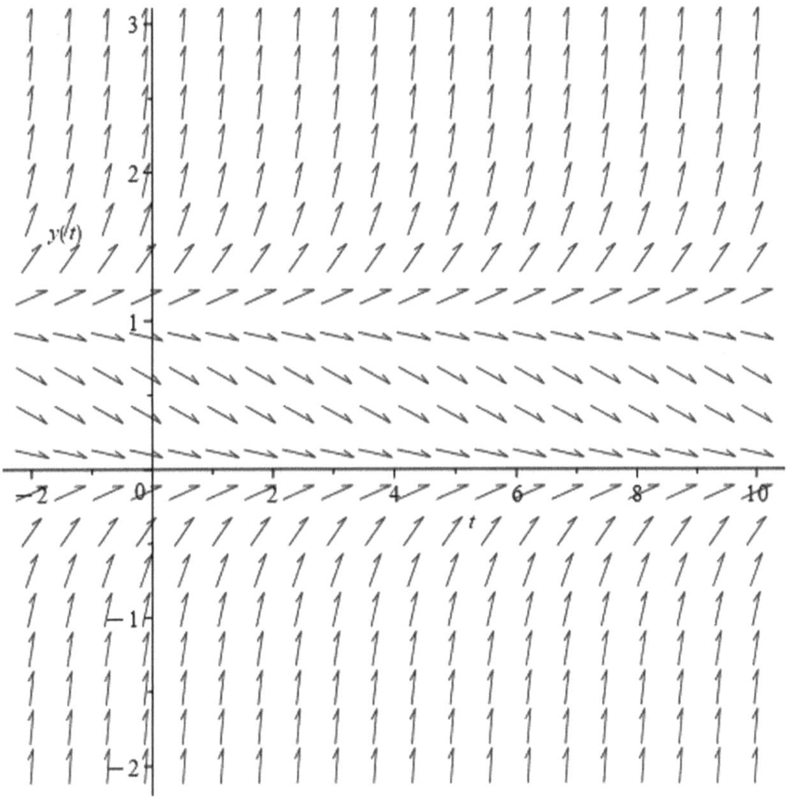

FIGURE 13.11
Solution for Exercise 9.1e.

f. $\frac{dy}{dx} = \frac{1}{y}$

No rest points as $1/y \neq 0$

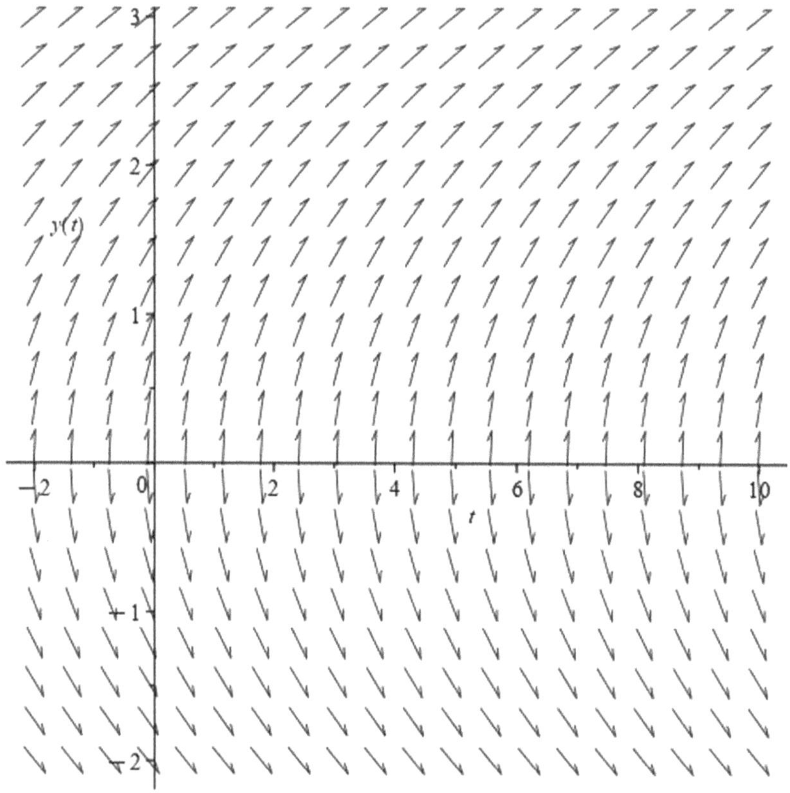

FIGURE 13.12
Solution for Exercise 9.1f.

9.2 Sketch a number of solutions to the following equations showing the correct slope, concavity, and any points of inflection.

a. $\frac{dy}{dx} = (y+2)(y-3)$

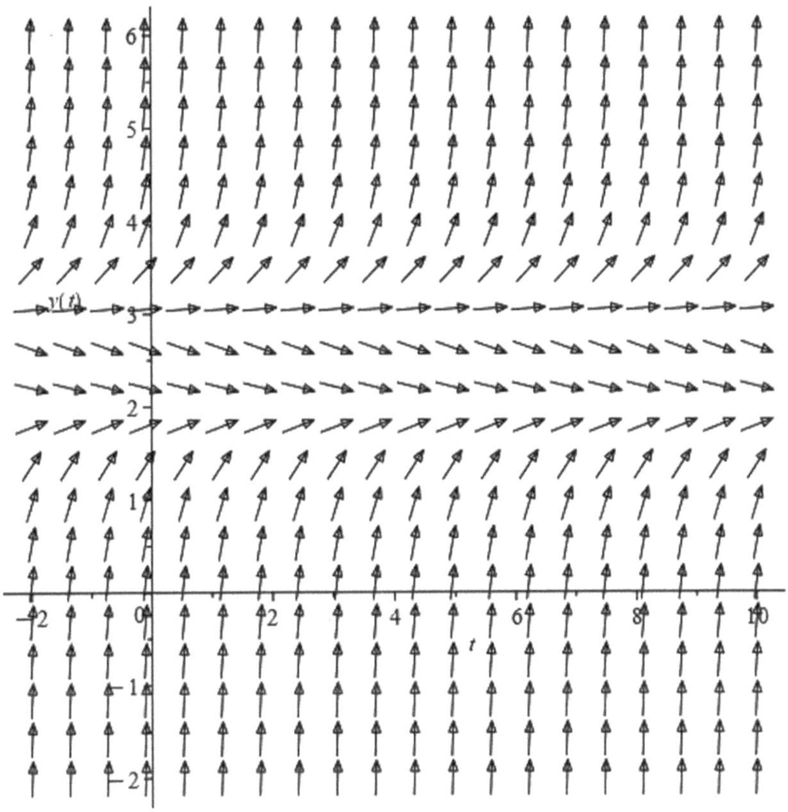

FIGURE 13.13
Solution for Exercise 9.2a.

b. $\dfrac{dy}{dx} = y^2 - 4$

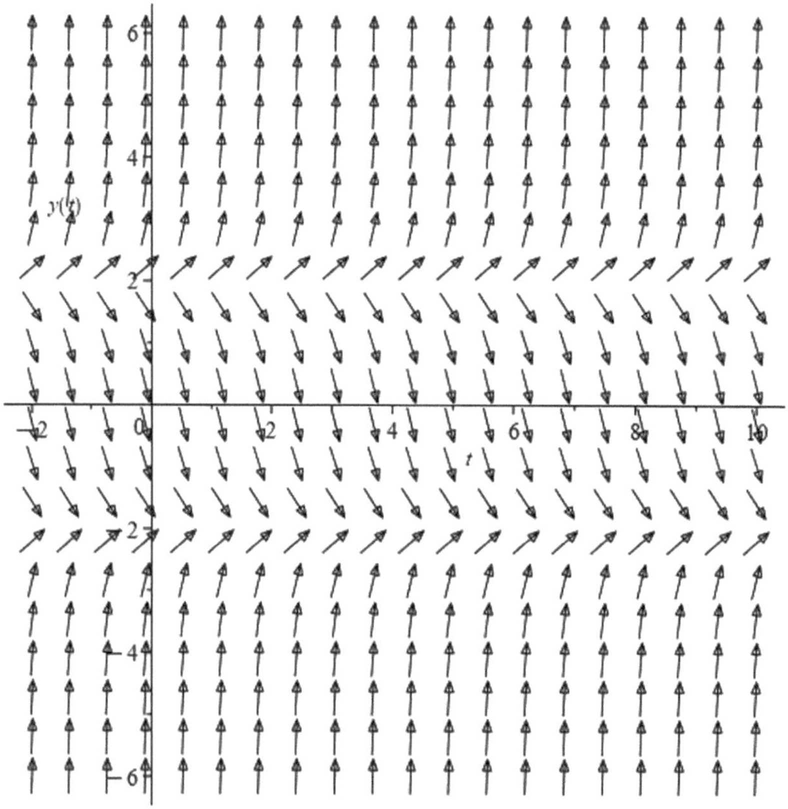

FIGURE 13.14
Solution for Exercise 9.2b.

c. $\frac{dy}{dx} = y^3 - 4$

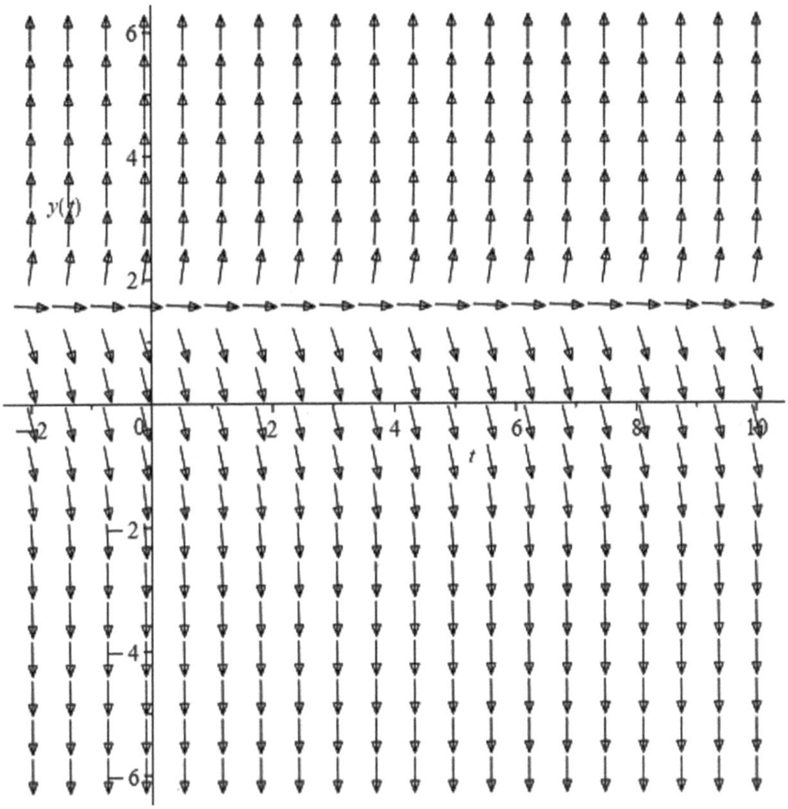

FIGURE 13.15
Solution for Exercise 9.2c.

d. $\frac{dy}{dx} = 6 - 2y$

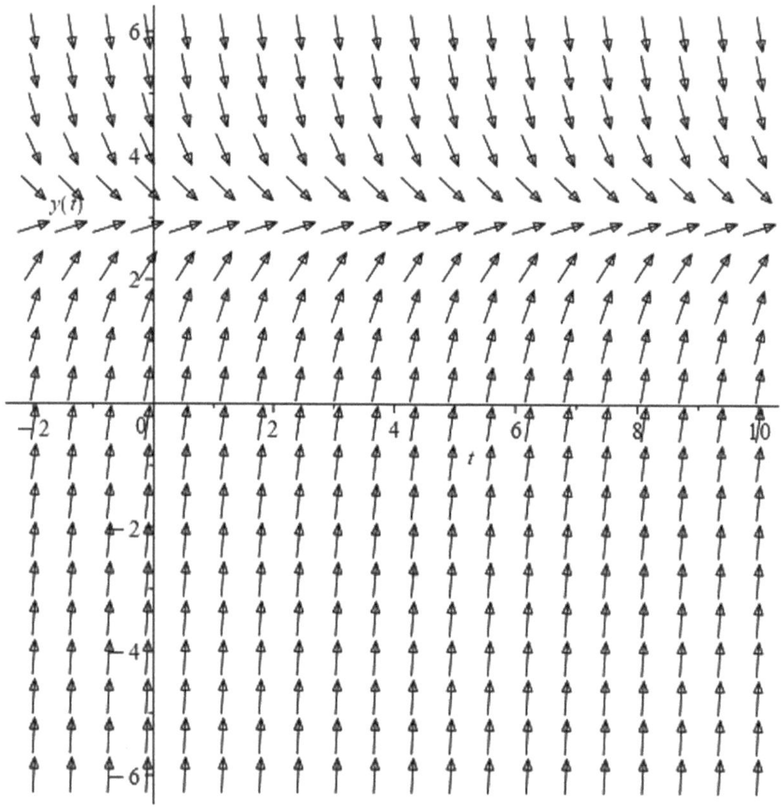

FIGURE 13.16
Solution for Exercise 9.2d.

9.3 Rest point is 0 and all trajectories move toward 0 so it is table.

Chapter 10 Exercises

Find and classify all the rest points. Then sketch a few trajectories to indicate the motion.

10.1 $dx/dt = x, dy/dt = y$

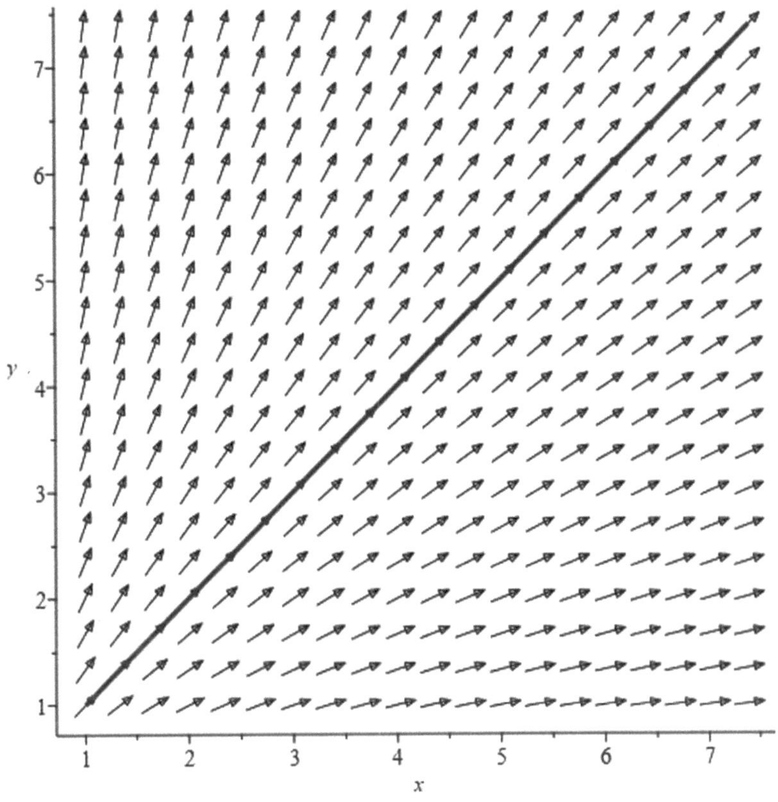

FIGURE 13.17
Solution for Exercise 10.1.

10.3 $dx/dt = y$, $dy/dt = -2x$
Rest point is (0, 0)
Asymptotically stable

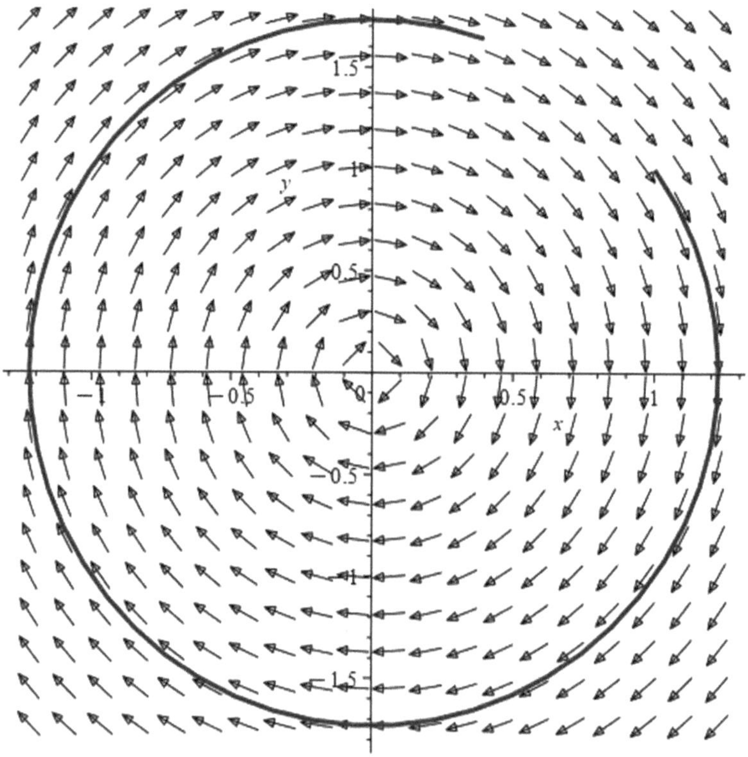

FIGURE 13.18
Solution for Exercise 10.3.

Chapter 11 Exercises

11.1 *Solve* $x' = 3x - 2y$, $y' = 5x - 4y$, *with* $x(0) = 3$, $y(0) = 6$

```
> dsys := {diff(x(t), t) = 3 x(t) − 2 y(t), diff(y(t), t) = 5 x(t − 4 y(t)), x(0) = 3, y(0) = 6};
        dsys := {d/dt x(t) = 3 x(t) − 2 y(t), d/dt y(t) = 5 x(t − 4 y(t)), x(0) = 3, y(0) = 6}
> dsn1 := dsolve(dsys, numeric, delaymax = 1);
        dsn1 := proc(x_rkf45_dae) ... end proc
> dsn1(0.5)
        [t = 0.5, x(t) = −7.08731813341968, y(t) = 13.5000000000000]
```

FIGURE 13.19
Maple solution for Exercise 11.1.

11.2 $x' = 3x + 4y, y' = 3x + 2y, x(0) = 1, y(0) = 1$

```
> dsys2 := {diff(x(t), t) = 3·x(t) + 4·y(t), diff(y(t), t) = 3·x(t) + 2·y(t), x(0) = 1, y(0) = 1};
        dsys2 := {d/dt x(t) = 3 x(t) + 4 y(t), d/dt y(t) = 3 x(t) + 2 y(t), x(0) = 1, y(0) = 1}
> dsn2 := dsolve(dsys2, numeric, delaymax = 1);
        dsn2 := proc(x_rkf45) ... end proc
> dsn2(0.5)
        [t = 0.5, x(t) = 22.8682401003791, y(t) = 17.3028127402100]
>
```

FIGURE 13.20
Maple solution for Exercise 11.2.

11.3 $x' = 2x - 3y + 5, y' = 3x - y + 7, x(0) = 1, y(0) = 1$

```
> dsys3 := {diff(x(t), t) = 2·x(t) − 3·y(t) + 5, diff(y(t), t) = 3·x(t) − y(t) + 7, x(0) = 1, y(0) = 1};
        dsys3 := {d/dt x(t) = 2 x(t) − 3 y(t) + 5, d/dt y(t) = 3 x(t) − y(t) + 7, x(0) = 1, y(0) = 1}
> dsn3 := dsolve(dsys3, numeric, delaymax = 1);
        dsn3 := proc(x_rkf45) ... end proc
> dsn3(0.5)
        [t = 0.5, x(t) = −0.0310445507697177, y(t) = 4.51900379007647]
>
```

FIGURE 13.21
Maple solution for Exercise 11.3.

Chapter 12 Exercises

Evaluate the following problems 12.1–12.4

12.1 $\mathcal{L}\{e^{-3t}\} = 1/(s + 3)$

12.2 $\mathcal{L}\{t\} = 1/s^2$

12.3 $\mathcal{L}\{\sin 2t\} = 2/(s^2 + 4)$

12.4 $\mathcal{L}\{3t - 5\sin 2t\} = (-7\,s^2 + 12)/(s^2(s^2 + 4)$

Evaluate the following inverse transforms for problems 12.5–12.9

12.5 $\mathcal{L}^{-1}\left\{\frac{1}{s^5}\right\} = 1/24\ t^4$

12.6 $\mathcal{L}^{-1}\left\{\frac{3s+5}{s^2+3}\right\} = 3\ \cos(\sqrt{3t} + \frac{5}{\sqrt{3}}\sin\left(\sqrt{3t}\right)$

12.7 $\mathcal{L}^{-1}\left\{\frac{1}{(s-1)(s+2)}\right\} = \text{A}/(\text{s}-1) + \text{B}/(\text{s}+2)/(\text{s}-1)*(\text{s}+2) = 1$

1 = A(s + 2) + B(s – 1), 2A – B = 1 and As + Bs = 0

Set s = 1, 1 = A(3) and s = –2 so 1 = B(–3) so A = 1/3 and B = –1/3

$= 1/3\ e^{t} - 1/3\ e^{-2t}$

12.9 $\mathcal{L}^{-1}\left\{\frac{s-4}{(s-1)(s+2)}\right\} = (s+4)/(s-1)(s+2) = A/(s-1) + B/(s+2) = s-4$

s-4 = A(s + 2) + B(s – 1)

Let s = 1, –3 = 3A so A = 1

Let s = –2, –6 = –3B, B = 1/2

$= \frac{1}{2} e^{-2t} + e^{t}$

12.11 Find $\mathcal{L}(f(t))$ for $f(t) = \begin{cases} 0 & 0 \le t < 3 \\ 2 & t \ge 3 \end{cases}$ $= 2\ e^{(-3s)}/s$

12.13 Evaluate $\mathcal{L}\{\mathcal{U}(t-5)\} = \begin{cases} 0 & 0 \le t < 5 \\ 1 & t \ge 5 \end{cases}$

Index

FSC
www.fsc.org
MIX
Papier | Fördert
gute Waldnutzung
FSC® C083411

Kolibri 360 GmbH
Ferdinand-Jühlke-Straße 7
99095 Erfurt, Deutschland
produktsicherheit@kolibri360.de